CHILTON'S
REPAIR & TUNE-UP GUIDE
JEEP WAGON
COMMANDO,
CHEROKEE
1966-79
All Models

Oil Filter. FRAM. PH.25.
Air Cleaner 32311 87

Howes Motors 0525 220508 Don.

Andrew Motors B.Ham.
021 778 1295.
Hyde Car Sales Stockport.
American Auto Sales 021 351 7655.

Plugs
RN 13L
GAP 30
·030

Managing Editor KERRY A. FREEMAN, S.A.E.
Senior Editor RICHARD J. RIVELE, S.A.E.
Editor RICHARD J. RIVELE, S.A.E.

President WILLIAM A. BARBOUR
Executive Vice President JAMES MIADES
Vice President and General Manager JOHN P. KUSHNERICK

CHILTON BOOK COMPANY
Radnor, Pennsylvania
19089

25871

SAFETY NOTICE

Proper service and repair procedures are vital to the safe, reliable operation of all motor vehicles, as well as the personal safety of those performing repairs. This book outlines procedures for servicing and repairing vehicles using safe, effective methods. The procedures contain many NOTES, CAUTIONS and WARNINGS which should be followed along with standard safety procedures to eliminate the possibility of personal injury or improper service which could damage the vehicle or compromise its safety.

It is important to note that repair procedures and techniques, tools and parts for servicing motor vehicles, as well as the skill and experience of the individual performing the work vary widely. It is not possible to anticipate all of the conceivable ways or conditions under which vehicles may be serviced, or to provide cautions as to all of the possible hazards that may result. Standard and accepted safety precautions and equipment should be used when handling toxic or flammable fluids, and safety goggles or other protection should be used during cutting, grinding, chiseling, prying, or any other process that can cause material removal or projectiles.

Some procedures require the use of tools specially designed for a specific purpose. Before substituting another tool or procedure, you must be completely satisfied that neither your personal safety, nor the performance of the vehicle will be endangered.

Although information in this guide is based on industry sources and is as complete as possible at the time of publication, the possibility exists that the manufacturer made later changes which could not be included here. While striving for total accuracy, Chilton Book Company cannot assume responsibility for any errors, changes, or omissions that may occur in the compilation of this data.

PART NUMBERS

Part numbers listed in this reference are not recommendations by Chilton for any product by brand name. They are references that can be used with interchange manuals and aftermarket supplier catalogs to locate each brand supplier's discrete part number.

ACKNOWLEDGMENTS

The Chilton Book Company expresses its appreciation to the Jeep Corporation, a Division of American Motors Corporation, Detroit, Michigan.

Manufactured in the United States of America
   4567890   87654321

Chilton's Repair & Tune-Up Guide: Jeep Wagoneer/Commando/Cherokee 1966–79
ISBN 0-8019-6739-2 pbk.
Library of Congress Catalog Card No. 79-65200

Contents

Quick Reference Specifications

For quick and easy reference, complete this page with the most commonly used specifications for your vehicle. The specifications can be found in Chapters 1 through 3 or on the tune-up decal under the hood of the vehicle.

TUNE-UP

Firing Order _____

Spark Plugs:

    Type _____

    Gap (in.) _____

Point Gap (in.) _____

Dwell Angle (°) _____

Ignition Timing (°) _____

    Vacuum (Connected/Disconnected) _____

Valve Clearance (in.)

    Intake _____   **Exhaust** _____

CAPACITIES

Engine Oil (qts)

    With Filter Change _____

    Without Filter Change _____

Cooling System (qts) _____

Manual Transmission (pts) _____

    Type _____

Automatic Transmission (pts) _____

    Type _____

Differential (pts) _____

    Type _____

COMMONLY FORGOTTEN PART NUMBERS

Use these spaces to record the part numbers of frequently replaced parts.

PCV VALVE	OIL FILTER	AIR FILTER
Manufacturer _____	Manufacturer _____	Manufacturer _____
Part No. _____	Part No. _____	Part No. _____

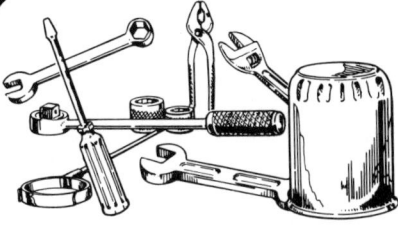

General Information and Maintenance

HOW TO USE THIS BOOK

This book covers all Wagoneer, Cherokee and Commando models from 1966 through 1979. Service procedures are specifically labelled as to model, engine and year, when it makes a difference. The "Troubleshooting" section is specially designed to help owners diagnose and fix minor ills before they result in major repair jobs. Each chapter contains the maintenance, repair, and overhaul procedures related to a particular component or system.

The purpose of this book is to cover maintenance or repair procedures that the owner will be able to perform without the aid of special tools or equipment. A lot of attention is given to the type of jobs on which the owner can save labor charges and time by doing it him-or-herself. Jobs that require special tools, such as automatic transmission overhaul, or those that the beginner is very unlikely to get right the first time, are purposely not covered.

A secondary purpose of this book is as a reference for owners who want to understand their vehicles or their mechanics, better. In this case, no tools are required.

TOOLS AND EQUIPMENT

Certain tools, plus a basic ability to handle tools are required to get started. A basic mechanics tool set, a torque wrench, and, for 1976 and later models, a Torx bits set. Torx bits are hexlobular drivers which fit both inside and outside on special Torx head fasteners used in various places on Jeep vehicles.

A special wheel bearing nut socket would be helpful when removing the front wheel bearings on 4x4 models.

When approaching a tune-up, the following items will be necessary:

    a. A spark plug socket.

    b. A spark plug gauge and gapping tool.

    c. A set of feeler gauges for the breaker points.

    d. A dwell meter/tachometer, and a timing light.

HISTORY AND MODEL IDENTIFICATION

The first Wagoneer-type Jeep was introduced in 1962. There were six versions of this body style available: 2- and 4-door, 2WD and 4WD station wagons and 2WD and 4WD panel deliveries. All of these vehicles came equipped with the 6 cylinder, 230 cu in., overhead cam engine, with the 327 cu in. V8 an option in 1965. The panel delivery was also available with the 232 cu in. 6 cylinder engine in 1966 and 1967. All vehicles were equipped with a standard 3-speed manual transmission or with the Turbo Hydra-Matic® automatic transmission, available in the Wagoneer station wagon only. The most distinguishing characteristic of these vehicles is the rather large, square radiator grille. The square-grilled Wagoneer station wagon was made until 1965; the panel delivery until 1967.

In 1965, the Wagoneer station wagon underwent a styling change. The large, square grille was changed to a low, wide grille that extends the full width of the front of the vehicle. The vehicle was available in either 2WD or 4WD with a 232 cu in. 6 cylinder engine or 327 cu in. V8 engine up until 1967. In 1968, a 350 cu in. Buick V8 engine became available with 4WD only. This engine was used until 1971. A Super Wagoneer with 4WD and the 327 cu in. V8 was available from 1965 to 1968. The Super Wagoneer package included deluxe interior and exterior trim including bucket seats, console shift and tilt steering wheel. All of the Wagoneers from 1965-on, were available with either a 3-speed manual transmission or a Turbo Hydra-Matic 3-speed automatic transmission.

Since 1965, the Wagoneer has remained basically the same in physical appearance. Different power train combinations have become available since the Kaiser Jeep Corporation was taken over by the American Motors Corporation in 1971; the 304 cu in. V8 being made available in 1971 and 1972; also, the 360 cu in. V8 and 258 cu in. 6 cylinder engines were made available. In 1973, Quadra-Trac® full-time 4WD was added to the Wagoneer.

In 1974, a 2-door version of the Wagoneer was introduced called the Cherokee. The Cherokee has all the equipment and options of the Wagoneer, including automatic transmission, 401 cu in. V8 and Quadra-Trac.

The first Jeepster Commando was introduced in 1966. Pre-1971 Commandos were equipped with either the F-Head 4 cylinder or the 225 cu in. Buick V6 engine. In 1972, all Commandos came with either the 232 cu in., 258 cu in. inline sixes or the 304 cu in. V8. A standard or automatic transmission coupled with 4WD together with convertible, station wagon, roadster or pickup body styles, made the Commando a very versatile vehicle. All pre-1971 Commandos have the square, vertically-slotted grille, while Commandos manufactured after 1971 have the wide, bright metal grille that extends fully across the front of the vehicle.

Since its inception, the Commando has been called by three names: Jeepster, Jeepster Commando and Commando. For the purpose of simplification and clarity, all of these vehicles will be referred to as the Commando.

SERIAL NUMBER IDENTIFICATION

Vehicle

Wagoneer

The vehicle identification number plate is located either on the left front door pillar or under the hood on the left firewall or both places.

The vehicle identification number on 1966–70 Wagoneers, can be deciphered with the following chart:

Model Identification by Serial Number

Serial # Prefix	Drive Axles	Body Style
1312	2WD	Station Wagon, 2-door
1312C	2WD	Station Wagon, 2-door, Custom
1314	2WD	Station Wagon, 4-door
1314C	2WD	Station Wagon, 4-door, Custom
1313	2WD	Panel Delivery
1412	4WD	Station Wagon, 2-door
1412C	4WD	Station Wagon, 2-door, Custom
1414	4WD	Station Wagon, 4-door
1414C	4WD	Station Wagon, 4-door, Custom
1413	4WD	Panel Delivery
1414D	4WD	Station Wagon, Super Wagoneer
1414X	4WD	Station Wagon, 4-door, Custom Special

The vehicle serial number suffix identifies the engine used in that particular vehicle:

Engine Model Identification through the Vehicle Identification Number Suffix 1966–70

Suffix Range	Engine Model
50,001 to 199,999	V8-327 Vigilante
200,001 to 299,999	6-232 High Torque
300,001 to 399,999	V8-350 Dauntless

When American Motors Corporation took over the Jeep Corporation, the numbering system was changed to the American Motors Corporation's 13-digit, alpha-numerical Vehicle Identification Number (VIN).

The serial number is interpreted in the following manner: The first digit, "J", signifies that the vehicle was made by the Jeep Corporation. The second digit, a number, signifies the year of production. The third digit, a letter, tells you where the vehicle was assembled, the type of transmission, and whether it is right-handed or left-hand drive. The fourth and fifth digits, which are numbers, tell you what model you have. The sixth digit, a number, tells the body style. The seventh digit, a letter, tells you the type of vehicle and its gross vehicle weight (GVW). The eighth digit, a letter, explains which engine the vehicle uses. The last 5 digits are sequential numbers, indicating the vehicle's sequence in production.

Vehicle serial number plate location-1966–70 Commando

Commando

The vehicle identification serial number is stamped on a metal plate which is mounted on the left of the firewall, under the hood on all Commandos. The serial number on 1966–70 Commandos, can be deciphered through the use of the following chart:

Model Identification by Serial Number 1966–70

Serial Number Prefix	Model
8701014 8701016	Jeepster Convertible
8701015 8701017	Jeepster Convertible, emission control
8702014 8702016	Jeepster Commando Convertible
8702015 8702017	Jeepster Commando Convertible, emission control
8705014 8705016	Jeepster Commando Roadster
8705015 8705017	Jeepster Commando Roadster, emission control
8705H14 8705H16	Jeepster Commando Pickup
8705H15 8705H17	Jeepster Commando Pickup, emission control
8705F14 8705F16	Jeepster Commando Station Wagon
8705F15 8705F17	Jeepster Commando Station Wagon, emission control

All Commandos manufactured after 1971 use the American Motors Corporation 13-digit, alpha-numerical Vehicle Identification Number (VIN). Look under "Wagoneer" for an example and explanation of this numbering system.

Engine

From 1966 to 1979 the Wagoneer, Cherokee and the Commando use 9 different engines; 4-134 F-head; V6-225; 232 and 258 inline sixes; 304, 327, 350, 360 and 401 V8s. The Commando uses the F-Head, the 225 V6, 232 and 258 inline sixes and the 304 V8. The Wagoneer and Cherokee use the 232 and 258 inline sixes, and the 304, 360 and 401 V8s.

The engine serial number for the F-Head 4 cylinder engine is located on the water pump boss at the front of the engine. It consists of a

PLANT	TRANS.	DRIVE
A = Toledo	AUTO.	LHD
B = CKD	AUTO.	LHD
F = Toledo	3-Speed	LHD
G = Toledo	3-Speed	RHD
J = CKD	3-Speed	LHD
K = CKD	3-Speed	RHD
M = Toledo	4-Speed	LHD
N = Toledo	4-Speed	RHD
O = CKD	4-Speed	LHD
P = CKD	4-Speed	RHD

BODY STYLE

1 = Thriftside (Truck)
2 = Townside (Truck)
3 = Platform Stake (Truck)
4 = 4-Dr. Wagon
5 = Open Body
6 = Cab & Chassis (Truck)
8 = Stripped Chassis
F = Full Cab-Metal
H = Half Cab-Metal

Sequential Serial Number (Five Digit Number)

Built by Jeep Corporation

J 2 A 1 4 4 C N 00001

1972 Model

TYPE		GVW
C = Custom Wagon		5600
O = Standard Wagon		5600
R = CJ-6	(Max.)	4750
S = CJ-5	(Max.)	4500
T = CJ-5	(Std.)	3750
U = Commando	(Max.)	4700
V = Commando/CJ-6	(Std.)	3900
W = Truck		5000
X = Truck		6000
Y = Truck		7000
Z = Truck		8000

VEHICLE LINE

14 = Wagoneer	—110″ W.B.	
24 = Truck	—120″ W.B.	
34 = Truck	—132″ W.B.	
63 = CJ-5	— 81″ W.B. CKD	
64 = CJ-6	—101″ W.B. CKD	
71 = MD		
72 = MDA	Government	
78 = M-606	Vehicles	
83 = CJ-5	— 84″ W.B.	
84 = CJ-6	—104″ W.B.	
87 = Commando	—104″ W.B.	

ENGINE

Code	CID	Cyl.	Comp.
A	258	6	Reg.
B	258	6	L/C
E	232	6	Reg.
F	232	6	L/C
H	304	V-8	Reg.
N	360	V-8	Reg.
R	134	4	Reg.
T	134	4	L/C

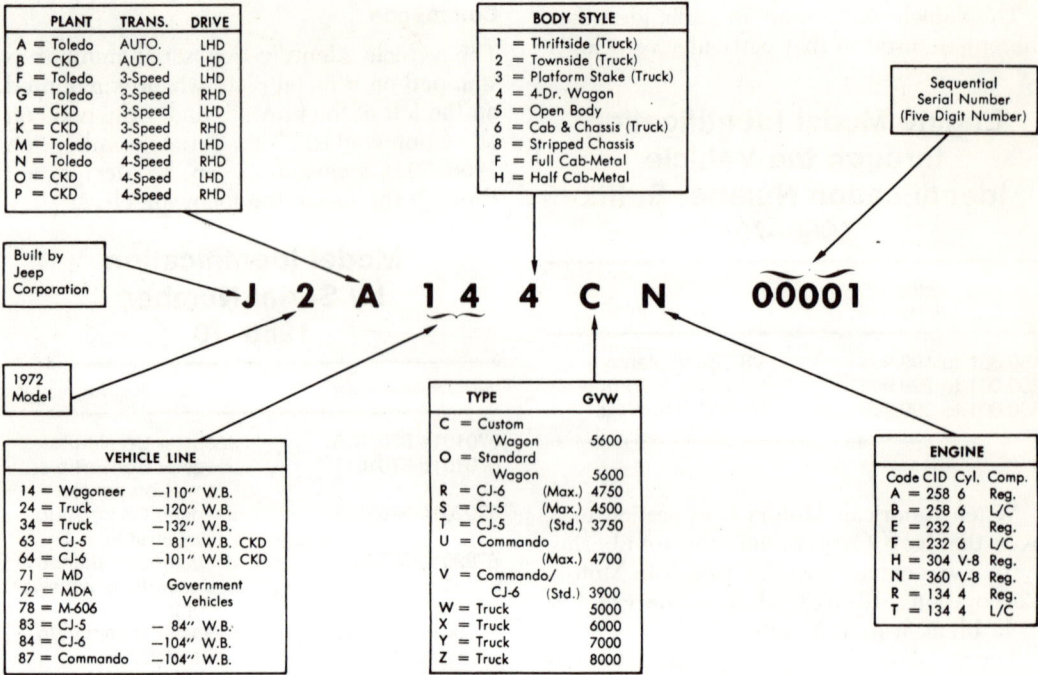

VIN decoding chart for all 1971–74 vehicles

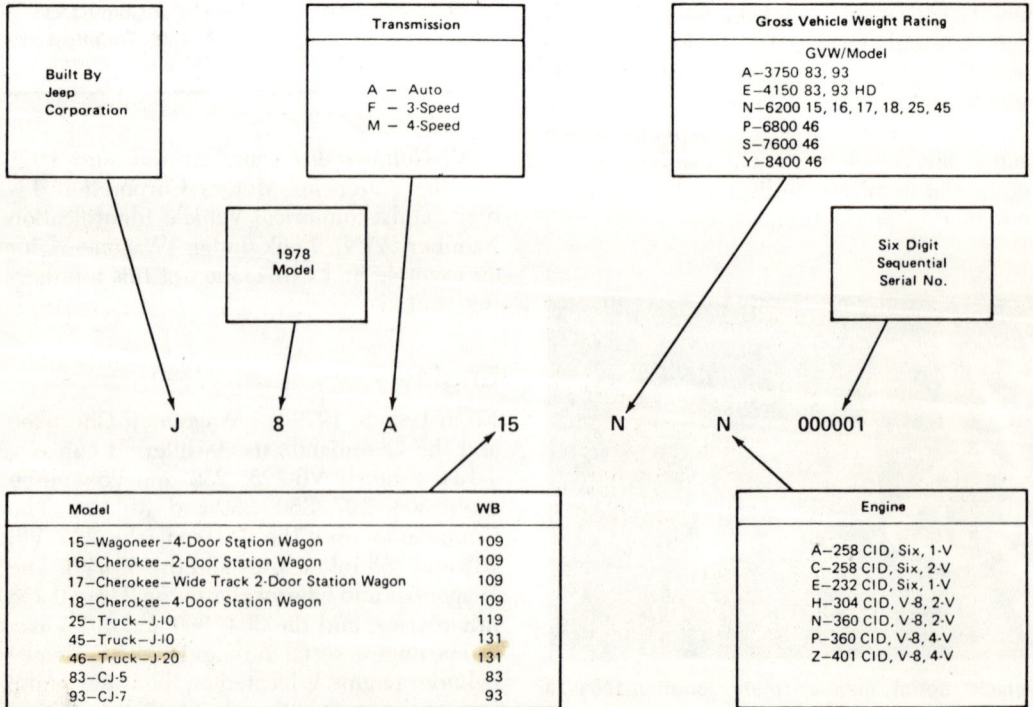

Built By Jeep Corporation

Transmission

A — Auto
F — 3-Speed
M — 4-Speed

Gross Vehicle Weight Rating

GVW/Model
A—3750 83, 93
E—4150 83, 93 HD
N—6200 15, 16, 17, 18, 25, 45
P—6800 46
S—7600 46
Y—8400 46

1978 Model

Six Digit Sequential Serial No.

J 8 A 15 N N 000001

Model	WB
15—Wagoneer—4-Door Station Wagon	109
16—Cherokee—2-Door Station Wagon	109
17—Cherokee—Wide Track 2-Door Station Wagon	109
18—Cherokee—4-Door Station Wagon	109
25—Truck—J-10	119
45—Truck—J-10	131
46—Truck—J-20	131
83—CJ-5	83
93—CJ-7	93

Engine

A—258 CID, Six, 1-V
C—258 CID, Six, 2-V
E—232 CID, Six, 1-V
H—304 CID, V-8, 2-V
N—360 CID, V-8, 2-V
P—360 CID, V-8, 4-V
Z—401 CID, V-8, 4-V

VIN decoding chart for all 1975–79 vehicles

Engine serial number location—4-134

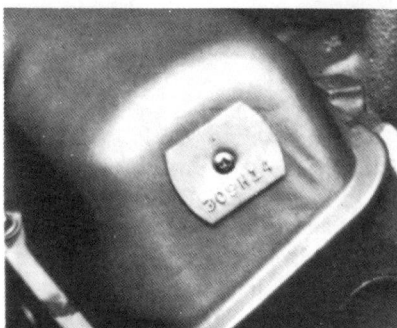

Engine serial number location—8-304, 360, 401

5 or 6 digit number. The engine code prefix for the F-Head is "4J".

The engine number for the V6 engine is located on the right side of the engine, on the crankcase, just below the head. The code is "KLH". The codes "RU" and "RV", included in the engine number of 1965 and 1966 engines, indicate manual or automatic transmission, respectively.

Engine serial number location—6-225

The American Motors engine code is, of course, found on the identification plate on the firewall. The second location is on the engine itself: on a machined surface of the block between number 2 and 3 spark plugs

Engine serial number location—6-232, 258 1971–79

on the six-cylinder engines. On the 304, 360 and 401 V8 engines, the number is located on a tag attached to the right valve cover. (For further identification, the displacement is cast into the side of the block.) The letter in the code identifies the engine by displacement (cu in.), carburetor type and compression ratio.

On vehicles equipped with the 232 six and made prior to 1971, the engine code number is located on a machined surface, adjacent to the distributor. The letter contained in the *code number* denotes the cu in. displacement of the engine. The letter "L" denotes 232 cu in., 8.5:1 compression ratio. The engine *code letter* is located on a boss directly above the oil filter.

Engine serial number location—1966–71 6-232

On the 327 V8, the engine code number is located on a tag which is attached to the alternator mounting bracket. The top line of the tag contains the letters "KJC" (Kaiser Jeep Corporation) and two letters which will identify the type of transmission and clutch used. The bottom line of the tag has a six-digit number, in which the first digit indicates the year that the engine was built; 7-1965; 8-1966; 1-1968. The second and third digit is the month in which the engine was made. The fourth digit indicates the compression ratio: E-8.7:1; F-9.7:1. The fifth and sixth digits indicate the day of the month in which

Engine serial number location—8-327

the engine was constructed. The lower half of the tag designates the specific engine group. The engine *code letter* is stamped directly under the date line on the tag and also on the front face, left bank, on the cylinder block.

The engine code number on the Buick 350 V8 is located on the left-hand side (looking from the rear) of the cylinder block deck, between the front two spark plugs. The first letter "K" indicates the engine was built for the Kaiser Jeep Corporation. The second letter in the engine code number indicates the year in which the engine was built: P-1968; R-1969; S-1970; T-1971. The third digit indicates the compression ratio and the fourth and fifth digits indicate the day of the year plus any change. The Buick 350 V8 has no engine *letter code*.

It is sometimes necessary to machine oversize or undersize clearances for cylinder blocks and crankshafts. If your engine is equipped with oversized or undersized parts, it is necessary to order parts that will match the old parts. To find out if your engine is one with odd-sized parts, check the engine code letter or the engine code number itself, which in some cases, is followed by a letter or a series of letters. The following chart explains just what the letters indicate on the F-Head and the 225 V6 engines.

Letter A (10001-A) indicates 0.010 in. undersized main and connecting rod bearings.
Letter B (10001-B) indicates 0.010 in. oversized cylinder bore.
Letter AB (10001-AB) indicates the combination of A and B above.
Letter C (10001-C) indicates 0.002 in. undersized piston pin.
Letter D (1001-D) indicates 0.010 in. undersized main bearing journals.
Letter E (10001-E) indicates 0.010 in. undersized connecting rod bearing journals.

On the 327 V8 and 232 sixes built prior to 1971, the undersize and oversize code is the same. On the 327 V8, the size code letters are stamped below the build date on the plate on the alternator bracket. On the 232 six, the size code is stamped on a tag located on the left front side of the baffle above the intake manifold below the build date, and on the boss above the oil filter. The following chart explains just what the letters indicate on the 327 V8 and 232 sixes made prior to 1971:

First Digit—Size of the bore:
  A, B, or C
Second Digit—Size of the main bearings:
  A, B, or C
Third Digit—Size of the connecting rod bearings:
  A, B, or C

A-Standard; B-0.010 in. Undersized; C-0.010 in. Oversized

All of the vehicles made after 1971 are equipped with American Motors Corporation engines, all of which have the same undersize/oversize letter codes, located on the boss directly above the oil filter on straight sixes and on the tag adjacent to the engine number on V8s. The parts size code is as follows:

Letter B indicates 0.010 in. oversized cylinder bore.
Letter M indicates 0.010 in. undersized main bearings.
Letter P indicates 0.010 in. undersized connecting rod bearings.
Letters PM indicates a combination of the above specification for P and M.
Letter C indicates 0.010 in. oversized camshaft block bores.

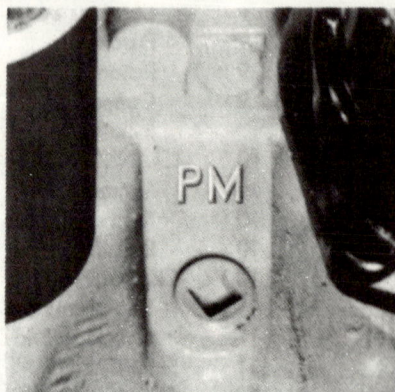

Parts letter size code, 6-232, 258

Transmission

There is a tag attached to the transmission case that identifies the manufacturer and model of the transmission. It is necessary to have the information on this tag before ordering parts. When reassembling the transmission, be sure that this tag is replaced on the transmission case so identification can be made in the future. In some cases, the transmission identification number may be cast into the transmission case.

ROUTINE MAINTENANCE

Air Cleaner

Three types of air cleaners have been used: the oil bath type, the dry cartridge type and the dry cartridge with polyurethane wrap type.

4-134 engines should have the oil bath type cleaned and refilled every 2,000 miles. All other engines with oil bath units should have them cleaned every 6,000 miles. Under

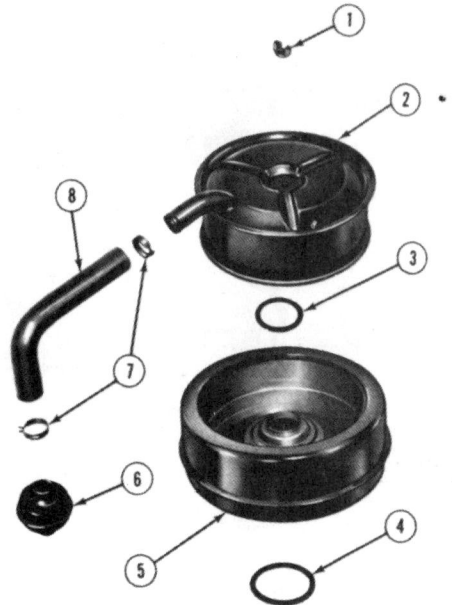

1. Wing nut	5. Oil cup
2. Cover	6. Breather
3. Rubber gasket	7. Clamps
4. Cork gasket	8. Vent tube

Exploded view of an oil bath air cleaner for the 6-225

dusty conditions, the units should be checked every week, or more frequently, if conditions warrant. To clean and refill an oil bath type, remove it from the engine, remove the oil cup from the unit, empty the oil, clean out the dirt from the oil cup and wash the element thoroughly in a detergent solvent. Dry the element with compressed air. In the summer, refill the cup with 40 or 50 weight engine oil. In the winter, use 20 weight.

Engines with the dry paper type filter should have the filter replaced every 12,000 miles. Under dusty conditions, the element

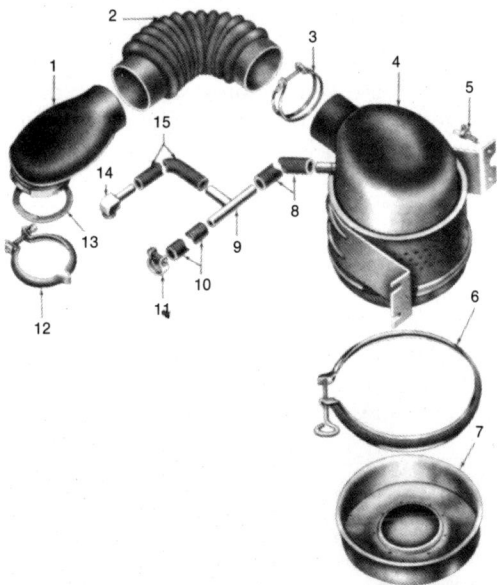

1. Horn	9. Hose tee
2. Flexible connector	10. Hose
3. Hose clamp	11. Hose clamp
4. Body	12. Clamp
5. Wing nut	13. Gasket
6. Clamp	14. Elbow
7. Oil cup	15. Hose
8. Hose	

Exploded view of an oil bath air cleaner for the 4-134

POLYURETHANE ELEMENT   PAPER CARTRIDGE

Polyurethane and paper air cleaner element

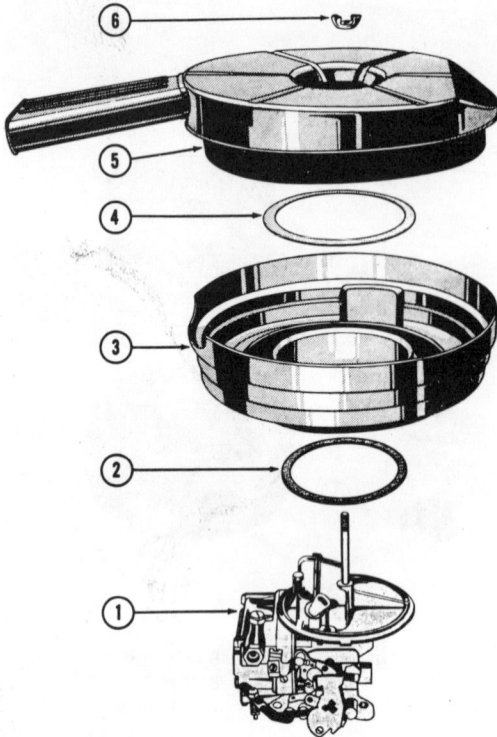

1. Carburetor
2. Cork gasket
3. Oil cup
4. Rubber gasket
5. Cover
6. Wing nut

Exploded view of an oil bath air cleaner for the 8-327

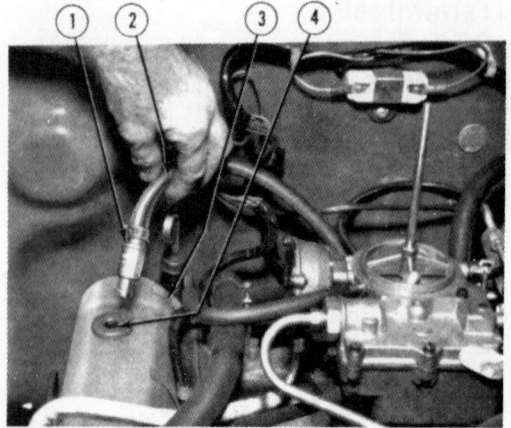

1. Ventilation valve
2. Hose to Carburetor Inlet
3. Right rocker cover
4. Grommet

Removing the PCV valve on 6-225 engines

Squeeze it flat to remove excess oil. At the same time, direct compressed air at the inside of the paper element to remove dirt. Replace the paper element every 15,000 miles, or sooner if necessary.

PCV Valve

The PCV valve, which is the heart of the positive crankcase ventilation system, should be changed every 15,000 miles. The main thing to keep in mind is that the valve should be free of dirt and residue and should be in working order. As long as the valve is kept clean and is not showing signs of becoming damaged or gummed up, it should perform its function properly. When the valve cannot be cleaned sufficiently or becomes sticky and will not operate freely, it should be replaced.

The PCV filter, which is located at the air filter housing on the six-cylinder models,

should be checked weekly, or more often if conditions warrant, and should be replaced at the first signs of clogging.

On engines using a dry paper filter with the polyurethane wrap, the polyurethane wrap should be carefully removed every 6,000 miles. Shake the dirt from the wrap, DO NOT WASH IT, squeeze the old oil out by pressing it flat between two rags then liberally soak it with SAE lOW-30 engine oil.

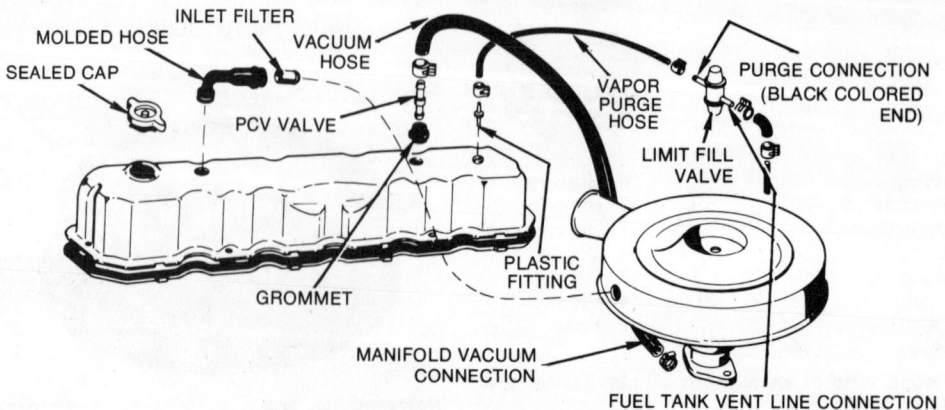

PCV valve location on the 6-232, 258

PURGE CONNECTION
(BLACK COLORED END)   LIMIT FILL VALVE   FUEL TANK VENT LINE   PURGE HOSE (ALL 6 CYL.
& V-8 MAN.)

CONNECTING HOSE
TO AIR CLEANER

PCV VALVE   VACUUM
HOSE

GROMMET

SEALED FILLER CAP

PURGE HOSE
(V-8 AUTO.)

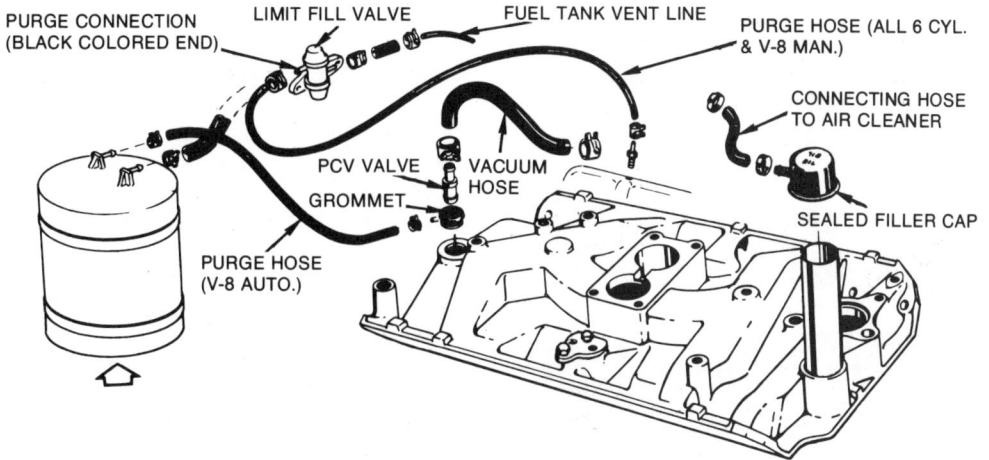

PCV valve location on the 8-304, 360, 401

AIR FILTER

6-232, 258 PCV air filter

should be checked and cleaned every 15,000 miles. Just blow out the screen with compressed air in the reverse direction of the normal air flow. Check to see that the screen forms a good seal around the edges of the air cleaner housing so no dirt can pass. If the screen is torn or clogged, or if it is seated improperly and cannot be repaired, replace it.

The PCV valve is located in the right valve cover on the 225 V6; in the intake manifold between number 2 and 3 cylinders on the F-Head; in the intake manifold behind the carburetor on the 350 V8 and all American Motors V8s; and on the valve rocker cover on the 232 and 258 sixes.

Evaporative Canister

All Wagoneers and Commandos equipped with American Motors V8s, and 1973 and later 6 cyl., have fuel evaporative emission control systems which include an evaporative storage canister. The purpose of this charcoal canister is to store gasoline vapors until they can be drawn into the engine and burned along with the air/fuel mixture. The air filter in the bottom of the canister should be replaced every 15,000 miles.

Fluid Level Checks

Engine Oil

First, it is necessary to make sure that your vehicle is on a level surface to ensure an accurate reading. Then, raise the hood, position the hold-up rod, if so equipped, and measure the oil with the dipstick which is on the right side of four-cylinder and V8 engines and on the left of six-cylinder engines. Add oil through the filler pipe on the right side of F head engines, through the valve cover filler hole on six-cylinder engines, and through the filler pipe at the front of the engine on American Motors V8s. The oil filler hole is located in the left side valve cover on the 350 V8.

If the oil is below the half mark, add a quart of oil, then recheck the level. If the level is still not reading full, add only a half of a quart at a time, until the dipstick reads full. Do not overfill the engine. When you check the oil in *any* engine, make sure that you allow sufficient time for all of the oil to drain back into the crankcase after stopping the engine or else you will only measure a fraction of the actual amount. A minute or so should be enough time.

Transmissions

The level of lubricant in the transmission should be maintained at the filler hole on all

The filler hole (arrow) and drain hole on a manual transmission

manual transmissions. This hole is located on the right side of the transmission.

Even though the transfer case and the transmission are linked together (on all three-speed transmissions) by passages that allow the oil from each unit to flow into the other, check each separately at their fill holes. When you check the level in the transmission, make sure that the vehicle is level so that you get a true reading. When you remove the filler plug, the lubricant should run out of the hole. If there is lubricant present at the hole, you know that the transmission is filled to the proper level. Replace the plug quickly for a minimum loss of lubricant. If lubricant does not run out of the hole when the plug is removed, lubricant should be added until it does. Replace the plug as soon as the lubricant reaches the level of the hole.

The fluid level in automatic transmissions is checked with a dipstick which is located in the filler pipe at the right rear of the engine. The fluid level should be maintained between the ADD and Full marks on the end of the dipstick with the automatic transmission fluid at normal operating temperatures. To raise the level from the ADD mark to the FULL mark, requires the addition of one pint of fluid. The fluid level with the fluid at room temperature (75° F) should be approximately ¼ in. below the ADD mark.

NOTE: *In checking the automatic transmission fluid, insert the dipstick in the filler tube with the markings toward the center of the car. Also, remember that the FULL mark on the dipstick is the indication of the level of the automatic transmis-*

sion *fluid when it is at operating temperature. This temperature is only obtained after at least 15 miles of expressway driving or the equivalent of city driving.*

To check the automatic transmission fluid level, follow the procedure given below. This procedure is applicable either when the fluid is at room temperature or at operating temperature.

1. With the transmission in Park, the engine running at idle speed, the foot brake applied and the vehicle resting on level ground, move the transmission gear selector through each of the gear positions, including Reverse, allowing time for the transmission to engage. Return the shift selector to the Park position and apply the parking brake. Do not turn the engine off, but leave it running at idle speed.

2. Clean all dirt from around the transmission dipstick cap and the end of the filler tube.

3. Pull the dipstick out of the tube, wipe it off with a clean cloth, and push it back into the tube all the way, making sure that it seats completely.

4. Pull the dipstick out of the tube again and read the level of the fluid on the stick. The level should be between the ADD and FULL marks. Do not overfill the transmission because this will cause foaming and loss of fluid through the vent and malfunctioning of the transmission. Use only DEXRON® or DEXRON® II transmission fluid.

BRAKE MASTER CYLINDER

The master cylinder reservoir is located under the hood, on the left side of the firewall. Before removing the master cylinder reservoir cap, make sure the vehicle is resting on level ground and clean all dirt away from the top of the master cylinder. Pry off the retaining clip or unscrew the hold-down bolt and remove the cap. The brake fluid level should be within ¼ in. of the top of the reservoir on both single and dual master cylinders.

If the level of the brake fluid is less than half the volume of the reservoir, it is advised that you check the brake system for leaks. Leaks in a hydraulic brake system most commonly occur at the wheel cylinders.

There is a rubber diaphragm in the top of the master cylinder cap. As the fluid level lowers in the reservoir due to normal brake shoe wear or leakage, the diaphragm takes up the space. This acts to prevent the loss of

Master cylinder location

brake fluid out of the vented cap and to pre-vent contamination of the brake fluid by dirt. After filling the master cylinder to the proper level with brake fluid, but before replacing the cap, fold the rubber diaphragm up into the cap, then replace the cap on the reservoir and tighten the retaining bolt or snap the re-taining clip into place.

COOLANT

The engine coolant level should be main-tained 2 in. below the bottom of the radiator filler neck when the engine is at air tempera-ture and 1 in. below the bottom of the filler neck when the engine is hot.

For best protection against freezing and overheating, maintain an approximate 50% water and 50% antifreeze mixture in the cool-ing system. Do not mix different brands of antifreeze to avoid possible chemical damage to the cooling system.

Avoid using water that is known to have a high alkaline content or is very hard, except in emergency situations. Drain and flush the cooling system as soon as possible after using such water.

CAUTION: *Cover the radiator cap with a thick cloth before removing it from a radia-tor in a vehicle that is hot. Turn the cap counterclockwise slowly until pressure can be heard escaping. Allow all pressure to es-cape from the radiator before completely removing the radiator cap. It is best to allow the engine to cool if possible, before removing the radiator cap.*

FRONT AND REAR AXLE

The standard front and rear axle differentials use SAE 80 or SAE 90 gear oil. Either is ac-ceptable for use in the differential housing. Traction-Lok differentials use only Jeep Trac-Lok Lubricant or its equivalent. Check the level of the oil in the differential housing every 5,000 miles under normal driving con-ditions and every 3,000 miles if the vehicle is used in severe driving conditions. The level should be up to the filler hole. When you remove the filler plug, the oil should start to run out. If it does not, replenish the supply until it does.

The lubricant should be changed every 30,000 miles. If running in deep water, change the lubricant daily.

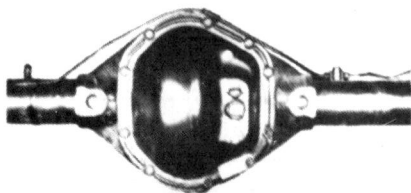

Rear axle fill hole location

STEERING GEAR

Check the steering gear lubricant level every 1,000 miles on 1966–71 vehicles and every 5,000 miles on 1972 and later vehicles. the level should be at the filler hole. Use SAE 80 gear oil to replenish the supply if it is needed. Don't forget to replace the filler bolt.

NOTE: *To fill the Saginaw manual steering gear, which was optional on pre-1971 Commandos, remove the side cover bolt (opposite the adjuster screw) and fill the gear with SAE 80 lubricant to the level of the bolt hole.*

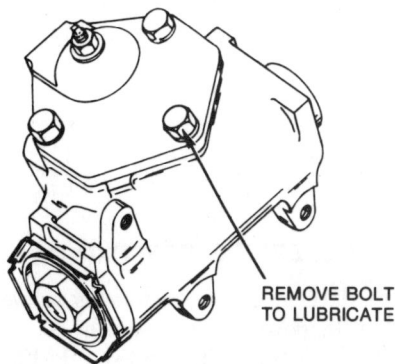

REMOVE BOLT TO LUBRICATE

Manual steering gear fill hole

Capacities Chart

Engine	Crankcase, Includes Filter	Transmission			Transfer Case		Drive Axle, Pts.		Fuel Tank Gal.	Cooling System, Qts.	
		Pints to Drain and Refill									
		3-sp	4-sp	Auto	Model 20	QT*	Front	Rear		without A/C	with A/C
Jeepster/Commando											
4-134	5	2.5	—	10	3.25	—	2.5	2.5	15	12.0	—
6-225	5	2.5	—	10	3.25	—	2.5	2.5	15	10.0	10.0
6-232	6	2.5	6.5	—	3.25	—	2.5	3.0	16	10.5	10.5
6-258	6	2.5	6.5	10	3.25	—	2.5	3.0	16	10.5	10.5
8-304	5	2.75	6.5	10	3.25	—	2.5	3.0	16	14.0	14.0
Wagoneer 1966–69											
6-232	6	2.5	—	10	3.25	—	2.5	3.0	20	10.5	10.5
8-327	6	2.75	—	10	3.25	—	2.5	3.0	20	19.5	19.5
8-350	5	2.5	—	10	3.25	—	2.5	3.0	20	15.0	15.0
Wagoneer 1970–73											
6-232	6	2.5	6.5	10	3.25	—	2.5	3.0	22	10.5	10.5
6-258	6	2.5	6.5	10	3.25	4.0	2.5	3.0	22	10.5	10.5
8-304	5	2.75	6.5	10	3.25	4.0	2.5	3.0	22	14.0	14.0
8-360	5	2.75	6.5	10	3.25	4.0	2.5	3.0	22	13.0	13.0
Wagoneer/Cherokee 1974–75											
6-258	6	2.5	—	10	3.25	4.0	2.5	3.0	22	10.5	10.5
8-360	5	2.75	6.5	10	3.25	4.0	2.5	3.0	22	14.0	14.0
8-401	5	—	—	10	—	4.0	2.5	3.0	22	14.0	14.0
Cherokee/Wagoneer 1976–79											
6-258	6	2.8	—	10	3.25	4.0	3.0	3.0	22	10.5	10.5
8-360	5	2.8	6.5	10	3.25	4.0	3.0	3.0	22	14.0	14.0
8-401	5	—	—	10	—	4.0	3.0	3.0	22	14.0	14.0

*Use Jeep Quadra-Trac Lubricant *only*. For models with a reduction unit, capacity is 5 pints.

POWER STEERING RESERVOIR

On models with power steering, check the fluid in the power steering pump every 1,000 miles. The level of the fluid should be at the correct point on the dipstick attached to the inside of the lid of the power steering pump. Fill the unit with DEXRON or DEXRON® II automatic transmission fluid. If the pump is abnormally low on fluid, check all the power steering hoses and connections, and the hydraulic cylinder for possible leaks.

STANDARD TRANSFER CASE

The transfer case should be checked in the same manner and frequency as the manual transmissions. The level should be up to the filler hole. Use the same viscosity oil as is being used in the transmission. The filler hole is located on the right side. Check the oil level at the top hole; the bottom one is for draining.

STEERING KNUCKLE

The axle shaft universal joints are located in the steering knuckle and are bathed in oil as they turn. To check the fluid level in the steering knuckle, remove the filler plug from the inside of the knuckle. The fluid should be at the level of the hole. If it is not, replenish the supply. Examine the knuckle for leaks if the level is abnormally low. A leak should be readily visible.

NOTE: *This applies to 1966–73 Wagoneers and Commandos thru 1970. Later models have open-type steering knuckles.*

Front axle steering knuckle fill plug

Wheel Bearings Lubrication and Adjustment

It is recommended that the front wheel bearings be cleaned, inspected and repacked every 12,000 miles, or as soon as possible

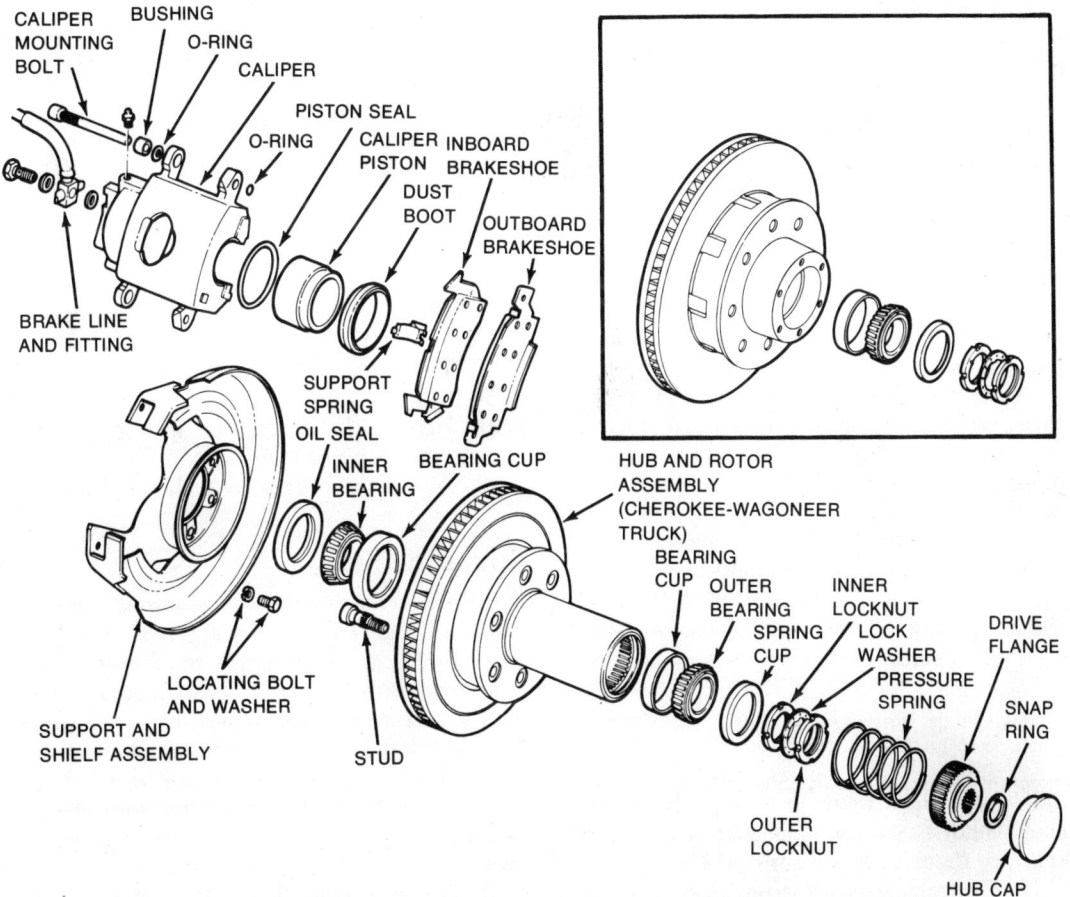

CALIPER MOUNTING BOLT
BUSHING
O-RING
CALIPER
PISTON SEAL
O-RING
CALIPER PISTON
INBOARD BRAKESHOE
DUST BOOT
OUTBOARD BRAKESHOE
BRAKE LINE AND FITTING
SUPPORT SPRING
OIL SEAL
INNER BEARING
BEARING CUP
LOCATING BOLT AND WASHER
SUPPORT AND SHIELF ASSEMBLY
STUD
HUB AND ROTOR ASSEMBLY (CHEROKEE-WAGONEER TRUCK)
BEARING CUP
OUTER BEARING
SPRING CUP
INNER LOCKNUT
LOCK WASHER
PRESSURE SPRING
OUTER LOCKNUT
DRIVE FLANGE
SNAP RING
HUB CAP

Front hub and brake assembly with disc brakes; inset shows assembly with locking hubs

after they have been run submerged in water for any appreciable length of time.

NOTE: *Sodium based grease is not compatible with lithium based grease. Read the package labels and be careful not to mix the two types. If there is any doubt as to the type of grease used, completely clean the old grease from the bearing and hub before replacing.*

1. Raise the front of the vehicle and place jackstands under the axle.

2. Remove the wheel.

3. Remove the front hub grease cap and driving hub snap-ring. On models equipped with locking hubs, remove the retainer knob hub ring, acitator knob, snap-ring, outer clutch retaining ring and actuating cam body.

4. Remove the splined driving hub and the pressure spring. This may require slight prying with a screwdriver.

5. Remove the wheel bearing locknut, lockring and adjusting nut.

6. On vehicle with drum brakes, remove the hub and drum assembly. This may require that the brake adjusting wheel be backed off a few turns. The outer wheel bearing and spring retainer will come off with the hub.

7. On vehicles with disc brakes, remove the caliper and suspend it out of the way by hanging it from a suspension or frame member with a length of wire. Do not disconnect the brake hose, and be careful to avoid stretching the hose. Remove the rotor and hub assembly. The outer wheel bearing and spring will come off with the hub.

8. Carefully drive out the inner bearing and seal from the hub, using a wood block.

9. Inspect the bearing races for excessive wear, pitting or grooves. If they are cracked or grooved, or if pitting and excess wear is present, drive them out with a drift or punch.

10. Check the bearing for excess wear, pitting or cracks, or excess looseness.

NOTE: *If it is necessary to replace either the bearing or the race, replace both. Never replace just a bearing or a race. These parts wear in a mating pattern. If just one is replaced, premature failure of the new part will result.*

11. If the old parts are retained, thoroughly clean them in a safe solvent and allow them to dry on a clean towel. Never spin dry them with compressed air.

12. On vehicles with drum brakes, cover the spindle with a cloth and thoroughly brush all dirt from the brakes. Never blow the dirt off the brakes due to the presence of asbestos in the dirt, which is harmful to your health when inhaled.

13. Remove the cloth and thoroughly clean the spindle.

14. Thoroughly clean the inside of the hub.

15. Pack the inside of the hub with EP wheel bearing grease. Add grease to the hub until it is flush with the inside diameter of the bearing cup.

16. Pack the bearing with the same grease. A needle shaped wheel bearing packer is best for this operation. If one is not available, place a large amount of grease in the palm of your hand and slide the edge of the bearing cage through the grease to pick up as much as possible, then work the grease in as best you can with your fingers.

17. If a new race is being installed, very carefully drive it into position until it bottoms all around, using a brass drift. Be careful to avoid scratching the surface.

18. Place the inner bearing in the race and install a new grease seal.

19. Place the hub assembly onto the spindle and install the outer bearing. Install the wheel bearing nut and torque it to 50 ft lb while turning the wheel back and forth to seat the bearings. Back off the nut about ¼ turn maximum.

20. Install the lockwasher with the inner tab aligned with the keyway in the spindle and turn the inner wheel bearing adjusting nut until the peg engages the nearest hole in the lockwasher.

21. Install the outer locknut and torque it to 50 ft lb.

22. Install the pressure spring, drive flange, snap ring and hub cap.

23. Install the caliper over the rotor.

Locking Hubs

Service procedures may vary among manufacturers. The following is a general service procedure that should apply to all types.

Locking hubs should be lubricated at least once a year and as soon as possible if running for extended periods submerged in water. The same type of grease should be used in the locking hubs as is used on the wheel bearings. EP lithium based chassis lube is preferred.

1. Remove the lock-out screws and washers.

2. Remove the hub ring and knob.

3. Remove the internal snap-ring from the groove in the hub.

4. Remove the cam body ring and clutch retainer from the hub and disassemble the parts.

5. Remove the axle shaft snap-ring. It may be necessary to push in on the gear and pull out on the axle with a bolt to make the snap-ring removal easier.

6. Remove the drive gear and clutch gear. A slight rocking of the hub may make them slide out easier.

7. Remove the coil spring and spring retainer.

8. Clean all the components in a safe solvent. Wipe out the hub with a clean cloth.

9. Grease the inside of the hub liberally.

10. Install the spring retainer ring with the undercut area facing inwards. Be sure it seats against the bearing.

11. Install the coil spring with the large end going in first.

12. Install the axle shaft sleeve and ring and the inner clutch ring with the teeth of both components meshed together in a locked position. It may be necessary to rock the hub to mesh the splines of the axle with those of the axle shaft sleeve and ring. Keep the two gears locked in position.

13. Install the axle shaft snap-ring. Push in on the gear and pull out on the axle with a bolt to allow the snap-ring to go into the groove.

14. Install the actuating cam body ring into the outer clutch retaining ring and install them in the hub.

15. Install the internal snap-ring.

16. Apply a small amount of Lubriplate® grease to the ears of the cam.

17. Assemble the knob in the hub ring and assemble them to the axle with the knob in the locked position. Tighten the screws and washers evenly and alternately, making sure the retainer ring is not cocked in the hub.

18. Torque the screws to 40 in. lb.

CAUTION: *Do not drive the vehicle until you are sure that both of the hubs are in the same position.*

Tires and Wheels

TIRE ROTATION

Tires should be rotated every 6,000 miles. If no spare is used, follow the "rotating four tires" diagram. If you have a spare and are including it in your tire rotating sequence, follow the "rotating five tires" diagram.

If uneven tire wear occurs before 6,000 miles, rotate the tires sooner. If the tires show abnormal wear patterns, have the axle alignment checked. Inflation pressures should be adjusted whenever tires are rotated.

TIRE LIFE AND SAFETY

Common sense and good driving habits will afford maximum tire life. Fast starts and stops, and hard cornering are hard on tires and will shorten their useful life span. If you start at normal speeds, allow yourself sufficient time to stop, and take corners at a reasonable speed, the life of your tires will increase greatly. Also make sure that you don't overload your vehicle or run with incorrect

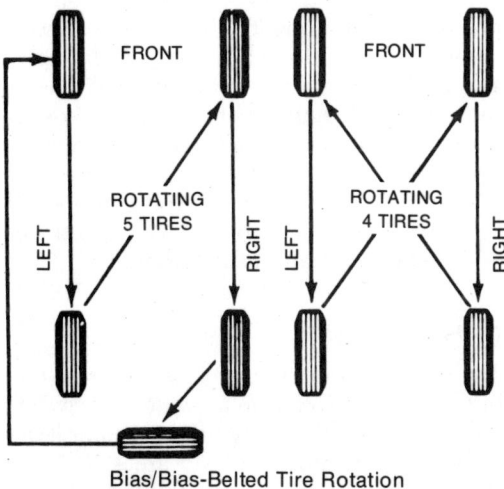

Bias/Bias-Belted Tire Rotation

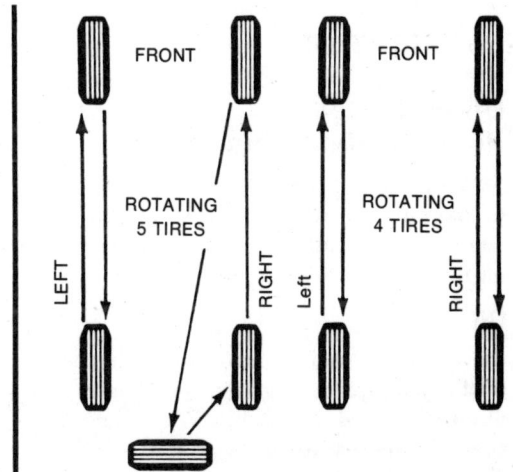

Radial Tire Rotation

Tire rotation

pressure in the tires. Both of these practices increase tread wear.

Inspect your tires frequently. Be especially careful to watch for bubbles in the tread or side wall, deep cuts, or underinflation. Remove any tires with bubbles. If the cuts are so deep they penetrate to the cords, discard the tire. Also look for uneven tread wear patterns that indicate that the front end is out of alignment or that the tires are out of balance.

Fuel Filter

There is an inline fuel filter located between the fuel pump and the carburetor on most models of Wagoneers and Commandos. On the 1966–70 232 sixes, the fuel filter consists of a sediment bowl mounted on the fuel pump. The inline fuel filter should be cleaned or replaced every 15,000 miles. If the vehicle is driven in abnormally dirty conditions or if contaminated gasoline was put in the gas tank, the filter could become clogged before 15,000 miles. The fuel sediment bowl type filter need not be serviced unless there is evidence of foreign matter (e.g., water, dirt) visible in the bowl. If there is, remove and empty the bowl, wipe it dry with a clean cloth and replace it.

Battery

Check the level of the water in the battery every time you fill the gas tank. The water level should meet the bottom of the filler hole. It should at least cover the plates.

LUBRICATION

Oil and Fuel Recommendations

All Jeep engines prior to 1975, are designed to operate on regular grade gasoline. All engines, 1975 and later, are designed to run on unleaded gasoline.

OIL TYPE

Many factors help to determine the proper oil for your Jeep. The big question is what viscosity to use and when. The whole question of viscosity revolves around the lowest anticipated ambient temperature to be encountered before your next oil change. The recommended viscosity ratings for tempera-

tures ranging from below 0° F to above 32° F are listed below. They are broken down into multiviscosities and single-viscosities. Multiviscosity oils are recommended because of their wider range of acceptable temperatures and driving conditions.

Engine Oil Viscosity Chart

Lowest Temperature Anticipated	Recommended Single Viscosity	Recommended Multi-Viscosity
Above 40° F	SAE 30 or 40	SAE 10W-30, 10W-40, 20W-40, 10W-50, 15W-50 or 20W-50
Above 0 F	SAE 20	SAE 10W-30, 10W-40, 10W-50
Below 0 F	SAE 10	SAE 5W-20 or 5W-30

Oil Changes

ENGINE

The engine oil is to be changed every 2,000 miles for the F-Head engine. For all of the other engines, the recommended interval is 6,000 miles. The oil should be changed more frequently, however, if the vehicle is being used in very dusty areas. Before draining the oil, make sure that the engine is at operating temperature. Hot oil will hold more impurities in suspension and will flow better, allowing it to remove more oil and dirt.

Drain the oil into a suitable receptacle. After the drain plug is loosened, unscrew the plug with your fingers, using a rag to shield your fingers from the heat. Push in on the plug as you unscrew it so you can feel when all of the screw threads are out of the hole. You can then remove the plug quickly with the minimum amount of oil running down your arm and you will also have the plug in your hand and not in the bottom of a pan of hot oil. Be careful of the oil. If it is at operating temperatures it is hot enough to burn you or at least make you uncomfortable.

Oil Filter Changes

The oil filter should be changed every 2,000 miles on the F-Head engines and every 6,000 miles on all of the other engines; sooner if the vehicle is operated in dusty areas. Change the oil filter when you change the oil. The engine should be at operating temperatures when the filter and oil are changed. The oil filter on all of the engines is the spin-on type.

On the F-Head engines, the filter is located on the right-hand side of the engine. On the V6 engine, the filter is located on the right side of the engine, just below the alternator. On the inline sixes, the filter is located on the lower, center right side of the engine. On all of the V8 engines, except the 327 V8, the oil filter is located on the right side of the engine. On the 327 V8 engine, the oil filter is on the rear left-hand side.

To remove the filter, you may need an oil filter wrench since the filter may have been screwed on too tightly and the heat from the engine may have made it even tighter. A filter wrench can be obtained at an auto parts store and is well worth the investment since it will save you a lot of grief. Loosen the filter with the filter wrench. With a rag wrapped around the filter, unscrew the filter from the oil pump housing. Be careful of hot oil that might run down the side of the filter, especially on the inline sixes and V8s. On the F-Head 4 cylinder engines, the filter is mounted with the open side facing downward so you won't have to worry about oil

An oil filter wrench

running down on your hand. Make sure that you have a pan under the filter before you start to remove it from the engine to avoid a mess and, if some of the hot oil does happen to get on you, you will have a place to dump the filter in a hurry. Wipe the base of the mounting plate with a clean, dry cloth. When you install the new filter, smear a small amount of oil on the gasket with your finger, just enough to coat the entire surface where it comes in contact with the mounting plate. When you tighten the filter, turn it ¼ to ¾ turn more after it comes in contact with the mounting plate.

TRANSMISSION

The lubricant in the manual transmission should be changed every 18,000 miles. All you have to do is remove the drain plug which is located at the bottom of the transmission or else on the side near the bottom. Allow all the lubricant to run out before replacing the plug. Replace the oil with the correct oil, usually SAE 80 or SAE 90 gear lubricant. Many brands of 80 and 90 weight now come in handy squeeze bottles which make filling easy and clean. See the section on level checks.

The transmission fluid in an automatic transmission should be changed every 24,000 miles of normal driving or every 12,000 miles of driving under abnormal or severe conditions. The fluid should be drained immediately after the vehicle has been driven for at least 20 minutes at expressway speeds or the equivalent of city driving, before it has had the chance to cool. Follow the procedure given below:

1. Drain the automatic transmission fluid from the transmission into an appropriate container, by removing the transmission bottom pan screws, pan and gasket. See "Automatic Transmission."

2. Remove the oil strainer and discard it.

3. Remove the O-ring seal from the pickup pipe and discard it.

4. Install a new O-ring seal on the pickup pipe and install the new oil strainer and pipe assembly.

5. Thoroughly clean the bottom pan and position a new gasket on the pan mating surface. Install the bottom pan and secure it with the attaching screws, torqued to 10–13 ft lbs.

6. Pour about 5 quarts of Dexron® or Dexron II® automatic transmission fluid in the filler pipe. 1966–70 vehicles use about 4 quarts of fluid. Make sure that the funnel, container, hose or any other item used to assist in filling the transmission is clean.

7. Start the engine. Do NOT race it. Allow the engine to idle for a few minutes.

8. Place the selector lever in Park and apply the parking brake. With the transmission fluid at operating temperatures, check the fluid level; add fluid to bring the level to the FULL mark.

FRONT AND REAR AXLE

The lubricant in the front and rear axle differentials should be checked about every 6,000 miles or sooner, if the vehicle is operated under severe conditions and should be changed every 12,000 miles. Follow the procedure given below for changing the lubricant in the front and rear axle differentials.

1. Remove the axle differential housing cover and allow the lubricant to drain out into a proper container.

2. Install the differential housing cover and a new gasket.

3. Tighten the cover attaching bolts to 15–25 ft lbs.

4. Remove the fill plug and add new lubricant to the fill hole level.

5. Replace the fill plug.

NOTE:*Trac-Lok (limited-slip) differentials may be cleaned only by disassembling the unit and wiping with clean rags. Refill Trac-Lok differentials with special limited-slip differential lubricant.*

MANUAL TRANSFER CASE

All manual transfer cases are to be serviced at the same time and in the same manner as the manual transmissions. The transfer case has its own drain plug which should be opened; do not rely on the transmission drain plug to completely drain the transfer case even if they are interconnected. Once the transfer case has been drained, replace the drain plug, remove the fill plug and fill the transfer case with the same type lubricant used in the manual transmissions. Replace the fill plug.

QUADR-TRAC TRANSFER CASE

Lubricant levels in the transfer case and low range reduction unit should be checked every 5,000 miles. Every 30,000 miles, the fluid in both units should be changed.

When checking the fluid levels, remove the filler plugs and add sufficient lubricant to bring the level up to the opening. Only Jeep Quadra-Trac® lubricant should be used. Allow all excess lubricant to run out of the hole before replacing the plug.

When changing the lubricant, remove the filler plugs from both the transfer case and the reduction unit. Remove the transfer case drain plug, and, after it has completely drained, install the drain plug. Loosen the five bolts on the reduction unit housing (the reduction unit has no drain plug), so that the unit can be pulled far enough for drainage. After it has completely drained, position the unit and tighten the bolts to 10–20 ft lb. Add one pint of Quadra-Trac® lubricant only, to the reduction unit. Install the filler plug. Fill the transfer case to the filler hole level with Quadra-Trac® lubricant ONLY. Allow excess lubricant to run out the hole. Install the filler plug.

CAUTION: *Filler and drain plugs must not be overtightened. Torque the plugs to 15–25 ft lb.*

After changing the fluid, it may be necessary to drive the vehicle in figure 8's for about fifteen minutes to allow fresh lubricant to enter the differential unit and force the clutches to operate.

CAUTION: *Do not hold the steering wheel at full stop during these manuevers, for more than five seconds at a time. Power steering fluid can overheat and damage to the pump or gear will result. Hold the wheel about ½ turn of full stop and avoid the overheating problem.*

PUSHING, TOWING, AND JUMP STARTING

To push-start your vehicle, (manual transmissions only) follow the procedures below. Check to make sure that the bumpers of both vehicles are aligned so neither will be damaged. Be sure that all electrical system components are turned off (headlights, heater, blower, etc.). Turn on the ignition switch. Place the shift lever in first or second and push in the clutch pedal. At about 15 mph, signal the driver of the pushing vehicle to fall back, depress the accelerator pedal, and release the clutch pedal slowly. The engine should start.

When you are doing the pushing or pulling, make sure that the two bumpers match so you won't damage the vehicle you are to push. Another good idea is to put an old tire in between the two vehicles. If the bumpers don't match, perhaps you should tow the other vehicle. If the other vehicle is just stuck, use first gear to slowly push it out. Tell the driver of the other vehicle to go slowly too. Try to keep your Jeep right up against the other vehicle while you are pushing. If the two vehicles do separate, stop and start over again instead of trying to catch up and ramming the other vehicle. Also try, as much as possible, to avoid riding or slipping the clutch. Low range makes this easy. When the other vehicle gains enough traction, it should pull away from your vehicle.

If you have to tow the other vehicle, make sure that the tow chain or rope is sufficiently long and strong, and that it is attached securely to both vehicles at a strong place. Attach the chain at a point on the frame or as close to it as possible. Once again, go slowly and tell the other driver to do the same.

Warn the other driver not to allow too much slack in the line when he gains traction and can move under his own power. Otherwise he may run over the tow line and damage both vehicles. If your Jeep has to be towed by a tow truck, it can be towed forward for any distance just as long as it is done fairly slowly. If your Jeep has to be towed backward, remove the front axle drive flanges to prevent the front differential from rotating. If the drive flanges are removed, improvise a cover to keep out dust and dirt.

JACKING AND HOISTING

Scissors jacks or hydraulic jacks are recommended for all Jeep vehicles. To change a tire, place the jack beneath the spring plate, below the axle, near the wheel to be changed.

Make sure that you are on level ground, that the transmission is in Reverse or with automatic transmissions, Park; the parking brake is set, and the tire diagonally opposite to the one to be changed is blocked so that it will not roll. Loosen the lug nuts before you jack the wheel to be changed completely free of the ground.

If you use a hoist, make sure that the pads of the hoist are located in such a way as to lift on the Jeep's frame and not on a shock absorber mount, floor boards, oil pan, or any other part that cannot support the full weight of the vehicle.

Preventive Maintenance Chart

1966–69 Jeepster/Commando

Interval	Item	Service
Every 2,000 miles	4-134 change oil and filter Complete chassis lube Steering gear Manual trans. Transfer case Differentials 4-134 distributor oiler, wick and cam 4-134 air cleaner	See oil viscosity chart EP grease Check level SAE 90 Check level SAE 90 Check level SAE 80 or 90 Check levels SAE 80 or 90* A few drops of engine oil Clean and refill
Every 6,000 miles	Auto. Trans. Air cleaner V6 V6 change oil and filter	Check level, refill Dexron II Clean and refill See oil viscosity chart
Every 12,000 miles	Steering knuckles Rear wheel bearings without lube fitting Front wheel bearings Manual trans. Transfer case Differentials Speedometer cable	Change lubricant Disassemble and repack— EP grease Repack—EP grease Change fluid—SAE 80 or 90 Change fluid—SAE 80 or 90 Change fluid—SAE 80 or 90* Lubricate—graphite grease
Every 24,000 miles	Auto. Trans.	Change fluid and filter—Dexron II

NOTE: *Cut intervals in half when operating under severe conditions.*
*With Power-Lok or Trac-Lok, use only Jeep Differential oil part #94557.

Preventive Maintenance Chart

1966–69 Wagoneer

Interval	Item	Service
Every 4,000 miles	6-232 crankcase 8-327 crankcase	Change oil and filter Change oil and filter
Every 6,000 miles	Steering gear Differentials	Check—Jeep lube #940657 Check—SAE 80 or 90*

Preventive Maintenance Chart (cont.)

1966–69 Wagoneer

Interval	Item	Service
Every 6,000 miles	Front axle king pin Manual trans. Transfer case Auto trans. Clutch cross shaft 8-350 crankcase Air filter	Check—EP chassis lube Check—SAE 80 or 90 Check—SAE 80 or 90 Check—Dexron II Lube—EP chassis lube Change oil and filter Change dry type; clean and refill oil bath type
Every 12,000 miles	Ball joints Front and rear wheel bearings All U-joints 8-350 distributor	Lube with part #934571 Lube with EP chassis grease Lube with EP chassis grease Replace cam lubricator
Every 30,000 miles	Differentials Front axle king pin Manual trans. Transfer case Auto trans.	Change fluid—SAE 80 or 90* Change lubricant—EP chassis lube Change lubricant—SAE 80 or 90 Change lubricant—SAE 80 or 90 Change fluid and filter—Dexron II

1970–73 Wagoneer, Commando

Interval	Item	Service
Every 6,000 miles or 6 mos.	Crankcase	Change oil and filter
Every 6,000 miles	Complete chassis lube Manual trans. Transfer case Steering gear Master cyl. Differentials Tires	EP chassis grease Check—SAE 80 or 90 Check—SAE 80 or 90 Check—Man.: SAE 80; Pwr.: Dexron II Check Check—SAE 80 or 90 Rotate
Every 12,000 miles	Fuel filter PCV valve Oil filter cap Drive belts Air cleaner—dry type Air cleaner—oil bath Distributor cam lubricator Timing and dwell Heat riser PCV filter-6 cyl. Points, condenser, roter Spark plugs V8 with auto trans.	Replace Replace Clean Adjust Replace Clean and refill—SAE engine oil Rotate Check Lubricate Clean Replace Replace Replace canister inlet filter
Every 24,000 miles	Auto trans.	Change fluid and filter—Dexron II
Every 30,000 miles	Manual trans. Transfer case Differentials Steering knuckle housing	Change fluid—SAE 80 or 90 Change fluid—SAE 80 or 90 Change fluid—SAE 80 or 90 Change lubricant—EP grease

1974–76 Wagoneer/Cherokee

Interval	Item	Service
Every 5,000 miles or 5 mos.	Crankcase	Change oil and filter
Every 5,000 miles	Manual trans. Transfer case Differentials	Check—SAE 80 or 90 Check—Model 20: SAE 80 or 90 Check—SAE 80

Preventive Maintenance Chart (cont.)

Interval	Item	Service
Every 5,000 miles	Steering gear	Check—SAE 80
	Power steering	Check—Dexron II
	Master cyl.	Check
	Radiator coolant	Check
	Complete chassis lube	EP chassis grease
	Point type distributor	Change points, condenser, rotor
	Point type ignition	Change spark plugs
	Heat riser	Lubricate
Every 10,000 miles	Drive shaft splines	Lubricate—EP chassis grease
Every 15,000 miles	Air cleaner	Replace
	Coil and plug wires	Replace
	Distributor cap and rotor	Replace
	Drive belts	Adjust
	Oil filler cap	Clean
	EGR valve port	Clean
	Fuel filter	Replace
	Canister filter	Replace
	Timing and dwell	Check
	PCV filter	Clean
	PCV valve	Replace
Every 25,000 miles	Automatic trans.	Change fluid and filter—Dexron II
Every 30,000 miles	Manual trans.	Change fluid—SAE 80 or 90
	Model 20 transfer case	Change fluid—SAE 80 or 90
	Differentials	Change fluid—SAE 80 or 90

1977–79 Wagoneer/Cherokee

Interval	Item	Service
Every 5,000 miles	Crankcase	Change oil and filter
	Complete chassis lube	EP chassis grease
	Coolant	Check
	Master cyl.	Check
	Steering gear	Check—SAE 80
	Power steering reservoir	Check—Dexron II
	Trans., man. or auto	Check—SAE 80; Dexron II
	Model 20 transfer case	Check—SAE 80
	Quadr-Trac	Check—Quadra-Trac fluid only
	Differentials	Check—SAE 80
	Drive belts	Adjust
	Catalytic converter	Check for bulging or heat damage
Every 25,000 miles	Coolant	Change
Every 30,000 miles	Plug and coil wires	Inspect and replace if necessary
	Spark plugs	Replace
	Air cleaner	Replace
	Timing	Check and adjust
	V8 oil filler cap	Clean
	Idle speed and mixture	Adjust
	Manual trans.	Drain and refill—SAE 80 or 90
	Transfer case—Model 20	Drain and refill—SAE 80 or 90
	Transfer case—Quadra-Trac	Drain and refill—Quadra-Trac fluid only
	Differentials	Drain and refill—SAE 80 or 90
	Front wheel bearings	Repack—EP chassis lube

NOTE: *When vehicle is subjected to severe use, cut the above intervals in half.*

2

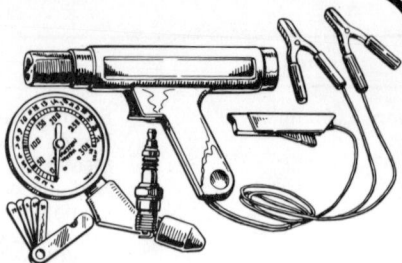

Tune-Up and
Troubleshooting

TUNE-UP PROCEDURES

Spark Plugs

Spark plugs ignite the air and fuel mixture in the cylinder as the piston reaches the top of the compression stroke. The controlled explosion that results forces the piston down, turning the crankshaft and the rest of the drive train.

The average life of a spark plug is dependent on a number of factors: the mechanical condition of the engine; the type of engine; the type of fuel; driving conditions; and the driver.

When you remove the spark plugs, check their condition. They are a good indicator of the condition of the engine. It is a good idea to remove the spark plugs at regular intervals, such as every 2,000 or 3,000 miles, just so you can keep an eye on the mechanical state of your engine.

A small deposit of light tan or gray material on a spark plug that has been used for any period of time is to be considered normal.

The gap between the center electrode and the side or ground electrode can be expected to increase not more than 0.001 in. every 1,000 miles under normal conditions.

When a spark plug is functioning normally or, more accurately, when the plug is in-

stalled in an engine that is functioning properly, the plugs can be taken out, cleaned, regapped, and reinstalled in the engine without doing the engine any harm.

When, and if, a plug fouls and begins to misfire, you will have to investigate, correct the cause of the fouling, and either clean or replace the plug.

There are several reasons why a spark plug will foul and you can learn which is at fault by just looking at the plug. A few of the most common reasons for plug fouling, and a description of the fouled plug's appearance, are listed in the "Troubleshooting" section, which also offers solutions to the problems.

Distributor Wiring Sequence and Firing Orders
REMOVAL

1. Remove the wires one at a time and number them so you won't cross them when you replace them.

2. Remove the wire from the end of the spark plug by grasping the wire by the rubber boot. If the boot sticks to the plug, remove it by twisting and pulling at the same time. Do not pull the wire itself or you will most certainly damage the core, or tear the connector.

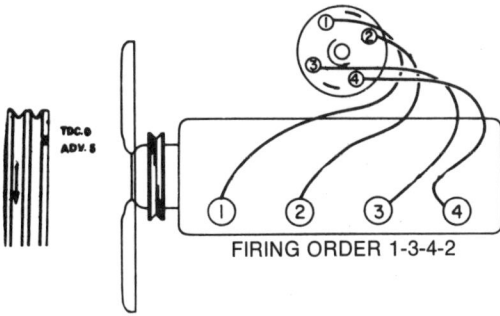

Distributor wiring and firing order: 4-134

FIRING ORDER 1-3-4-2

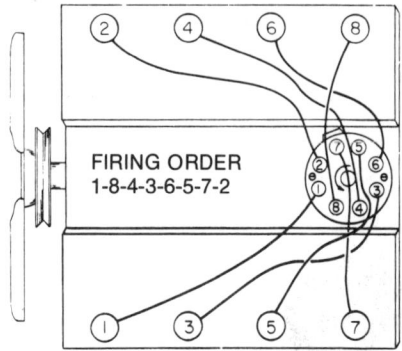

FIRING ORDER
1-8-4-3-6-5-7-2

Distributor wiring and firing order: 8-327

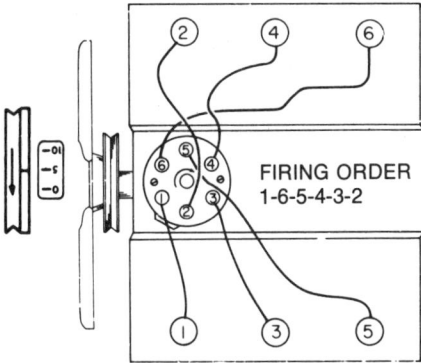

FIRING ORDER
1-6-5-4-3-2

Distributor wiring and firing order: 6-225

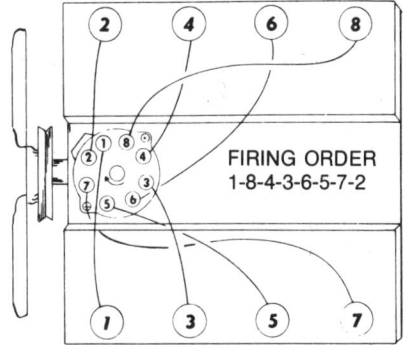

FIRING ORDER
1-8-4-3-6-5-7-2

Distributor wiring and firing order: 8-350

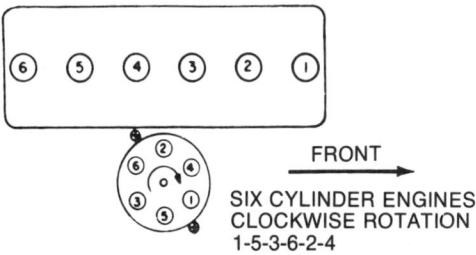

FRONT

SIX CYLINDER ENGINES
CLOCKWISE ROTATION
1-5-3-6-2-4

Distributor wiring and firing order: 1966–74
6-232, 258

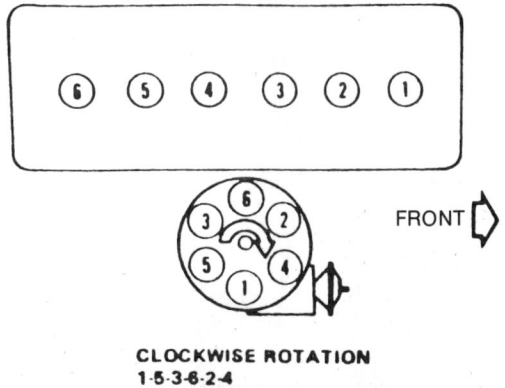

FRONT

CLOCKWISE ROTATION
1-5-3-6-2-4

Distributor wiring and firing order: 1975–79
6-232, 258

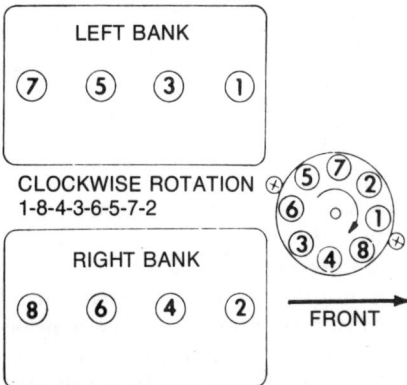

LEFT BANK

CLOCKWISE ROTATION
1-8-4-3-6-5-7-2

RIGHT BANK

FRONT

Distributor wiring and firing order: 1971–74
8-304, 360, 401

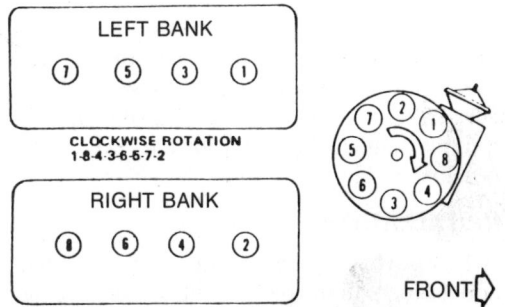

LEFT BANK

CLOCKWISE ROTATION
1-8-4-3-6-5-7-2

RIGHT BANK

FRONT

Distributor wiring and firing order: 1975–79
8-360, 401

3. Use a $^{13}/_{16}$ in. or $^5/_8$ in. on some 350 cu. in. V8's, spark plug socket to loosen all of the plugs about two turns.

4. If compressed air is available, blow off the area around the spark plug holes. Otherwise, use a rag or a brush to clean the area. Be careful not to allow any foreign material to drop into the spark plug holes.

5. Remove the plugs by unscrewing them the rest of the way from the engine.

INSPECTION

Check the plugs for deposits and wear. If they are not going to be replaced, clean the plugs thoroughly. Remember that any kind of deposit will decrease the efficiency of the plug. Plugs can be cleaned on a spark plug cleaning machine, which can sometimes be found in service stations, or you can do an acceptable job of cleaning with a stiff brush.

Measuring a spark plug electrode gap

Check spark plug gap before installation. The ground electrode must be aligned with the center electrode and the specified size wire gauge should pass through the gap with a slight drag. If the electrodes are worn, it is possible to file them level.

INSTALLATION

1. Insert the plugs in the spark plug hole and tighten them hand-tight. Take care not to cross-thread them.

2. Tighten the plugs to the torque figure specified in the "Tune-Up" section at the end of this chapter.

3. Install the spark plug wires on their plugs. Make sure that each wire is firmly connected to each plug.

Breaker Points

The points function as a circuit breaker for the primary circuit of the ignition system. The ignition coil must boost the 12 volts of electrical pressure supplied by the battery to as much as 25,000 volts in order to fire the plugs. To do this, the coil depends on the points and the condenser to make a clean break in the primary circuit.

The coil has both primary and secondary circuits. When the ignition is turned on, the battery supplies voltage through the coil and onto the points. The points are connected to ground, completing the primary circuit. As the current passes through the coil, a magnetic field is created in the iron center core of the coil. As the cam in the distributor turns, the points open and the primary circuit collapses. The magnetic field in the primary circuit of the coil also collapses and cuts through the secondary circuit windings around the iron core. Because of the scientific phenomenon called "electromagnetic induction," the battery voltage is increased to a level sufficient to fire the spark plugs.

When the points open, the electrical charge in the primary circuit jumps the gap created between the two open contacts of the points. If this electrical charge were not transferred elsewhere, the metal contacts of the points would melt and the gap between the points would start to change rapidly. If this gap is not maintained, the points will not break the primary circuit. If the primary circuit is not broken, the secondary circuit will not have enough voltage to fire the spark plugs.

Condenser

The function of the condenser is to absorb excessive voltage from the points when they open and thus prevent the points from becoming pitted or burned.

It is interesting to note that the above cycle must be completed by the ignition system every time a spark plug fires. In a V8 engine, all of the spark plugs fire once for every two revolutions of the crankshaft. That means that in one revolution, 4 spark plugs

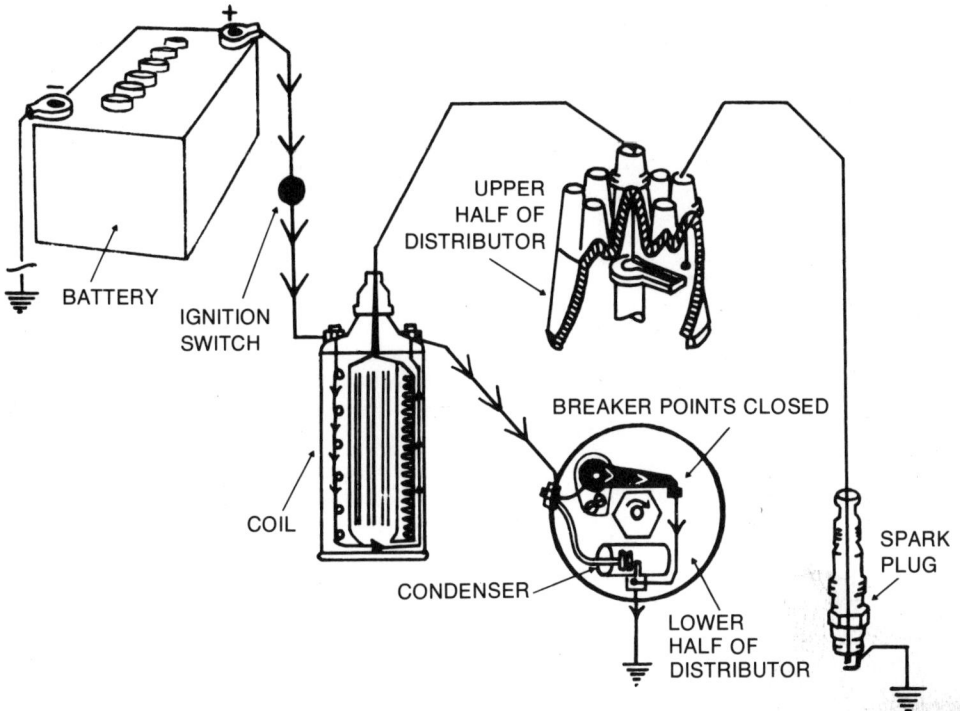

Diagram of a primary ignition circuit

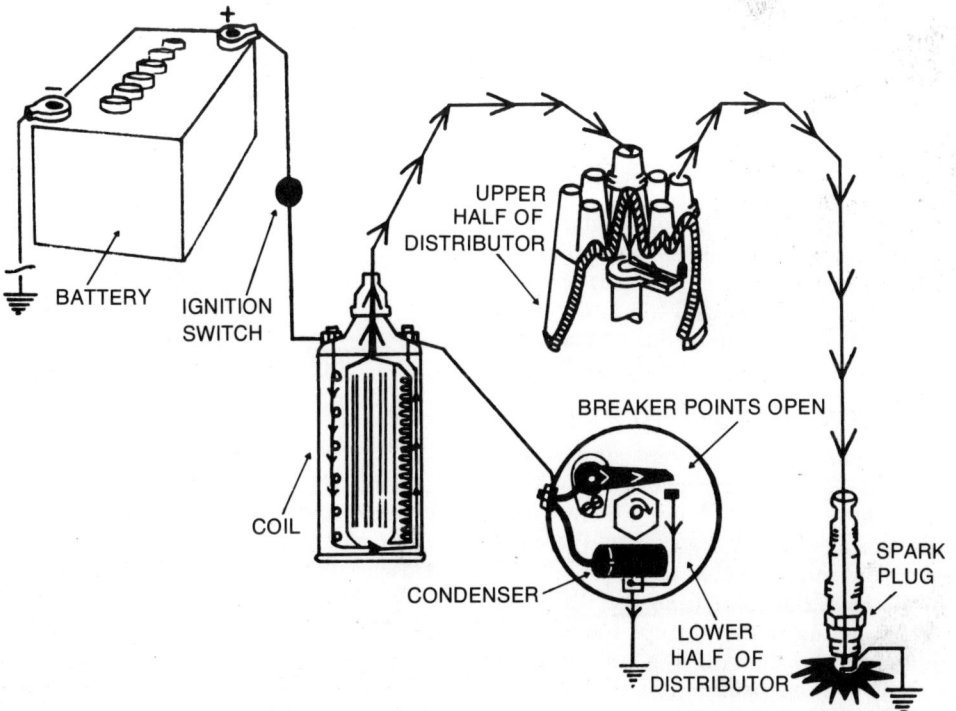

Diagram of a secondary ignition circuit

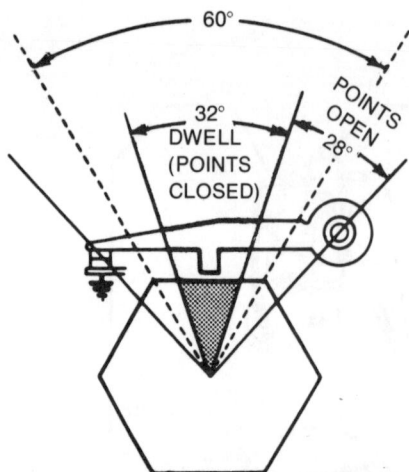

Contact breaker point dwell—6-232, 258

Contact point material transfer

REPLACE CONTACT SET
WHEN TRANSFER HAS
EXCEEDED .020"

fire. So, when the engine is at an idle speed of 800 rpm, the points are opening and closing 3,200 times a minute.

There are two ways to check the breaker point gap: It can be done with a feeler gauge or a dwell meter. Either way you set the points; you are basically adjusting the amount of time that the points remain open. The time is measured in degrees of distributor rotation. When you measure the gap between the breaker points with a feeler gauge, you are setting the maximum amount the points will open when the rubbing block on the points is on a high point of the distributor cam. When you adjust the points with a dwell meter, you are adjusting the number of degrees that the points will remain closed before they start to open as a high point of the distributor cam approaches the rubbing block of the points.

When you replace a set of points, always replace the condenser at the same time.

When you change the point gap or dwell, you will also have changed the ignition timing. So, if the point gap or dwell is changed, the ignition timing must be adjusted also.

INSPECTION OF THE POINTS

1. Disconnect the high-tension wire from the top of the distributor and the coil.

2. Remove the distributor cap by prying off the spring clips on the sides of the cap or turning the screw-headed fasteners.

3. Remove the rotor from the distributor shaft by pulling it straight up. Examine the condition of the rotor. If it is cracked or the metal tip is excessively worn or burned, it should be replaced.

4. Pry open the contacts of the points with

a screwdriver and check the condition of the contacts. If they are excessively worn, burned or pitted, they should be replaced.

5. If the points are in good condition, adjust them and replace the rotor and the distributor cap. If the points need to be replaced, follow the replacement procedure given below.

REPLACEMENT OF THE BRAKER POINTS AND CONSENSER

1. Remove the coil high-tension wire from top of the distributor cap. Remove the dis-

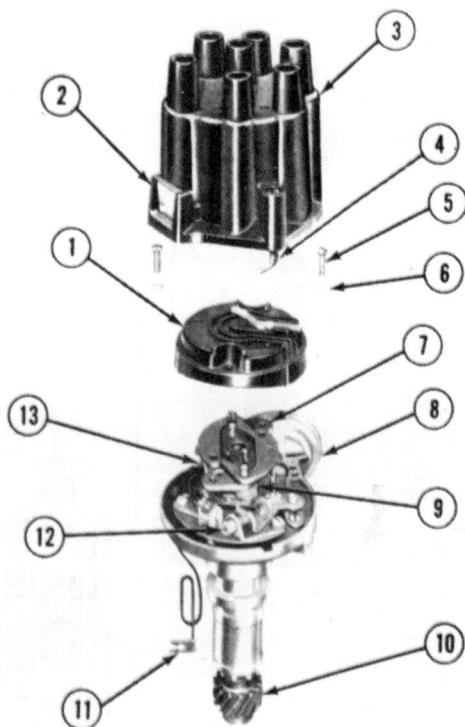

1. Rotor	8. Vacuum unit
2. Window	9. Breaker cam
3. Distributor cap	10. Drive gear
4. Cap retainer	11. Primary lead
5. Rotor mounting screw	12. Contact set
6. Lock washer	13. Condenser
7. Advance mechanism	

Exploded view of the 6-225 distributor

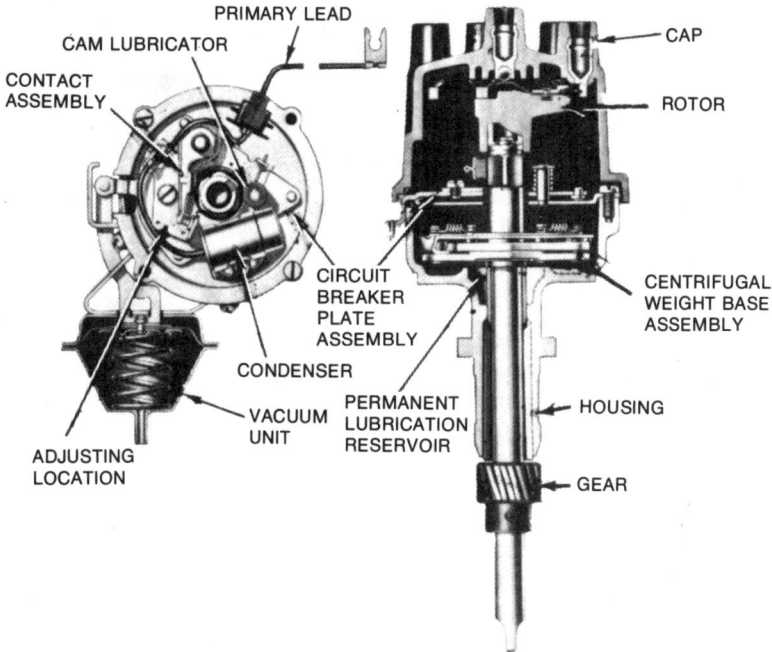

Cutaway view of the distributor on the 6-232, 258

tributor cap from the distributor and place it out of the way. Remove the rotor from the distributor shaft.

2. Loosen the screw that holds the condenser lead to the body of the breaker points and remove the condenser lead from the points.

3. Remove the screw that holds and grounds the condenser to the distributor body. Remove the condenser from the distributor and discard it.

4. Remove the points assembly attaching screws and adjustment lockscrews. A screw-driver with a holding mechanism will come in handy here, so that you don't drop a screw into the distributor and have to remove the entire distributor to retrieve it.

5. Remove the points by lifting them straight up and off the locating dowel on the plate. Wipe off the cam and apply new cam lubricant. Discard the old set of points.

6. Slip the new set of points onto the locating dowel and install the screws that hold the assembly onto the plate. Do not tighten them all the way.

7. Attach the new condenser to the plate with the ground screw.

8. Attach the condenser lead to the points at the proper place.

9. Apply a small amount of cam lubricant

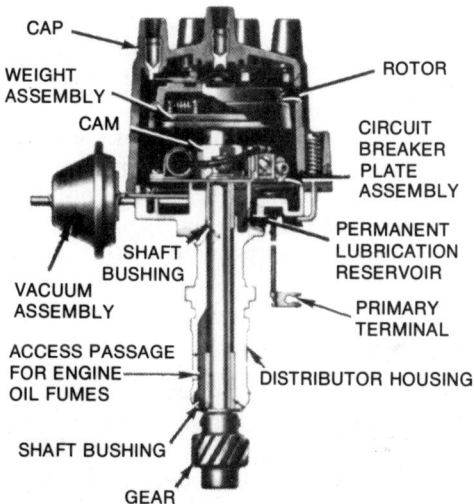

Cutaway view of the distributor on all V8's

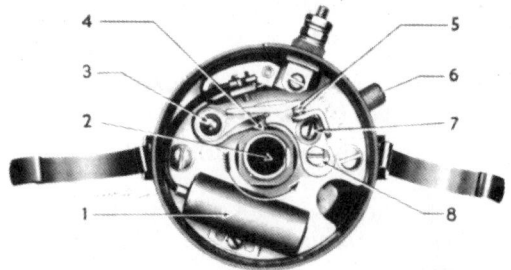

1. Condenser
2. Wick
3. Breaker pivot
4. Distributor Cam
5. Point contacts
6. Oiler
7. Lock screw
8. Adjusting screw

4-134 distributor

to the shaft where the rubbing block of the points touches.

ADJUSTMENT OF THE BREAKER POINTS WITH A FEELER GAUGE

1. If the contact points of the assembly are not parallel, bend the stationary contact so that they make contact across the entire surface of the contacts. Bend only the stationary bracket part of the point assembly; not the moveable contact.

2. Turn the engine until the rubbing block of the points is on one of the high points of the distributor cam. You can do this by either turning the ignition switch to the start position and releasing it quickly ("bumping" the engine) or by using a wrench on the bolt that holds the crankshaft pulley to the crankshaft.

3. Place the correct size feeler gauge between the contacts. Make sure it is parallel with the contact surfaces.

4. With your free hand, insert a screwdriver into the notch provided for adjustment or into the eccentric adjusting screw, then twist the screwdriver to either increase or decrease the gap to the proper setting.

5. Tighten the adjustment lockscrew and recheck the contact gap to make sure that it didn't change when the lockscrew was tightened.

6. Replace the rotor and distributor cap, and the high-tension wire that connects the top of the distributor and the coil. Make sure that the rotor is firmly seated all the way onto the distributor shaft and that the tab of the rotor is aligned with notch in the shaft. Align the tab in the base of the distributor cap with the notch in the distributor body. Make sure that the cap is firmly seated on the distributor and that the retainer springs or "L"-shaped, screw-headed clips are in place. Make sure that the end of the high-tension wire is firmly placed in the top of the distributor and the coil.

ADJUSTMENT OF THE BREAKER POINTS WITH A DWELL METER

1. Adjust the points with a feeler gauge as previously described.

2. Connect the dwell meter to the ignition circuit as according to the manufacturer's instructions.

3. If the dwell meter has a set line on it, adjust the meter to zero the indicator.

4. Start the engine.

NOTE: *Be careful when working on any*

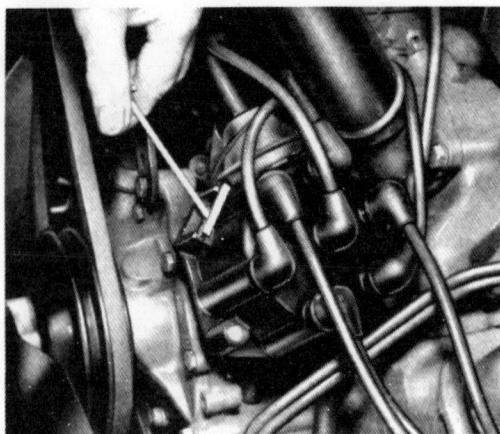

Adjusting the dwell on the V6 or V8

vehicle while the engine is running. Make sure that the transmission is in Neutral and that the parking brake is applied. Keep hands, clothing, tools and the wires of the test instruments clear of the rotating fan blades.

5. Observe the reading on the dwell meter. If the reading is within the specified range, turn off the engine and remove the dwell meter.

NOTE: *If the meter does not have a scale for 4 cylinder engines, multiply the 8 cylinder reading by two.*

6. If the reading is above the specified range, the breaker point gap is too small. If the reading is below the specified range, the gap is too large. In either case, the engine must be stopped and the gap adjusted in the manner previously covered.

NOTE: *On the V6 engine and all of the V8 engines, it is possible to adjust the dwell while the engine is running. There is a little window in the side of the distributor that can be raised, so that the points can be adjusted with an allen wrench.*

After making the adjustment, start the engine and check the reading on the dwell meter. When the correct reading is obtained, disconnect the dwell meter.

7. Check the adjustment of the ignition timing.

Ignition Timing

Ignition timing is the measurement, in degrees of crankshaft rotation, of the point at which the spark plugs fire in each of the cylinders. It is measured in degrees before or after Top Dead Center (TDC) of the compression stroke. Ignition timing is controlled by turning the distributor body in the engine.

Ideally, the air/fuel mixture in the cylinder will be ignited by the spark plug just as the piston passes TDC of the compression stroke. If this happens, the piston will be beginning its downward motion of the power stroke just as the compressed and ignited air/fuel mixture starts to expand. The expansion of the air/fuel mixture then forces the piston down on the power stroke and turns the crankshaft.

Because it takes a fraction of a second for the spark plug to ignite the mixture in the cylinder, the spark plug must fire a little before the piston reaches TDC. Otherwise, the mixture will not be completely ignited as the piston passes TDC and the full power of the explosion will not be used by the engine.

The timing measurement is given in degrees of crankshaft rotation before the piston reaches TDC (BTDC). If the setting for the ignition timing is 5° BTDC, the spark plug must fire 5° before each piston reaches TDC. This only holds true, however, when the engine is at idle speed.

As the engine speed increases, the pistons go faster. The spark plugs have to ignite the fuel even sooner if it is to be completely ignited when the piston reaches TDC. To do this, the distributor has a means to advance the timing of the spark as the engine speed increases. This is accomplished by centrifugal weights within the distributor and a vac-

uum diaphragm, mounted on the side of the distributor. It is necessary to disconnect the vacuum line from the diaphragm when the ignition timing is being set.

If the ignition is set too far advanced (BTDC), the ignition and expansion of the fuel in the cylinder will occur too soon and tend to force the piston down while it is still traveling up. This causes engine ping. If the ignition spark is set too far retarded, after TDC (ATDC), the piston will have already passed TDC and started on its way down when the fuel is ignited. This will cause the piston to be forced down for only a portion of its travel. This will result in poor engine performance and lack of power.

The timing is best checked with a timing light. This device is connected in series with the No. 1 spark plug. The current that fires the spark plug also causes the timing light to flash.

There is a notch on the front of the crankshaft pulley on the F-Head engine. There are also marks to indicate TDC and 5° BTDC on the timing gear cover that will assist you in setting ignition timing.

On the 232 and 258 sixes, there is a mark on the crankshaft pulley and a scale divided into degrees. The 304, 350, 360 and 401 V8s have the same mark and scale arrangement.

The 327 V8 has the scale on the crankshaft pulley and the pointer mark on the engine.

When the engine is running, the timing light is aimed at the marks on the engine and crankshaft pulley.

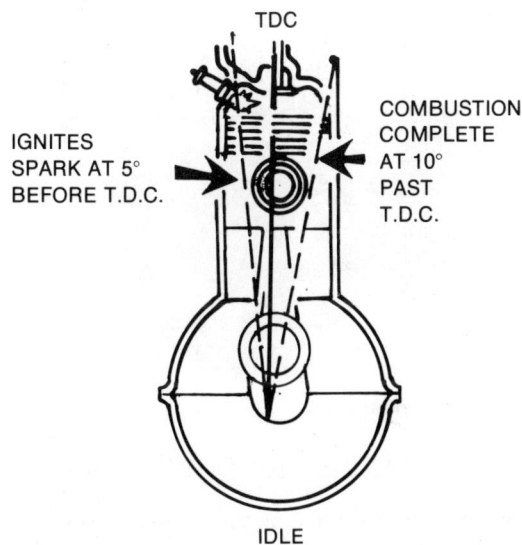

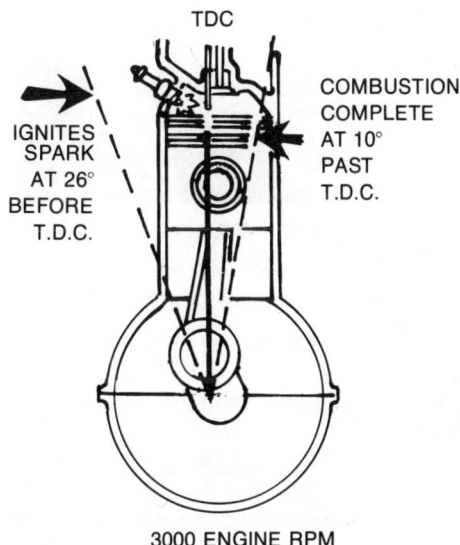

TDC		TDC	
IGNITES SPARK AT 5° BEFORE T.D.C.	COMBUSTION COMPLETE AT 10° PAST T.D.C.	IGNITES SPARK AT 26° BEFORE T.D.C.	COMBUSTION COMPLETE AT 10° PAST T.D.C.
IDLE		3000 ENGINE RPM	

Ignition timing at idle and at 3,000 rpm

Timing marks: 4-134

Timing marks: 6-225

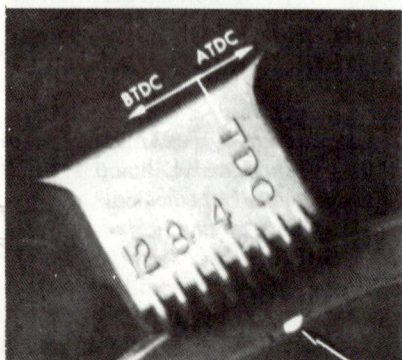

NOTCH

Timing marks: 6-232, 258

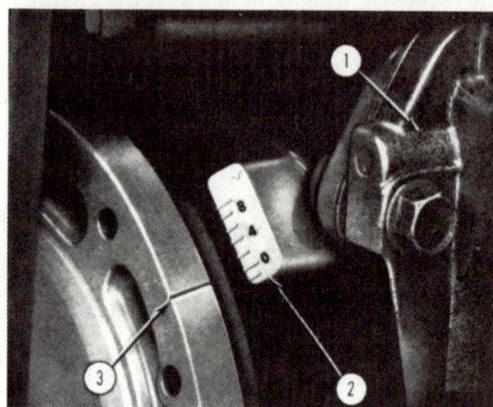

Timing marks: 8-350

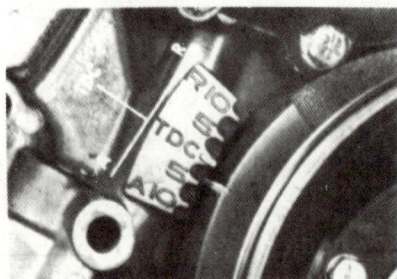

Timing marks: 8-304, 360, 401

T.D.C.   5°   10° B.T.D.C.

Timing marks: 8-327

IGNITION TIMING ADJUSTMENT

1. Locate the timing marks on the crankshaft pulley and the front of the engine.

2. Clean off the timing marks, so that you can see them.

3. Use chalk or white paint to color the mark on the scale that will indicate the correct timing, when aligned with the mark on the pulley or the pointer. It is also helpful to mark the notch in the pulley or the tip of the pointer with a small dab of color.

4. Attach a tachometer to the engine.

5. Attach a timing light to the engine, according to the manufacturer's instructions.

6. Disconnect the vacuum line to the distributor at the distributor and plug the vacuum line.

7. Check to make sure that all of the wires clear the fan and then start the engine.

8. Adjust the idle to the correct setting.

9. Aim the timing light at the timing marks. If the marks that you put on the pulley and the engine are aligned when the light flashes, the timing is correct. Turn off the engine and remove the tachometer and the timing light. If the marks are not in alignment, proceed with the following steps.

10. Turn off the engine.

11. Loosen the distributor lockbolt just enough so that the distributor can be turned with a little effort.

12. Start the engine. Keep the wires of the timing light clear of the fan.

13. With the timing light aimed at the pulley and the marks on the engine, turn the distributor in the direction of rotor rotation to retard the spark, and in the opposite direction of rotor rotation to advance the spark. Align the marks on the pulley and the engine with the flashes of the timing light.

14. When the marks are aligned, tighten the distributor lockbolt and recheck the timing with the timing light to make sure that the distributor did not move when you tightened the lockbolt.

15. Turn off the engine and remove the timing light.

Valve Lash

Valve adjustment determines how far the valves enter the cylinder and how long they stay open and closed.

If the valve clearance is too large, part of the lift of the camshaft will be used in removing the excessive clearance. Consequently, the valve will not be opening as far as it should. This condition has two effects: the valve train components will emit a tapping sound as they take up the excessive clearance and the engine will perform poorly because the valves don't open fully and allow the proper amount of gases to flow into and out of the engine.

If the valve clearance is too small, the intake valves and the exhaust valves will open too far and they will not fully seat on the cylinder head when they close. When a valve seats itself on the cylinder head, it does two things: it seals the combustion chamber so that none of the gases in the cylinder escape and it cools itself by transferring some of the heat it absorbs from the combustion in the cylinder to the cylinder head and to the

engine's cooling system. If the valve clearance is too small, the engine will run poorly because of the gases escaping from the combustion chamber. The valves will also become overheated and will warp, since they cannot transfer heat unless they are touching the valve seat in the cylinder head.

NOTE: *While all valve adjustments must be made as accurately as possible, it is better to have the valve adjustment slightly loose than slightly tight, as a burned valve may result from overly tight adjustments.*

The only engine in which the valves can be adjusted is the F-Head 4 cylinder. The V6-225, 232 and 258 sixes, 304, 327, 350, 360 and 401 V8's all have hydraulic lifters and are not adjustable in service.

F-HEAD ENGINE

NOTE: *The engine must be cold (at air temperature) when the valves are adjusted on the F-Head engine.*

1. Remove the side valve spring cover.

2. After the engine has cooled to ambient temperature, turn the engine until the lifter for the front intake valve is down as far as it will go. The lifter should be resting on the center of the heel (back) of the cam lobe for that valve. You can observe the position of the lifter by looking through the side valve spring cover opening. Put the correct size feeler gauge between the rocker arm and the valve stem for that particular intake valve. There should be a very slight drag on the feeler gauge when it is pulled through the gap. If there is a slight drag, you can assume that the valves are at the correct setting. If the feeler gauge cannot pass between the rocker arm and the valve stem, the gap between them is too small and must be increased. If the gauge can be passed through the gap without any drag, the gap is too large and must be decreased. Loosen the locknut on the top of the rocker arm (pushrod side) by turning it counterclockwise.

Turn the adjusting screw clockwise to

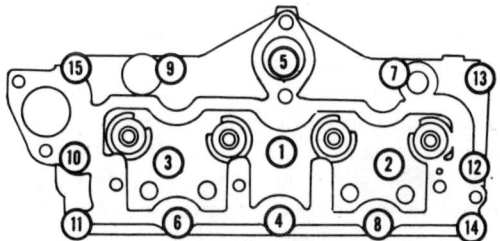

Head bolt tightening sequence: 4-134

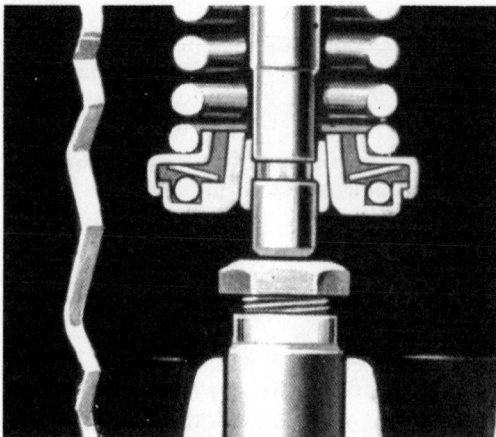

Exhaust valve adjustment screw: 4-134

Exhaust valve lash measurement: 4-134

lessen the gap and counterclockwise to increase the gap. When the gap is correct, turn the locknut clockwise to lock the adjusting screw. Follow this procedure for all of the intake valves, making sure that the lifter is all the way down for each adjustment.

3. Turn the engine so that the first exhaust valve is completely closed and the lifter that operates that particular valve is all the way down and on the heel of the cam lobe that operates it.

4. Insert the correct size feeler gauge between the valve stem of the exhaust valve and the adjusting screw. This is done through the side of the engine in the space that is exposed when the side valve spring cover is removed. If there is a slight drag on the feeler gauge, you can assume that the gap is correct. If there is too much drag or not enough, turn the adjusting screw clockwise to increase the gap and counterclockwise to decrease the gap.

5. When all of the valves have been adjusted to the proper clearance, replace the valve covers. Always use new valve cover gaskets when replacing the covers.

Carburetor

This section contains only tune-up adjustment procedures for carburetors. Descriptions, adjustments, and overhaul procedures for carburetors can be found in the "Fuel System" section.

When the engine in your Jeep is running, the air-fuel mixture from the carburetor is being drawn into the engine by a partial vacuum which is created by the movement of the pistons downward on the intake stroke. The amount of air-fuel mixture that enters into the engine is controlled by the throttle plate(s) in the bottom of the carburetor. When the engine is not running the throttle plate(s) is closed, completely blocking off the bottom of the carburetor from the inside of the engine. The throttle plates are connected by the throttle linkage to the accelerator pedal in the passenger compartment of the Jeep. When you depress the pedal, you open the throttle plates in the carburetor to admit more air-fuel mixture to the engine.

When the engine is not running, the throttle plates are closed. When the engine is idling, it is necessary to have the throttle plates open slightly. To prevent having to hold your foot on the pedal when the engine is idling, an idle speed adjusting screw was added to the carburetor linkage.

The idle adjusting screw contacts a lever (throttle lever) on the outside of the carburetor. When the screw is turned, it either opens or closes the throttle plates of the carburetor, raising or lowering the idle speed of the engine. This screw is called the curb idle adjusting screw.

IDLE SPEED ADJUSTMENT 1966–74

1. Start the engine and run it until it reaches operating temperature.

2. If it hasn't already been done, check and adjust the ignition timing. After you have set the timing, turn off the engine.

3. Attach a tachometer to the engine.

4. Remove the air cleaner, except on 1971 and later engines. Leave the air cleaner on these models. Turn on the headlights to high beam.

5. Start the engine and, with the transmission in Neutral or Park, check the idle speed on the tachometer. If the reading on the tachometer is correct, turn off the engine and remove the tachometer. If it is not correct, proceed to the following steps.

1. Fuel inlet
2. Choke plate
3. Manual choke cable bracket
4. Idle speed adjusting screw
5. Mixture adjusting screws

Rochester 2G

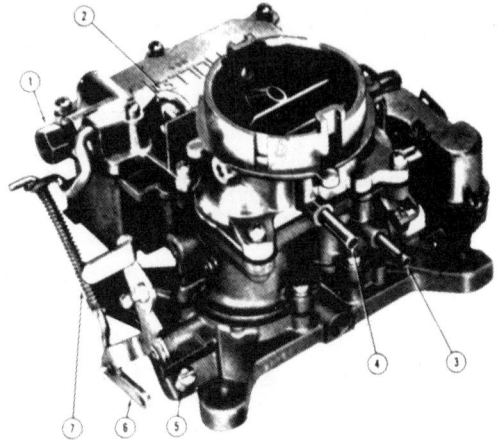

1. Fuel inlet
2. Pump rod
3. Distributor vacuum fitting
4. Choke clean air tube
5. Idle screw
6. Throttle lever
7. Pump link and spring

Holley 2209: 8-327

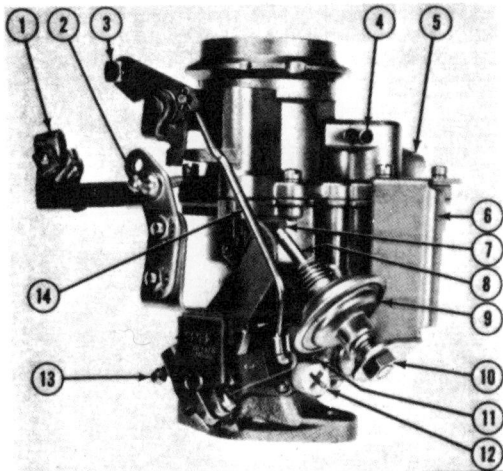

1.	Choke cable bracket	8,9.	Dashpot plunger
2,7.	Throttle lever	10.	Locknut
3.	Choke shaft	11.	Stop pin
4.	Bowl vent	12.	Idle mixture limiter
5.	Fuel inlet	13.	Idle speed screw
6.	Dashpot bracket	14.	Fast idle rod

Late model Carter YF: 4-134

6. Turn the idle adjusting screw at the bottom of the carburetor with a screwdriver—clockwise to increase idle speed and counterclockwise to decrease it.

MIXTURE ADJUSTMENT 1966–74

The idle mixture screw is located at the very bottom of the carburetor.

1. Turn the screw until it is all the way in. Do not force the screw in any further because it is very easy to damage the needle valve and its seat by screwing the adjusting screw in too tightly.

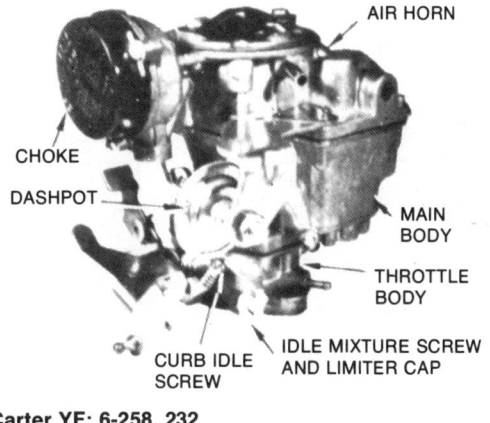

AIR HORN

CHOKE

DASHPOT

MAIN BODY

THROTTLE BODY

IDLE MIXTURE SCREW AND LIMITER CAP

CURB IDLE SCREW

Carter YF: 6-258, 232

2. Turn the screw out ¾–1¾ turns. This should be the normal adjustment setting. For a richer mixture, turn the screw out. The ideal setting for the mixture adjustment screw results in the maximum engine rpm.

NOTE: *Limiter caps are installed on all carburetors on all 1971 and later American Motors engines and some pre-1971 engines. These caps limit the amount of adjustment that can be made and should not be removed, if possible. If a satisfactory idle cannot be obtained however, they can be removed by installing a sheet metal screw in the center of the screw and turning clockwise. After removing the caps, adjust the carburetor in the same manner as was used without the caps. There are special service limiter caps available to replace the ones removed. Install the service limiter caps with the ears positioned against*

CHOKE COVER
AIR HORN
MAIN BODY
SOLENOID
SOLENOID ADJUSTER
IDLE LIMITER CAP
POWER VALVE
ACCELERATOR PUMP

Autolite 2100: 8-304, 360

1. Bowl vent gap   3. Accelerator pump screw
2. Idle speed screw 4. Idle mixture needle

Holley 4160: 8-327

1. Fuel inlet   3. Idle screw
2. Vacuum diaphragm   4. Mixture screw

Rochester 2GV: 8-350

the full rich stops. Be careful not to disturb the idle setting while installing the caps. Press the caps squarely and firmly into place.

Idle Speed and Mixture Adjustments

1975–79

CAUTION: *On vehicles equipped with a catalytic converter, do not idle the engine over three minutes at a time. If the adjustments are not completed within three minutes, run the engine at 2000 rpm for one minute.*

1. Turn the idle screw(s) to the full rich position. Note the position of the screw head slot inside the limiter cap slots.

2. Remove the limiter cap(s) carefully with a pair of needle nosed pliers. Reset the idle speed screws to the approximate position before cap removal.

3. Connect an accurate tachomter to the engine according to the manufacturer's instructions.

4. Run the engine to operating temperature.

5. Adjust the idle to 30 rpm above the recommended idle speed.

NOTE: *On V8 engines with automatic transmission, the throttle-stop solenoid is used to adjust the idle speed. Use the following procedure for these vehicles:*

a. With the solenoid wire connected, loosen the locknut and turn the solenoid in or out to obtain the specified rpm.

b. Tighten the solenoid bracket.

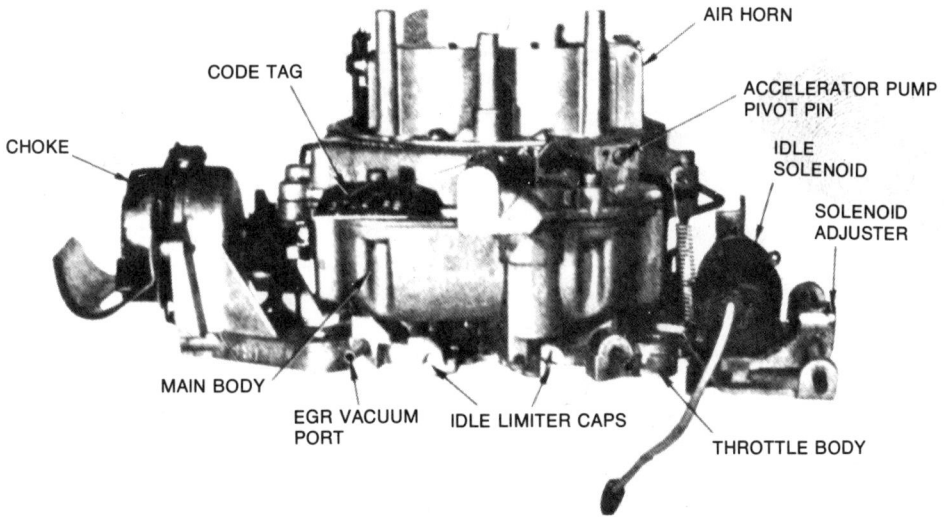

Autolite 4300: 8-360, 401

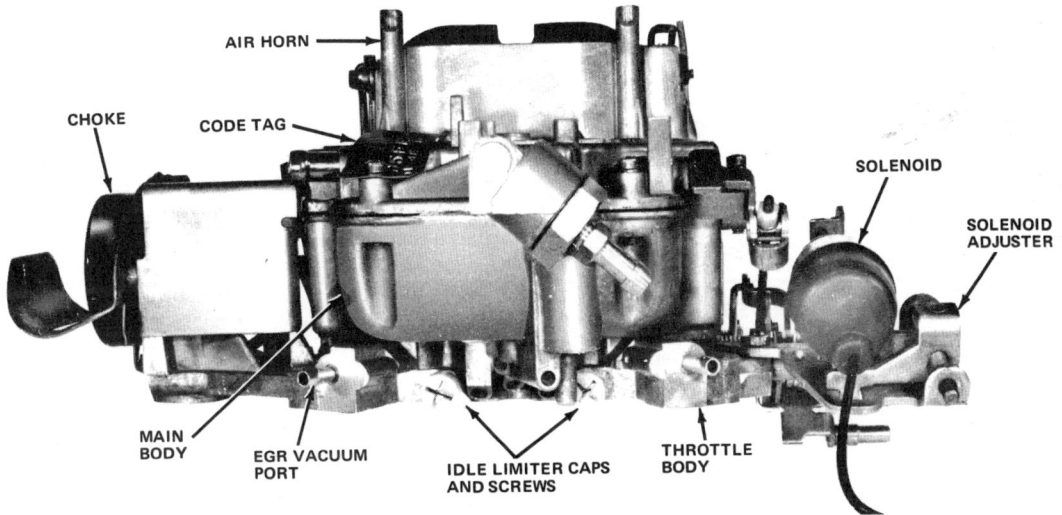

Carter BBD: 6-258

c. Disconnect the solenoid wire and adjust the idle speed screw to 500 rpm. Connect the wire.

6. Starting from the full rich stop position (established before the limiters were removed) turn the mixture screws clockwise (leaner) until a slight rpm drop is indicated.

7. Turn the mixture screws counterclockwise until the highest rpm reading is obtained at the best lean idle setting. On carburetors with two screws, turn them evenly in alternating equal increments.

NOTE: *If the idle speed changed more than 30 rpm during the adjustment, reset it to 30 rpm above the specified rpm and repeat the adjustment.*

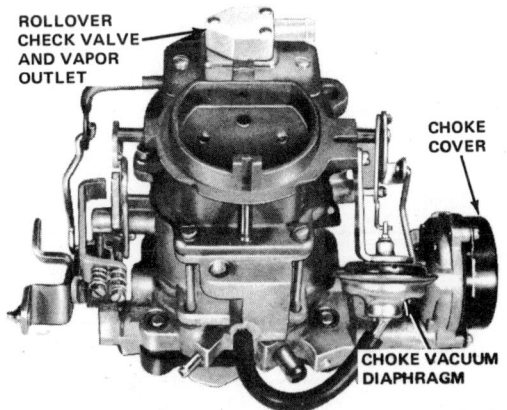

Holley 4350: 8-401

Tune-Up Specifications—Jeepster/Commando

| Engine | Spark Plugs | | Distributor | | Ignition Timing (deg.) | | Intake Valve Opens (deg.) | Fuel Pump Press. (psi) | Idle Speed | | Valve Clearance | |
	Type	Gap	Dwell (deg.)	Gap (in.)	Man. Trans.	Auto. Trans.			Man. Trans.	Auto. Trans.	Intake	Exhaust
1966–67 4-134	J-8	.030	42	.020	5B	5B	9B	2.5–3.5	600	600	.018	.016
6-225	J-12Y	.035	30	.016	5B	5B	24B	2–4	550	550	Hyd.	Hyd.
1968–69 4-134	J-8	.030	42	.020	0 ①	0 ①	9B	2–4	600 ②	600 ②	.018	.016
6-225	J-12Y	.035	26–32 ③	.016	5B ④	5B ④	24B	2–4	550 ⑤	550 ⑤	Hyd.	Hyd.
1970–72 6-232	N-12Y ⑥	.035	31–34	.016	5B	—	12.5B	3–5	650	—	Hyd.	Hyd.
6-258	N-12Y	.035	31–34	.016	3B	3B	12.5B	3–5	650	600	Hyd.	Hyd.
8-304	N-12Y	.035	29–31	.016	5B	5B	14.75B	4–6	700	650	Hyd.	Hyd.
1973 6-232	N-12Y	.035	31–34	.016	5B	—	12.5B	3–5	600	—	Hyd.	Hyd.
6-258	N-12Y	.035	31–34	.016	3B	3B	12.5	3–5	600	550	Hyd.	Hyd.
8-304	N-12Y	.035	29–31	.016	5B	5B	14.75B	4–6	750	700	Hyd.	Hyd.

① With distributor #IAY-4012: 5B.
② With distributor #IAY-4401: 650 solenoid off; 700 solenoid on.
   With distributor #IAY-4401B: 700 solenoid off; 750 solenoid on.
③ With Delco-Remy distributor: 30.
④ With Prestolite distributor #IAT-4502A: 0.
⑤ With emission controls: 650 solenoid off; 700 solenoid on.
⑥ 1970: N-14Y.

Tune-Up Specifications—Wagoneer/Cherokee

| Engine | Spark Plugs | | Distributor | | Ignition Timing (deg.) | | Intake Valve Opens (deg.) | Fuel Pump Press. (psi) | Idle Speed | |
	Type	Gap	Dwell (deg.)	Gap (in.)	Man. Trans.	Auto. Trans.*			Man. Trans.	Auto. Trans.*
1966–69 6-232	N-14Y	.035	31–34	.016	0 ①	0 ①	12.5B	3–5	550 ②	550 ②
8-327	H-14Y	.035	30	.016	5B ③	5B ③	24B	5–7	550 ②	500 ②
8-350	RBL-13Y	.030	30	.016	0 ④	0 ④	24B	5–7	650 ⑤	650 ⑤
1970–72 6-258	N-12Y	.035	31–34	.016	3B	3B	12.5B	3–5	650	600
8-304	N-12Y	.035	29–31	.016	5B	5B	14.75B	4–6	700	650
8-360	N-12Y	.035	29–31	.016	5B	5B	14.75	4–6	700	650

Tune-Up Specifications—Wagoneer/Cherokee (cont.)

| Engine | Spark Plugs | | Distributor | | Ignition Timing (deg.) | | Intake Valve Opens (deg.) | Fuel Pump Press. (psi) | Idle Speed | |
	Type	Gap	Dwell (deg.)	Gap (in.)	Man. Trans.	Auto. Trans.*			Man. Trans.	Auto. Trans.*
1973 6-258	N-12Y	.035	31–34	.016	3B	3B	12.5B	3–5	600	550
8-360	N-12Y	.035	29–31	.016	5B	5B	14.75	4–6	750	700
1974 6-258	N-12Y	.035	31–34	.016	3B	3B	12.5	3–5	600	550
8-360	N-12Y	.035	29–31	.016	5B	5B	14.75	4–6	750	700
8-401	N-12Y	.035	29–31	.016	—	5B	25.57	5–7	—	700
1975 6-258	N-12Y	.035	Electronic		3B	3B	12.5B	3–5	650	550
8-360	N-12Y	.035	Electronic		2.5B	2.5B	14.75B	4–6	750	700
8-401	N-12Y	.035	Electronic		—	2.5B	25.57B	5–7	—	700
1976 6-258	N-12Y	.035	Electronic		6B	8B	12.5B	3–5	600	550(700)
8-360	N-12Y	.035	Electronic		5B	8B(5B)	14.57B	4–6	750	700
8-401	N-12Y	.035	Electronic		—	8B(5B)	25.57B	5–7	—	700
1977 6-258	N-12Y	.035	Electronic		6B	6B	12.5B	3–5	650	550
8-360	RN-12Y	.035	Electronic		5B	8B	14.75B	4–6	750	700
8-401	RN-12Y	.035	Electronic		—	8B	25.57B	5–7	—	700
1978 6-258	N-13L	.035	Electronic		6B	6B	12.5B	3–5	650	550
8-360	N-12Y	.035	Electronic		5B	8B	14.75B	4–6	750	700
8-401	N-12Y	.035	Electronic		—	8B	25.57B	5–7	—	700
1979 6-258	N-14LY	.035	Electronic		8B	8B	12.5B	3–5	700	600
8-360	N-12Y	.035	Electronic		8B	10B	14.75B	4–6	700	700

*Figures in parentheses are for California.
① With distributor #1110340: 5B.
② With emission controls: 650 solenoid off; 700 solenoid on.
③ With emission controls: 0.
④ With distributor #1111964: 5B.
⑤ Solenoid on: 700.

Troubleshooting

The following section is designed to aid in the rapid diagnosis of engine problems. The systematic format is used to diagnose problems ranging from engine starting difficulties to the need for engine overhaul. It is assumed that the user is equipped with basic hand tools and test equipment (tachdwell meter, timing light, voltmeter, and ohmmeter).

Troubleshooting is divided into two sections. The first, *General Diagnosis*, is used to locate the problem area. In the second, *Specific Diagnosis*, the problem is systematically evaluated.

General Diagnosis

Problem: Symptom	Begin at Specific Diagnosis, Number _____
Engine Won't Start:	
Starter doesn't turn	1.1, 2.1
Starter turns, engine doesn't	2.1
Starter turns engine very slowly	1.1, 2.4
Starter turns engine normally	3.1, 4.1
Starter turns engine very quickly	6.1
Engine fires intermittently	4.1
Engine fires consistently	5.1, 6.1
Engine Runs Poorly:	
Hard starting	3.1, 4.1, 5.1, 8.1
Rough idle	4.1, 5.1, 8.1
Stalling	3.1, 4.1, 5.1, 8.1
Engine dies at high speeds	4.1, 5.1
Hesitation (on acceleration from standing stop)	5.1, 8.1
Poor pickup	4.1, 5.1, 8.1
Lack of power	3.1, 4.1, 5.1, 8.1
Backfire through the carburetor	4.1, 8.1, 9.1
Backfire through the exhaust	4.1, 8.1, 9.1
Blue exhaust gases	6.1, 7.1
Black exhaust gases	5.1
Running on (after the ignition is shut off)	3.1, 8.1
Susceptible to moisture	4.1
Engine misfires under load	4.1, 7.1, 8.4, 9.1
Engine misfires at speed	4.1, 8.4
Engine misfires at idle	3.1, 4.1, 5.1, 7.1, 8.4

Engine Noise Diagnosis

Problem: Symptom	Probable Cause
Engine Noises:①	
Metallic grind while starting	Starter drive not engaging completely
Constant grind or rumble	*Starter drive not releasing, worn main bearings
Constant knock	Worn connecting rod bearings
Knock under load	Fuel octane too low, worn connecting rod bearings
Double knock	Loose piston pin
Metallic tap	*Collapsed or sticky valve lifter, excessive valve clearance, excessive end play in a rotating shaft
Scrape	*Fan belt contacting a stationary surface
Tick while starting	S.U. electric fuel pump (normal), starter brushes
Constant tick	*Generator brushes, shreaded fan belt
Squeal	*Improperly tensioned fan belt
Hiss or roar	*Steam escaping through a leak in the cooling system or the radiator overflow vent
Whistle	*Vacuum leak
Wheeze	Loose or cracked spark plug

①—It is extremely difficult to evaluate vehicle noises. While the above are general definitions of engine noises, those starred (*) should be considered as possibly originating elsewhere in the car. To aid diagnosis, the following list considers other potential sources of these sounds.

Metallic grind:
Throwout bearing; transmission gears, bearings, or synchronizers; differential bearings, gears; something metallic in contact with brake drum or disc.

Metallic tap:
U-joints; fan-to-radiator (or shroud) contact.

Scrape:
Brake shoe or pad dragging; tire to body contact; suspension contacting undercarriage or exhaust; something non-metallic contacting brake shoe or drum.

Tick:
Transmission gears; differential gears; lack of radio suppression; resonant vibration of body panels; windshield wiper motor or transmission; heater motor and blower.

Squeal:
Brake shoe or pad not fully releasing; tires (excessive wear, uneven wear, improper inflation); front or rear wheel alignment (most commonly due to improper toe-in).

Hiss or whistle:
Wind leaks (body or window); heater motor and blower fan.

Roar:
Wheel bearings; wind leaks (body and window).

Index

Topic		Group
Battery	*	1
Cranking system	*	2
Primary electrical system	*	3
Secondary electrical system	*	4
Fuel system	*	5
Engine compression	*	6
Engine vaccuum	**	7
Secondary electrical system	**	8
Valve train	**	9
Exhaust system	**	10
Cooling system	**	11
Engine lubrication	**	12

   * The engine need not be running
   **The engine must be running

Sample Section

Test and Procedure	Results and Indications	Proceed to
4.1—Check for spark: Hold each spark plug wire approximately ¼″ from ground with gloves or a heavy, dry rag. Crank the engine and observe the spark.	If no spark is evident:	**4.2**
	If spark is good in some cases:	**4.3**
	If spark is good in all cases:	**4.6**

Specific Diagnosis

This section is arranged so that following each test, instructions are given to proceed to another, until a problem is diagnosed.

1.1—Inspect the battery visually for case condition (corrosion, cracks) and water level.	If case is cracked, replace battery:	**1.4**
	If the case is intact, remove corrosion with a solution of baking soda and water (**CAUTION:** *do not get the solution into the battery*), and fill with water:	**1.2**
1.2—Check the battery cable connections: Insert a screwdriver between the battery post and the cable clamp. Turn the headlights on high beam, and observe them as the screwdriver is gently twisted to ensure good metal to metal contact.	If the lights brighten, remove and clean the clamp and post; coat the post with petroleum jelly, install and tighten the clamp:	**1.4**
	If no improvement is noted:	**1.3**

Testing battery cable connections using a screwdriver

1.3—Test the state of charge of the battery using an individual cell tester or hydrometer.	If indicated, charge the battery. **NOTE:** *If no obvious reason exists for the low state of charge (i.e., battery age, prolonged storage), the charging system should be tested:*	**1.4**

Spec. Grav. Reading	Charged Condition
1.260–1.280	Fully Charged
1.230–1.250	Three Quarter Charged
1.200–1.220	One Half Charged
1.170–1.190	One Quarter Charged
1.140–1.160	Just About Flat
1.110–1.130	All The Way Down

State of battery charge

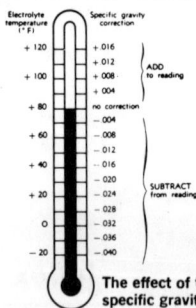

Electrolyte temperature (° F) Specific gravity correction

+ 120 / + 016
+ 100 / + 012 / + 008 ADD to reading
+ 80 / + 004 / no correction
+ 60 / – 004
+ 40 / – 008 / – 012 / – 016
+ 20 / – 020 / – 024 SUBTRACT from reading
0 / – 028 / – 032
– 20 / – 036 / – 040

The effect of temperature on the specific gravity of battery electrolyte

1.4—Visually inspect battery cables for cracking, bad connection to ground, or bad connection to starter.	If necessary, tighten connections or replace the cables:	**2.1**

Tests in Group 2 are performed with coil high tension lead disconnected to prevent accidental starting.

2.1—Test the starter motor and solenoid: Connect a jumper from the battery post of the solenoid (or relay) to the starter post of the solenoid (or relay).	If starter turns the engine normally:	**2.2**
	If the starter buzzes, or turns the engine very slowly:	**2.4**
	If no response, replace the solenoid (or relay).	**3.1**
	If the starter turns, but the engine doesn't, ensure that the flywheel ring gear is intact. If the gear is undamaged, replace the starter drive.	**3.1**

Test and Procedure	Results and Indications	Proceed to
2.2—Determine whether ignition override switches are functioning properly (clutch start switch, neutral safety switch), by connecting a jumper across the switch(es), and turning the ignition switch to "start".	If starter operates, adjust or replace switch:	**3.1**
	If the starter doesn't operate:	**2.3**
2.3—Check the ignition switch "start" position: Connect a 12V test lamp between the starter post of the solenoid (or relay) and ground. Turn the ignition switch to the "start" position, and jiggle the key.	If the lamp doesn't light when the switch is turned, check the ignition switch for loose connections, cracked insulation, or broken wires. Repair or replace as necessary:	**3.1**
	If the lamp flickers when the key is jiggled, replace the ignition switch.	**3.3**

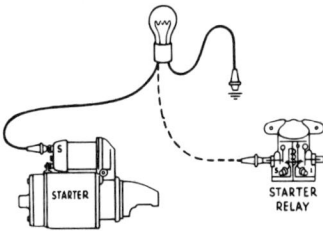

Checking the ignition switch "start" position

2.4—Remove and bench test the starter, according to specifications in the car section.	If the starter does not meet specifications, repair or replace as needed:	**3.1**
	If the starter is operating properly:	**2.5**
2.5—Determine whether the engine can turn freely: Remove the spark plugs, and check for water in the cylinders. Check for water on the dipstick, or oil in the radiator. Attempt to turn the engine using an 18″ flex drive and socket on the crankshaft pulley nut or bolt.	If the engine will turn freely only with the spark plugs out, and hydrostatic lock (water in the cylinders) is ruled out, check valve timing:	**9.2**
	If engine will not turn freely, and it is known that the clutch and transmission are free, the engine must be disassembled for further evaluation:	**Next Chapter**
3.1—Check the ignition switch "on" position: Connect a jumper wire between the distributor side of the coil and ground, and a 12V test lamp between the switch side of the coil and ground. Remove the high tension lead from the coil. Turn the ignition switch on, and jiggle the key.	If the lamp lights:	**3.2**
	If the lamp flickers when the key is jiggled, replace the ignition switch:	**3.3**
	If the lamp doesn't light, check for loose or open connections. If none are found, remove the ignition switch and check for continuity. If the switch is faulty, replace it:	**3.3**

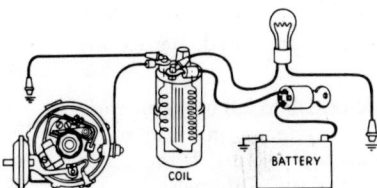

Checking the ignition switch "on" position

Test and Procedure	Results and Indications	Proceed to
3.2—Check the ballast resistor or resistance wire for an open circuit, using an ohmmeter.	On cars with point-type ignition systems, replace the resistor or resistance wire if the resistance is zero. On cars equipped with Solid-State Ignition, the resistance should be 1.35 ohms for 1975 and 1976 cars. The resistance should be 1.10 ohms for 1977 cars, except for California V8's, which are equipped with the Dura-Spark I system and have no ballast resistors. If resistance is zero, replace the resistor or resistance wiring.	3.3
3.3—On point-type ignition systems, visually inspect the breaker points for burning, pitting or excessive wear. Gray coloring of the point contact surfaces is normal. Rotate the crankshaft until the contact heel rests on a high point of the distributor cam and adjust the point gap to specifications. On electronic ignition models, remove the distributor cap and visually inspect the armature. Ensure that the armature pin is in place, and that the armature is on tight and rotates when the engine is cranked. Make sure there are no cracks, chips or rounded edges on the armature.	If the breaker points are intact, clean the contact surfaces with fine emery cloth, and adjust the point gap to specifications. If the points are worn, replace them. On electronic systems, replace any parts which appear defective. If condition persists:	3.4
3.4—On point-type ignition systems, connect a dwell-meter between the distributor primary lead and ground. Crank the engine and observe the point dwell angle. On electronic ignition systems, conduct a stator (magnetic pickup assembly) test. See "Troubleshooting the Solid-State Ignition System".	On point-type systems, adjust the dwell angle if necessary. **NOTE:** *Increasing the point gap decreases the dwell angle and vice-versa.* If the dwell meter shows little or no reading; On electronic ignition systems, if the stator is bad, replace the stator. If the stator is good, proceed to the other tests in the solid-state ignition troubleshooting section.	3.6 3.5
3.5—On point-type ignition systems, check the condenser for short: connect an ohmmeter across the condenser body and the pigtail lead. Checking the condenser for short OHMMETER	If any reading other than infinite is noted, replace the condenser:	3.6
3.6—Test the coil primary resistance: On point-type ignition systems, connect an ohmmeter across the coil primary terminals, and read the resistance on the low scale. Note whether an external ballast resistor or resistance wire is utilized. On electronic ignition systems, test the coil primary resistance. Connect an ohmmeter between the coil BAT terminal and socket #1 in the harness.	Coils utilizing ballast resistors or resistance wires should have approximately 1.0 ohms resistance. Coils with internal resistors should have approximately 4.0 ohms resistance. If values far from the above are noted, replace the coil. Resistance should be 1.0 to 2.0 ohms for early and Dura-Spark II systems, and 0.5 to 1.5 ohms for Dura-Spark I systems. If the coil is defective, replace the coil. Otherwise:	4.1 4.1

Test and Procedure	Results and Indications	Proceed to
4.1—Check for spark: Hold each spark plug wire approximately ¼″ from ground with gloves or a heavy, dry rag. Crank the engine, and observe the spark.	If no spark is evident:	**4.2**
	If spark is good in some cylinders:	**4.3**
	If spark is good in all cylinders:	**4.6**
4.2—Check for spark at the coil high tension lead: Remove the coil high tension lead from the distributor and position it approximately ¼″ from ground. Crank the engine and observe spark. **CAUTION:** *This test should not be performed on cars equipped with transistorized ignition.*	If the spark is good and consistent:	**4.3**
	If the spark is good but intermittent, test the primary electrical system starting at 3.3:	**3.3**
	If the spark is weak or non-existent, replace the coil high tension lead, clean and tighten all connections and retest. If no improvement is noted:	**4.4**
4.3—Visually inspect the distributor cap and rotor for burned or corroded contacts, cracks, carbon tracks, or moisture. Also check the fit of the rotor on the distributor shaft (where applicable).	If moisture is present, dry thoroughly, and re-test per 4.1:	**4.1**
	If burned or excessively corroded contacts, cracks, or carbon tracks are noted, replace the defective part(s) and retest per 4.1:	**4.1**
	If the rotor and cap appear intact, or are only slightly corroded, clean the contacts thoroughly (including the cap towers and spark plug wire ends) and retest per 4.1: If the spark is good in all cases:	**4.6**
	If the spark is poor in all cases:	**4.5**
4.4—Check the coil secondary resistance: On point-type systems, connect an ohmmeter across the distributor side of the coil and the coil tower. Read the resistance on the high scale of the ohmmeter. On electronic ignition systems, connect an ohmmeter between socket #4 and the coil tower (see the charts). Testing the coil secondary resistance	The resistance of a satisfactory coil should be between 4,000 and 10,000 ohms. If resistance is considerably higher (i.e. 40,000 ohms) replace the coil and retest per 4.1. **NOTE:** *this does not apply to high performance coils.* On electronic systems, resistance should be between 7,000 and 13,000 ohms. If not, replace the coil and retest per 4.1.	
4.5—Visually inspect the spark plug wires for cracking or brittleness. Ensure that no two wires are positioned so as to cause induction firing (adjacent and parallel). Remove each wire, one by one, and check resistance with an ohmmeter.	Replace any cracked or brittle wires. If any of the wires are defective, replace the entire set. Replace any wires with excessive resistance (over 8000Ω per foot for suppression wire), and separate any wires that might cause induction firing.	**4.6**
4.6—Remove the spark plugs, noting the cylinders from which they were removed, and evaluate according to the chart below.	See following.	**See following.**

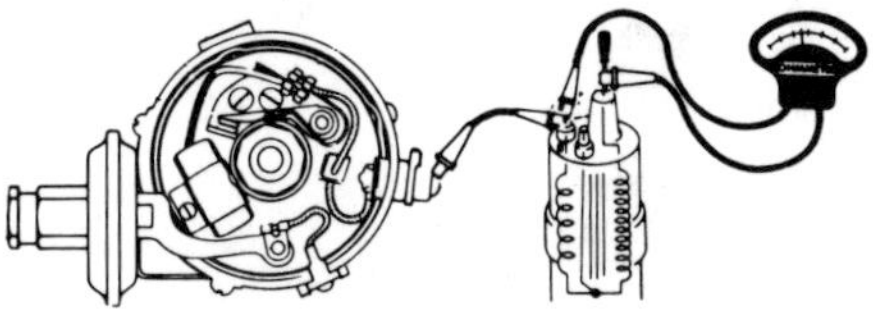

	Condition	Cause	Remedy	Proceed to
	Electrodes eroded, light brown deposits.	Normal wear. Normal wear is indicated by approximately .001″ wear per 1000 miles.	Clean and regap the spark plug if wear is not excessive: Replace the spark plug if excessively worn:	4.7
	Carbon fouling (black, dry, fluffy deposits).	If present on one or two plugs:		
		Faulty high tension lead(s).	Test the high tension leads:	4.5
		Burnt or sticking valve(s).	Check the valve train: (Clean and regap the plugs in either case.)	9.1
		If present on most or all plugs: Overly rich fuel mixture, due to restricted air filter, improper carburetor adjustment, improper choke or heat riser adjustment or operation.	Check the fuel system:	5.1
	Oil fouling (wet black deposits)	Worn engine components. **NOTE:** *Oil fouling may occur in new or recently rebuilt engines until broken in.*	Check engine vacuum and compression: Replace with new spark plug	6.1
	Lead fouling (gray, black, tan, or yellow deposits, which appear glazed or cinder-like).	Combustion by-products.	Clean and regap the plugs: (Use plugs of a different heat range if the problem recurs.)	4.7
	Gap bridging (deposits lodged between the electrodes).	Incomplete combustion, or transfer of deposits from the combustion chamber.	Replace the spark plugs:	4.7
	Overheating (burnt electrodes, and extremely white insulator with small black spots).	Ignition timing advanced too far.	Adjust timing to specifications:	8.2
		Overly lean fuel mixture.	Check the fuel system:	5.1
		Spark plugs not seated properly.	Clean spark plug seat and install a new gasket washer: (Replace the spark plugs in all cases.)	4.7

	Condition	Cause	Remedy	Proceed to
	Fused spot deposits on the insulator.	Combustion chamber blow-by.	Clean and regap the spark plugs:	**4.7**
	Pre-ignition (melted or severely burned electrodes, blistered or cracked insulators, or metallic deposits on the insulator).	Incorrect spark plug heat range.	Replace with plugs of the proper heat range:	**4.7**
		Ignition timing advanced too far.	Adjust timing to specifications:	**8.2**
		Spark plugs not being cooled efficiently.	Clean the spark plug seat, and check the cooling system:	**11.1**
		Fuel mixture too lean.	Check the fuel system:	**5.1**
		Poor compression.	Check compression:	**6.1**
		Fuel grade too low.	Use higher octane fuel:	**4.7**

Test and Procedure	Results and Indications	Proceed to
4.7—Determine the static ignition timing. Using the crankshaft pulley timing marks as a guide, locate top dead center on the compression stroke of the number one cylinder.	The rotor should be pointing toward the no. 1 tower in the distributor cap, and the armature spoke for that cylinder should be lined up with the stator.	**4.8**
4.8—Check coil polarity: Connect a voltmeter negative lead to the coil high tension lead, and the positive lead to ground (**NOTE:** *reverse the hook-up for positive ground cars*). Crank the engine momentarily. Checking coil polarity	If the voltmeter reads up-scale, the polarity is correct:	**5.1**
	If the voltmeter reads down-scale, reverse the coil polarity (switch the primary leads):	**5.1**
5.1—Determine that the air filter is functioning efficiently: Hold paper elements up to a strong light, and attempt to see light through the filter.	Clean permanent air filters in gasoline (or manufacturer's recommendation), and allow to dry. Replace paper elements through which light cannot be seen:	**5.2**
5.2—Determine whether a flooding condition exists: Flooding is identified by a strong gasoline odor, and excessive gasoline present in the throttle bore(s) of the carburetor.	If flooding is not evident:	**5.3**
	If flooding is evident, permit the gasoline to dry for a few moments and restart.	
	If flooding doesn't recur:	**5.6**
	If flooding is persistent:	**5.5**
5.3—Check that fuel is reaching the carburetor: Detach the fuel line at the carburetor inlet. Hold the end of the line in a cup (not styrofoam), and crank the engine.	If fuel flows smoothly:	**5.6**
	If fuel doesn't flow (**NOTE:** *Make sure that there is fuel in the tank*), or flows erratically:	**5.4**

Test and Procedure	Results and Indications	Proceed to
5.4—Test the fuel pump: Disconnect all fuel lines from the fuel pump. Hold a finger over the input fitting, crank the engine (with electric pump, turn the ignition or pump on); and feel for suction.	If suction is evident, blow out the fuel line to the tank with low pressure compressed air until bubbling is heard from the fuel filler neck. Also blow out the carburetor fuel line (both ends disconnected):	**5.6**
	If no suction is evident, replace or repair the fuel pump: NOTE: *Repeated oil fouling of the spark plugs, or a no-start condition, could be the result of a ruptured vacuum booster pump diaphragm, through which oil or gasoline is being drawn into the intake manifold (where applicable).*	**5.6**
5.5—Check the needle and seat: Tap the carburetor in the area of the needle and seat.	If flooding stops, a gasoline additive (e.g., Gumout) will often cure the problem:	**5.6**
	If flooding continues, check the fuel pump for excessive pressure at the carburetor (according to specifications). If the pressure is normal, the needle and seat must be removed and checked, and/or the float level adjusted:	**5.6**
5.6—Test the accelerator pump by looking into the throttle bores while operating the throttle.	If the accelerator pump appears to be operating normally:	**5.7**
	If the accelerator pump is not operating, the pump must be reconditioned. Where possible, service the pump with the carburetor(s) installed on the engine. If necessary, remove the carburetor. Prior to removal:	**5.7**
5.7—Determine whether the carburetor main fuel system is functioning: Spray a commercial starting fluid into the carburetor while attempting to start the engine.	If the engine starts, runs for a few seconds, and dies:	**5.8**
	If the engine doesn't start:	**6.1**
5.8—Uncommon fuel system malfunctions: See below:	If the problem is solved:	**6.1**
	If the problem remains, remove and recondition the carburetor.	

Condition	Indication	Test	Usual Weather Conditions	Remedy
Vapor lock	Car will not restart shortly after running.	Cool the components of the fuel system until the engine starts.	Hot to very hot	Ensure that the exhaust manifold heat control valve is operating. Check with the vehicle manufacturer for the recommended solution to vapor lock on the model in question.
Carburetor icing	Car will not idle, stalls at low speeds.	Visually inspect the throttle plate area of the throttle bores for frost.	High humidity, 32–40°F.	Ensure that the exhaust manifold heat control valve is operating, and that the intake manifold heat riser is not blocked.

Condition	Indication	Test	Usual Weather Conditions	Remedy
Water in the fuel	Engine sputters and stalls; may not start.	Pump a small amount of fuel into a glass jar. Allow to stand, and inspect for droplets or a layer of water.	High humidity, extreme temperature changes.	For droplets, use one or two cans of commercial gas dryer (Dry Gas) For a layer of water, the tank must be drained, and the fuel lines blown out with compressed air.

Test and Procedure	Results and Indications	Proceed to
6.1—Test engine compression: Remove all spark plugs. Insert a compression gauge into a spark plug port, crank the engine to obtain the maximum reading, and record.	If compression is within limits on all cylinders:	**7.1**
	If gauge reading is extremely low on all cylinders:	**6.2**
	If gauge reading is low on one or two cylinders: (If gauge readings are identical and low on two or more adjacent cylinders, the head gasket must be replaced.)	**6.2**

Testing compression
(© Chevrolet Div. G.M. Corp.)

Compression pressure limits
(© Buick Div. G.M. Corp.)

Maxi. Press. Lbs. Sq. In.	Min. Press. Lbs. Sq. In.	Maxi. Press. Lbs. Sq. In.	Min. Press. Lbs. Sq. In.	Max. Press. Lbs. Sq. In.	Min. Press. Lbs. Sq. In.	Max. Press. Lbs. Sq. In.	Min. Press. Lbs. Sq. In.
134	101	162	121	188	141	214	160
136	102	164	123	190	142	216	162
138	104	166	124	192	144	218	163
140	105	168	126	194	145	220	165
142	107	170	127	196	147	222	166
146	110	172	129	198	148	224	168
148	111	174	131	200	150	226	169
150	113	176	132	202	151	228	171
152	114	178	133	204	153	230	172
154	115	180	135	206	154	232	174
156	117	182	136	208	156	234	175
158	118	184	138	210	157	236	177
160	120	186	140	212	158	238	178

Test and Procedure	Results and Indications	Proceed to
6.2—Test engine compression (wet): Squirt approximately 30 cc. of engine oil into each cylinder, and retest per 6.1.	If the readings improve, worn or cracked rings or broken pistons are indicated:	**Next Chapter**
	If the readings do not improve, burned or excessively carboned valves or a jumped timing chain are indicated: **NOTE: A jumped timing chain is often indicated by difficult cranking.**	**7.1**
7.1—Perform a vacuum check of the engine: Attach a vacuum gauge to the intake manifold beyond the throttle plate. Start the engine, and observe the action of the needle over the range of engine speeds.	See below.	**See below**

	Reading	Indications	Proceed to
	Steady, from 17–22 in. Hg.	Normal:	**8.1**

	Reading	Indications	Proceed to
	Low and steady.	Late ignition or valve timing, or low compression:	6.1
	Very low.	Vacuum leak:	7.2
	Needle fluctuates as engine speed increases.	Ignition miss, blown cylinder head gasket, leaking valve or weak valve spring:	6.1, 8.3
	Gradual drop in reading at idle.	Excessive back pressure in the exhaust system:	10.1
	Intermittent fluctuation at idle.	Ignition miss, sticking valve:	8.3, 9.1
	Drifting needle.	Improper idle mixture adjustment, carburetors not synchronized (where applicable), or minor intake leak. Synchronize the carburetors, adjust the idle, and retest. If the condition persists:	7.2
	High and steady.	Early ignition timing:	8.2

Test and Procedure	Results and Indications	Proceed to
7.2—Attach a vacuum gauge per 7.1, and test for an intake manifold leak. Squirt a small amount of oil around the intake manifold gaskets, carburetor gaskets, plugs and fittings. Observe the action of the vacuum gauge.	If the reading improves, replace the indicated gasket, or seal the indicated fitting or plug:	8.1
	If the reading remains low:	7.3
7.3—Test all vacuum hoses and accessories for leaks as described in 7.2. Also check the carburetor body (dashpots, automatic choke mechanism, throttle shafts) for leaks in the same manner.	If the reading improves, service or replace the offending part(s):	8.1
	If the reading remains low:	6.1
8.1—Remove the distributor cap and check to make sure that the armature turns when the engine is cranked. Visually inspect the distributor components.	Clean, tighten or replace any components which appear defective.	8.2

Test and Procedure	Results and Indications	Proceed to
8.2—Connect a timing light (per manufacturer's recommendation) and check the dynamic ignition timing. Disconnect and plug the vacuum hose(s) to the distributor if specified, start the engine, and observe the timing marks at the specified engine speed.	If the timing is not correct, adjust to specifications by rotating the distributor in the engine: (Advance timing by rotating distributor opposite normal direction of rotor rotation, retard timing by rotating distributor in same direction as rotor rotation.)	**8.3**
8.3—Check the operation of the distributor advance mechanism(s): To test the mechanical advance, disconnect all but the mechanical advance, and observe the timing marks with a timing light as the engine speed is increased from idle. If the mark moves smoothly, without hesitation, it may be assumed that the mechanical advance is functioning properly. To test vacuum advance and/or retard systems, alternately crimp and release the vacuum line, and observe the timing mark for movement. If movement is noted, the system is operating.	If the systems are functioning: If the systems are not functioning, remove the distributor, and test on a distributor tester:	**8.4** **8.4**
8.4—Locate an ignition miss: With the engine running, remove each spark plug wire, one by one, until one is found that doesn't cause the engine to roughen and slow down.	When the missing cylinder is identified:	**4.1**
9.1—Evaluate the valve train: Remove the valve cover, and ensure that the valves are adjusted to specifications. A mechanic's stethoscope may be used to aid in the diagnosis of the valve train. By pushing the probe on or near push rods or rockers, valve noise often can be isolated. A timing light also may be used to diagnose valve problems. Connect the light according to manufacturer's recommendations, and start the engine. Vary the firing moment of the light by increasing the engine speed (and therefore the ignition advance), and moving the trigger from cylinder to cylinder. Observe the movement of each valve.	See below.	**See below**

Observation	Probable Cause	Remedy	Proceed to
Metallic tap heard through the stethoscope.	Sticking hydraulic lifter or excessive valve clearance.	Adjust valve. If tap persists, remove and replace the lifter:	**10.1**

Observation	Probable Cause	Remedy	Proceed to
Metallic tap through the stethoscope, able to push the rocker arm (lifter side) down by hand.	Collapsed valve lifter.	Remove and replace the lifter:	**10.1**
Erratic, irregular motion of the valve stem.*	Sticking valve, burned valve.	Recondition the valve and/or valve guide:	**Next Chapter**
Eccentric motion of the pushrod at the rocker arm.*	Bent pushrod.	Replace the pushrod:	**10.1**
Valve retainer bounces as the valve closes.*	Weak valve spring or damper.	Remove and test the spring and damper. Replace if necessary:	**10.1**

*—When observed with a timing light.

Test and Procedure	Results and Indications	Proceed to
9.2—Check the valve timing: Locate top dead center of the No. 1 piston, and install a degree wheel or tape on the crankshaft pulley or damper with zero corresponding to an index mark on the engine. Rotate the crankshaft in its direction of rotation, and observe the opening of the No. 1 cylinder intake valve. The opening should correspond with the correct mark on the degree wheel according to specifications.	If the timing is not correct, the timing cover must be removed for further investigation:	
10.1—Determine whether the exhaust manifold heat control valve is operating: Operate the valve by hand to determine whether it is free to move. If the valve is free, run the engine to operating temperature and observe the action of the valve, to ensure that it is opening.	If the valve sticks, spray it with a suitable solvent, open and close the valve to free it, and retest. If the valve functions properly:	**10.2**
	If the valve does not free, or does not operate, replace the valve:	**10.2**
10.2—Ensure that there are no exhaust restrictions: Visually inspect the exhaust system for kinks, dents, or crushing. Also note that gasses are flowing freely from the tailpipe at all engine speeds, indicating no restriction in the muffler or resonator.	Replace any damaged portion of the system:	**11.1**

Test and Procedure	Results and Indications	Proceed to
11.1—Visually inspect the fan belt for glazing, cracks, and fraying, and replace if necessary. Tighten the belt so that the longest span has approximately ½″ play at its midpoint under thumb pressure. Checking the fan belt tension	Replace or tighten the fan belt as necessary:	**11.2**
11.2—Check the fluid level of the cooling system.	If full or slightly low, fill as necessary: If extremely low:	**11.5** **11.3**
11.3—Visually inspect the external portions of the cooling system (radiator, radiator hoses, thermostat elbow, water pump seals, heater hoses, etc.) for leaks. If none are found, pressurize the cooling system to 14–15 psi.	If cooling system holds the pressure: If cooling system loses pressure rapidly, reinspect external parts of the system for leaks under pressure. If none are found, check dipstick for coolant in crankcase. If no coolant is present, but pressure loss continues: If coolant is evident in crankcase, remove cylinder head(s), and check gasket(s). If gaskets are intact, block and cylinder head(s) should be checked for cracks or holes. If the gasket(s) is blown, replace, and purge the crankcase of coolant: **NOTE:** *Occasionally, due to atmospheric and driving conditions, condensation of water can occur in the crankcase. This causes the oil to appear milky white. To remedy, run the engine until hot, and change the oil and oil - filter.*	**11.5** **11.4** **12.6**
11.4— Check for combustion leaks into the cooling system: Pressurize the cooling system as above. Start the engine, and observe the pressure gauge. If the needle fluctuates, remove each spark plug wire, one by one, noting which cylinder(s) reduce or eliminate the fluctuation. Radiator pressure tester	Cylinders which reduce or eliminate the fluctuation, when the spark plug wire is removed, are leaking into the cooling system. Replace the head gasket on the affected cylinder bank(s).	

Test and Procedure	Results and Indications	Proceed to
11.5—Check the radiator pressure cap: Attach a radiator pressure tester to the radiator cap (wet the seal prior to installation). Quickly pump up the pressure, noting the point at which the cap releases.	If the cap releases within ± 1 psi of the specified rating, it is operating properly:	**11.6**
	If the cap releases at more than ± 1 psi of the specified rating, it should be replaced:	**11.6**

Testing the radiator pressure cap

Test and Procedure	Results and Indications	Proceed to
11.6—Test the thermostat: Start the engine cold, remove the radiator cap, and insert a thermometer into the radiator. Allow the engine to idle. After a short while, there will be a sudden, rapid increase in coolant temperature. The temperature at which this sharp rise stops is the thermostat opening temperature.	If the thermostat opens at or about the specified temperature:	**11.7**
	If the temperature doesn't increase: (If the temperature increases slowly and gradually, replace the thermostat.)	**11.7**
11.7—Check the water pump: Remove the thermostat elbow and the thermostat, disconnect the coil high tension lead (to prevent starting), and crank the engine momentarily.	If coolant flows, replace the thermostat and re-test per 11.6:	**11.6**
	If coolant doesn't flow, reverse flush the cooling system to alleviate any blockage that might exist. If system is not blocked, and coolant will not flow, recondition the water pump.	—
12.1—Check the oil pressure gauge or warning light: If the gauge shows low pressure, or the light is on, for no obvious reason, remove the oil pressure sender. Install an accurate oil pressure gauge and run the engine momentarily.	If oil pressure builds normally, run engine for a few moments to determine that it is functioning normally, and replace the sender.	—
	If the pressure remains low:	**12.2**
	If the pressure surges:	**12.3**
	If the oil pressure is zero:	**12.3**
12.2—Visually inspect the oil: If the oil is watery or very thin, milky, or foamy, replace the oil and oil filter.	If the oil is normal:	**12.3**
	If after replacing oil the pressure remains low:	**12.3**
	If after replacing oil the pressure becomes normal:	—
12.3—Inspect the oil pressure relief valve and spring, to ensure that it is not sticking or stuck. Remove and thoroughly clean the valve, spring, and the valve body.	If the oil pressure improves:	—
	If no improvement is noted:	**12.4**

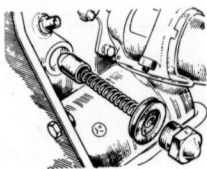

Oil pressure relief valve
(© British Leyland Motors)

Test and Procedure	Results and Indications	Proceed to
12.4—Check to ensure that the oil pump is not cavitating (sucking air instead of oil): See that the crankcase is neither over nor underfull, and that the pickup in the sump is in the proper position and free from sludge.	Fill or drain the crankcase to the proper capacity, and clean the pickup screen in solvent if necessary. If no improvement is noted:	**12.5**
12.5—Inspect the oil pump drive and the oil pump:	If the pump drive or the oil pump appear to be defective, service as necessary and retest per 12.1:	**12.1**
	If the pump drive and pump appear to be operating normally, the engine should be disassembled to determine where blockage exists:	**Next Chapter**
12.6—Purge the engine of ethylene glycol coolant: Completely drain the crankcase and the oil filter. Obtain a commercial butyl cellosolve base solvent, designated for this purpose, and follow the instructions precisely. Following this, install a new oil filter and refill the crankcase with the proper weight oil. The next oil and filter change should follow shortly thereafter (1000 miles).		

3

Engine and Engine Rebuilding

ENGINE ELECTRICAL

Distributor

REMOVAL

To remove the distributor assembly, follow the procedure below.

1. Remove the high-tension wires from the distributor cap terminal towers, noting their positions to assure correct reassembly. For diagrams of firing orders and distributor wiring, refer to the tune-up and troubleshooting section.

2. Remove the primary lead from the terminal post at the side of the distributor.

3. Disconnect the vacuum line if there is one.

4. Remove the two distributor cap retaining hooks or screws and remove the distributor cap.

5. Note the position of the rotor in relation to the base. Scribe a mark on the base of the distributor and on the engine block to facilitate reinstallation. Align the marks with the direction the metal tip of the rotor is pointing.

6. Remove the bolt that holds the distributor to the engine.

7. Lift the distributor assembly from the engine.

INSTALLATION

1. Insert the distributor shaft and assembly into the engine. Line up the mark on the distributor and the one on the engine with the metal tip of the rotor. Make sure that the vacuum advance diaphragm is pointed in the same direction as it was pointed originally. This will be done automatically if the marks on the engine and the distributor are lined up with the rotor.

NOTE: *On the F-Head, the distributor shaft fits into a slot in the end of the oil pump shaft.*

Marking the rotor-distributor base relationship

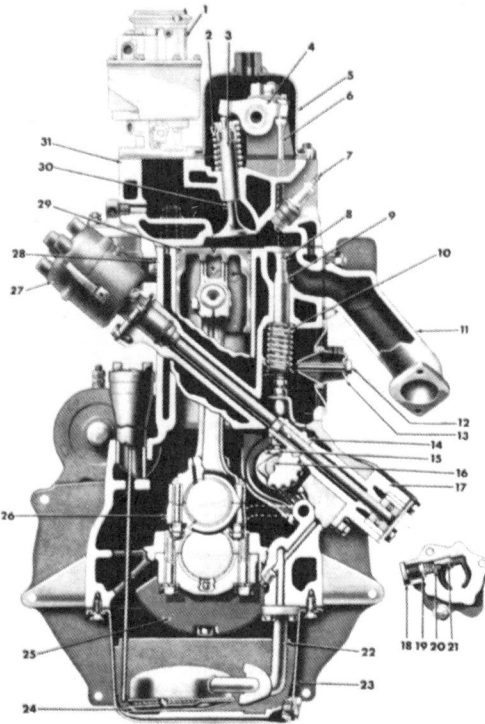

1. Carburetor
2. Intake valve spring
3. Intake valve stem guide
4. Rocker arm
5. Rocker arm cover
6. Intake valve pushrod
7. Spark plug
8. Exhaust valve
9. Exhaust valve stem guide
10. Exhaust valve spring
11. Exhaust manifold
12. Valve spring cover screw
13. Crankcase ventilator
14. Oil pump drive gear
15. Exhaust valve tappet
16. Camshaft
17. Oil pump
18. Oil pump relief spring retainer
19. Oil pump relief plunger gasket
20. Oil plunger relief spring
21. Oil pump relief plunger
22. Oil float support
23. Oil pan
24. Oil float
25. Crankshaft
26. Connecting rod
27. Distributor
28. Piston and pin
29. Cylinder block
30. Intake valve
31. Cylinder head

Frontal cross-section of the 4-134

2. Install the distributor hold-down bolt and clamp. Leave the screw loose enough so that you can move the distributor with heavy hand pressure.

3. Connect the primary wire to the distributor side of the coil. Install the distributor cap on the distributor housing. Secure the distributor cap with the spring clips or the screw type retainers, whichever is used.

4. Install the spark plug wires. Make sure that the wires are pressed all of the way into the top of the distributor cap and firmly onto the spark plugs.

NOTE: *Design of the V6 engine requires a special form of distributor cam. The distributor may be serviced in the regular way and should cause no more problems than any other distributor, if the firing plan is thoroughly understood. The distributor cam is not ground to standard six cylinder indexing intervals. This particular form requires that the original pattern of spark plug wiring be used. The engine will not run in balance if number one spark plug wire is inserted into number six distributor cap tower, even though each wire in the firing sequence is advanced to the next distributor tower. There is a difference between the firing intervals of each succeeding cylinder through the 720° engine cycle.*

5. Adjust the point cam dwell and set the ignition timing. Refer to the tune-up section.

If the engine has been turned while the distributor has been removed, or if the marks were not drawn, it will be necessary to initially time the engine. Follow the procedure below.

INSTALLATION, ENGINE DISTURBED

1. It is necessary to place the no. 1 cylinder in the firing position to correctly install the distributor. To locate this position, some engines have marks placed on the flywheel while other engines have marks placed on the timing gear covers and crankshaft pulleys. The flywheel marks may be viewed through a covered opening directly in back of the starting motor by loosening the hole cover and sliding it to one side.

2. Remove the no. 1 cylinder spark plug. Turn the engine until the piston in no. 1 cylinder is moving up on the compression stroke. This can be determined by placing your thumb over the spark plug hole and feeling the air being forced out of the cylinder. Stop turning F-head engines when either the 5° mark on the flywheel is in the middle of the flywheel inspection opening, or the marks on the crankshaft pulley and the timing gear cover are in alignment.

3. Install the distributor so that the rotor,

which is mounted on the shaft, points toward the no. 1 spark plug terminal tower position when the cap is installed. Of course you won't be able to see the direction in which the rotor is pointing if the cap is on the distributor. Lay the cap on the top of the distributor and make a mark on the side of the distributor housing just below the no. 1 spark plug terminal. Make sure the rotor points toward that mark when you install the distributor.

4. When the distributor shaft has reached the bottom of the hole, move the rotor back and forth slightly until the driving lug on the end of the shaft enters the slot, which is cut in the end of the oil pump gear on the F-Head, or when the drive gears of the distributor and cam mesh on the other engines, and until the distributor assembly slides down into place.

On models that have a gear on the end of the distributor shaft and a gear on the end of the oil pump drive, these gears have to mesh with the same teeth as originally installed when the distributor is inserted into the engine. Once again, the marks that were placed on the engine and the base of the distributor housing come into play. If the distributor shaft gear and the oil pump drive gear are but one tooth off from what they are supposed to be, the engine will not run correctly.

5. When the distributor is correctly installed, the breaker points should be in such a position that they are just ready to break contact with each other. This is accomplished by rotating the distributor body after it has been installed in the engine. Once again, line up the marks that you made before the distributor was removed from the engine.

6. Install the distributor hold-down screw and the hold-down bracket. Be sure that the models that have vacuum advance units are free to turn in the mounting socket. Note that the vacuum advance control of some distributors is connected directly to the plate on which the points are mounted. When this is the case, the plate must be free to turn rather than the distributor body.

7. Install the spark plug into the no. 1 spark plug hole and continue from step 3 of the distributor installation procedure.

Alternator

All J-series Wagoneers and Commandos made prior to 1971, use a 35-amp, 12-volt, negative ground alternator and a transis-

1. Auxiliary terminal
2. Output terminal
3. Auxiliary terminal
4. Field terminal
5. Ground terminal
6. Ground terminal

Early Motorola alternator

torized voltage regulator as standard equipment. Vehicles made in 1971 and after, have 37-amp alternators as standard.

An alternator differs from a conventional DC shunt generator in that the armature is stationary, and is called the stator, while the field rotates and is called the rotor. The higher current values in the alternator's stator are conducted to the external circuit through fixed leads and connections, rather than through a rotating commutator and brushes as in a DC generator. This eliminates a major point of maintenance.

The alternator employs a three-phase stator winding. The rotor consists of a field coil encased between six-poled, interleaved sections, producing a twelve-pole magnetic field with alternating north and south poles. By rotating the rotor inside the stator, an al-

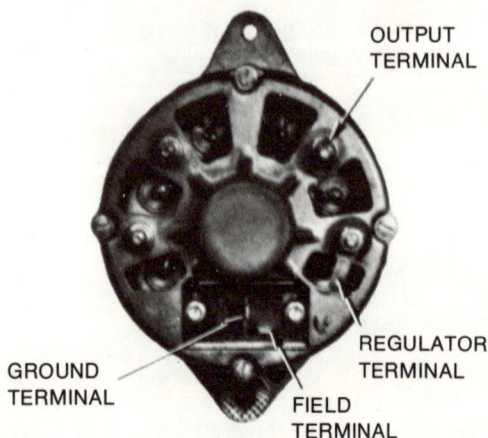

OUTPUT TERMINAL
REGULATOR TERMINAL
FIELD TERMINAL
GROUND TERMINAL

Typical alternator used after 1971

ternating current is induced in the stator windings. This alternating current is changed to direct current by diodes and is routed out of the alternator through the output terminal. Diode rectifiers act as one-way electrical valves. Half of the diodes have a negative polarity and are grounded. The other half of the diodes have a positive polarity and are connected to the output terminal.

Since the diodes have a high resistance to the flow of current in one direction, and a low resistance in the opposite direction, they are connected in a manner which allows current to flow from the alternator to the battery in the low-resistance direction.

The high resistance in the other direction prevents the flow of current from the battery to the alternator. Because of this feature, there is no need for a circuit breaker between the alternator and the battery.

Residual magnetism in the rotor field poles is minimal. The starting field current must, therefore, be supplied by the battery. It is connected to the field winding through the ignition switch and the charge indicator lamp of the ammeter in the dash.

As in the DC shunt generator, the alternator voltage is regulated by varying the field current. This is accomplished electronically in the transistorized voltage regulator. No current regulator is required because all alternators have self-limiting current characteristics.

An alternator is better than a conventional, DC shunt generator because it is lighter and more compact, because it is designed to supply the battery and accessory circuits through a wide range of engine speeds, and because it eliminates the necessary maintenance of replacing brushes and servicing commutators.

The transistorized voltage regulator is an electronic switching device. It senses the voltage at the auxiliary terminal of the alternator and supplies the necessary field current for maintaining the system voltage at the output terminal. The output current is determined by the battery electrical load—such as operating the headlights or heater blower.

The transistorized voltage regulator is a sealed unit that has no adjustments and must be replaced as a complete unit when it ceases to operate.

ALTERNATOR PRECAUTIONS

To prevent damage to the alternator and regulator, the following precautionary measures must be taken when working with the electrical system.

1. Never reverse battery connections. Always check the battery polarity visually. This is to be done before any connections are made to be sure that all of the connections correspond to the battery ground polarity of the Jeep.

2. Booster batteries for starting must be connected properly. Make sure that the positive cable of the booster battery is connected to the positive terminal of the battery that is getting the boost. This applies to both negative and ground cables.

3. Disconnect the battery cables before using a fast charger; the charger has a tendency to force current through the diodes in the opposite direction for which they were designed. This burns out the diodes.

4. Never use a fast charger as a booster for starting the vehicle.

5. Never disconnect the voltage regulator while the engine is running.

6. Do not ground the alternator output terminal.

7. Do not operate the alternator on an open circuit with the field energized.

8. Do not attempt to polarize an alternator.

REMOVAL AND INSTALLATION

1. Remove all of the electrical connections from the alternator. Label all of the wires so that you can install them correctly.

2. Remove all of the attaching nuts, bolts and washers noting different sized threads or nuts and bolts that go in certain holes.

3. Remove the alternator carefully. It is expensive to replace.

4. To install, reverse the above procedure.

BELT TENSION ADJUSTMENT

The fan belt drives the alternator and the water pump. If it is too loose, it will slip and the generator/alternator will not be able to produce the rated current. If the belt is too loose, the water pump would not operate efficiently and the engine could overheat. Check the tension of the fan belt by pushing your thumb down on the longest span of belt midway between the pulleys. If the belt flexes more than ½ in., adjust it. Loosen the bolt on the adjusting bracket and move the alternator or generator away from the engine to tighten the belt. Do not apply pressure to

PRY BAR   ADJUSTING HOLE IN BRACKET

Adjusting the drive belt tension

the rear of the cast aluminum housing of an alternator; it might break. Tighten the adjusting bolt when the proper tension is reached.

Regulator

The voltage regulators that are used with alternators are transistorized and cannot be serviced. If the voltage regulator is not operating properly, it must be replaced.

Starter Motor

The starter on the F-Head engine can be removed from the top of the engine. The starter motor on all other engines must be removed from beneath the vehicle.

REMOVAL AND INSTALLATION

1. Disconnect the positive battery cable from the battery. Disconnect the battery and solenoid leads from the starter. Mark them so you can replace them in the correct position when the starter is to be installed.

2. Remove all of the attachment bolts that attach the starter to the bellhousing.

3. Lift the starter from the engine.

4. Install the starter in the reverse order of the above.

NOTE: *It is necessary to remove the right side exhaust manifold from the engine (out the top) on the 327 V8 in order to remove the starter.*

STARTER DRIVE REPLACEMENT

Autolite

1. Remove the cover of the starter drive's actuating lever arm. Remove the thru-bolts, starter drive gear housing, and the return spring of the drive gear's actuating lever.

2. Remove the pivot pin which retains the starter gear actuating lever and remove the lever and armature.

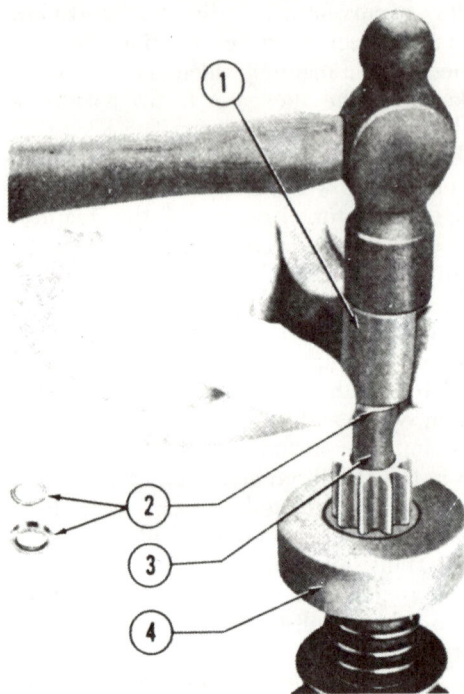

1. ½ in. pipe coupling   3. Armature shaft
2. Snap-ring and retainer 4. Drive assembly

Removing the starter drive

3. Remove the stop-ring retainer. Remove and discard the stop-ring which holds the drive gear to the armature shaft and then remove the drive gear assembly.

To install the unit:

1. Lightly Lubriplate® the armature shaft splines and install the starter drive gear assembly on the shaft. Install a new stop-ring and stop-ring retainer.

2. Position the starter drive gear actuating lever to the frame and starter drive assembly. Install the pivot pin.

3. Fill the starter drive gear housing one-quarter full of grease.

4. Position the drive actuating lever return spring and the drive gear housing to the frame, then install and tighten the thru-bolts. Be sure that the stop-ring retainer is properly seated in the drive housing.

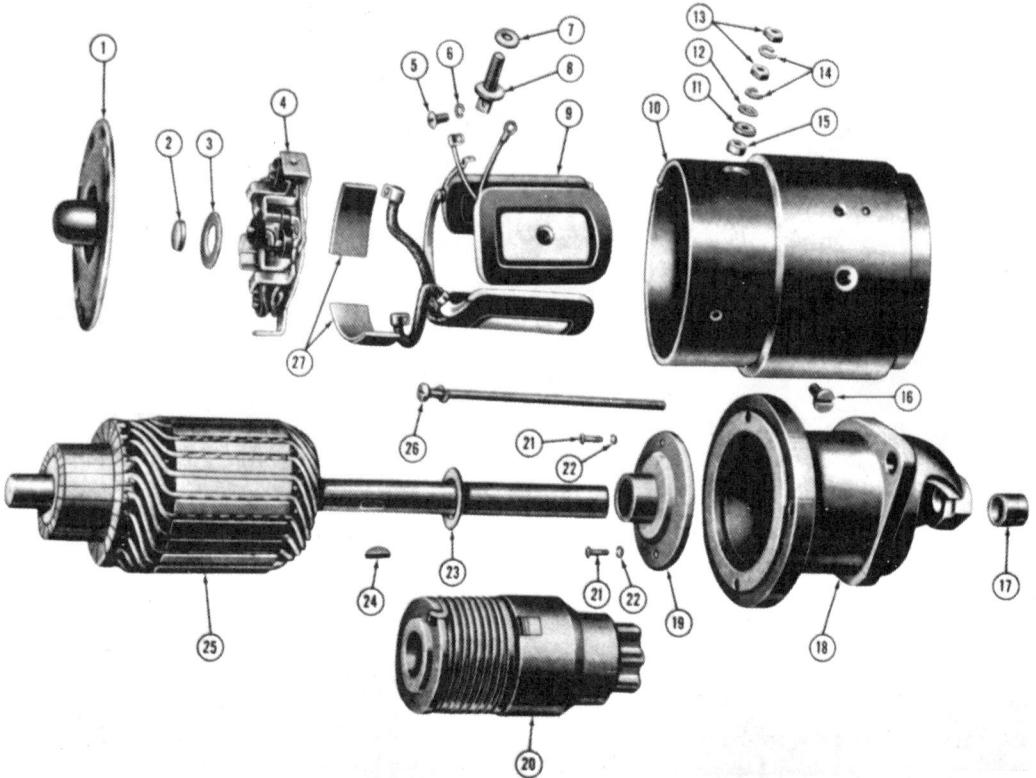

1. End plate	10. Frame	19. Intermediate bearing
2. Plug	11. Insulating washer	20. Bendix drive
3. Thrust washer	12. Washer	21. Screw
4. Brush plate assembly	13. Nut	22. Lockwasher
5. Screw	14. Lockwasher	23. Thrust washer
6. Lockwasher	15. Insulating bushing	24. Key
7. Insulating washer	16. Pole shoe screw	25. Armature
8. Terminal	17. Sleeve bearing	26. Thru-bolt
9. Field coil and pole shoe set	18. Drive end frame	27. Insulator

4-134 starter motor

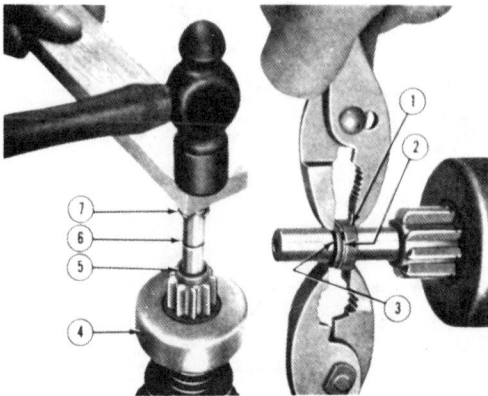

1. Retainer	5. Retainer
2. Snap-ring	6. Groove in the armature
3. Thrust collar	shaft
4. Drive assembly	7. Snap-ring

Installing the pinion stop retainer and thrust collar

Delco-Remy

1. Remove the three-bolts.

2. Remove the starter drive housing.

3. Slide the two-piece thrust collar off the end of the armature shaft.

4. Slide a standard ½ in. pipe coupling, or other spacer, onto the shaft so the end of the coupling butts against the edge of the retainer.

5. Tap the end of the coupling with a hammer, driving the retainer toward the armature end of the snap-ring.

6. Remove the snap-ring from its groove in the shaft with pliers. Slide the retainer and the starter drive from the armature.

To install the unit:

1. Lubricate the drive end of the shaft with silicone lubricant.

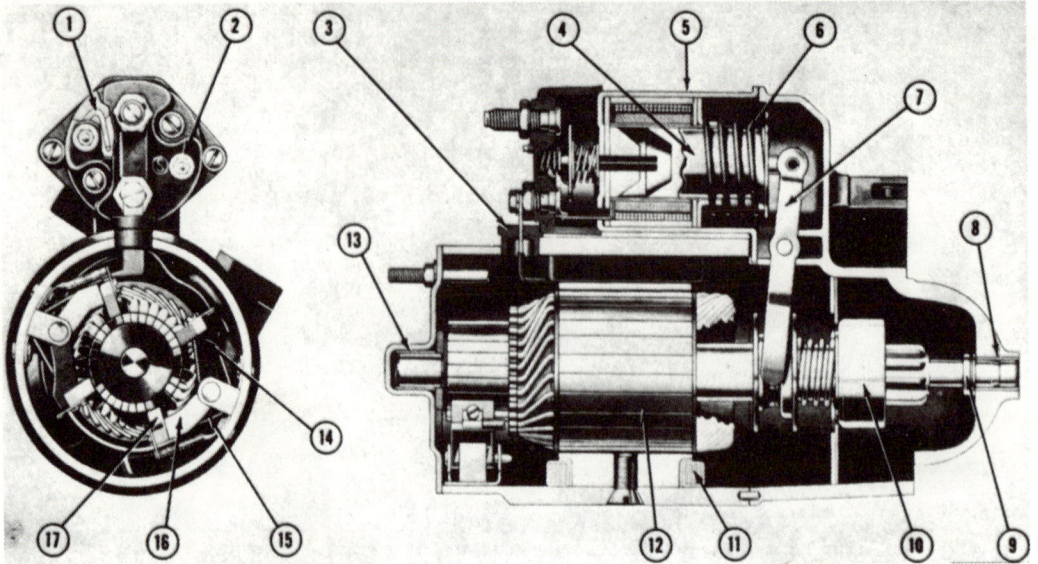

1. 'R' terminal contact
2. Switch terminal
3. Grommet
4. Plunger
5. Solenoid
6. Return spring
7. Shift lever
8. Bushing
9. Pinion stop
10. Overrunning clutch
11. Field coil
12. Armature
13. Bushing
14. Insulated brush holders
15. Brush spring
16. Grounded brush holder
17. Brush

6-225 and 8-350 starter motor

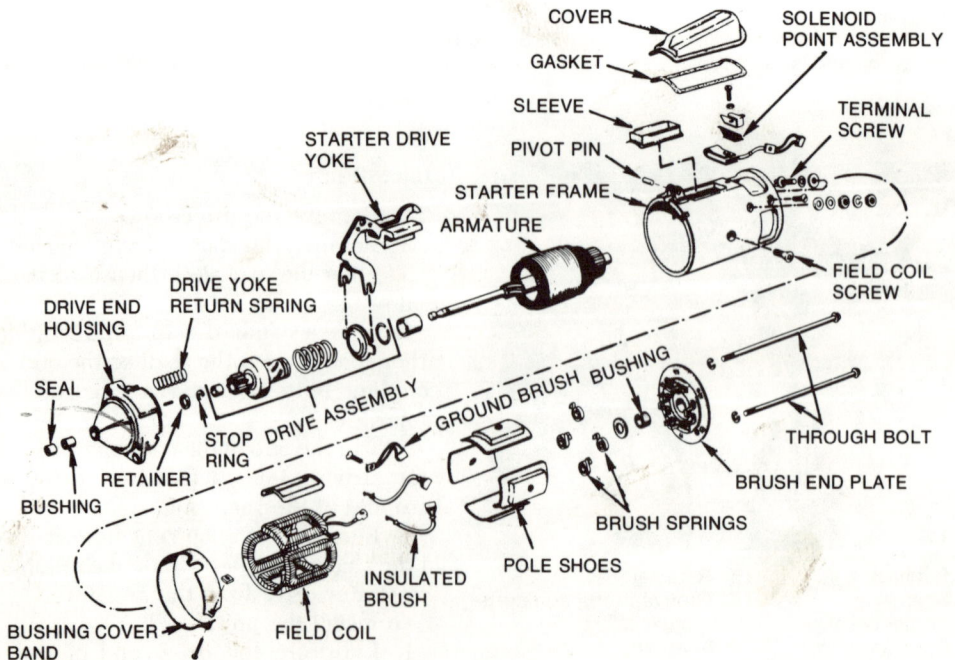

American motors starter motor

2. Slide the drive gear assembly onto the shaft, with the gear facing outward.

3. Slide the retainer onto the shaft with the cupped surface facing away from the gear.

4. Stand the whole starter assembly on a block of wood with the snap-ring positioned on the upper end of the shaft. Drive the snap-ring down with a small block of wood and a hammer. Slide the snap-ring into its groove.

5. Install the thrust collar onto the shaft with the shoulder next to the snap-ring.

6. With the retainer on one side of the snap-ring and the thrust collar on the other side, squeeze them together with a pair of pliers until the ring seats in the retainer. On models without a thrust collar, use a washer. Remember to remove the washer before installing the starter in the engine.

Prestolite

1. Slide the thrust collar off the armature shaft.

2. Using a standard ½ in. pipe connector, drive the snap-ring retainer off the shaft.

3. Remove the snap-ring from the groove, and then remove the drive assembly.

To install the unit:

1. Lubricate the drive end and splines with Lubriplate.®

2. Install the clutch assembly onto the shaft.

3. Install the snap-ring retainer with the cupped surface facing toward the end of the shaft.

4. Install the snap-ring into the groove. Use a new snap-ring if necessary.

5. Install the thrust collar onto the shaft with the shoulder against the snap-ring.

6. Force the retainer over the snap-ring in the same manner as was used for the Delco-Remy starters.

SOLENOID OR RELAY REPLACEMENT

Of the 3 makes of starters used on the various engines in the Wagoneer, Commando and Cherokee (Autolite, Delco-Remy and Presto-lite) there are 3 types of solenoids or relays.

The Delco-Remy starter, installed on the 225 V6 and 350 V8, has a solenoid externally mounted on top of the starter. The Autolite system, used in all 1971 and later models, has a remote relay mounted on the inside of the fender well. The Prestolite starters, used with the F-Head engine, have the solenoid externally mounted on top of the starter like the Delco-Remy units. The Prestolite starters, used in the 327 V8 and the pre-1971 232 Six have an internal solenoid, the solenoid coil being integral with the pinion housing.

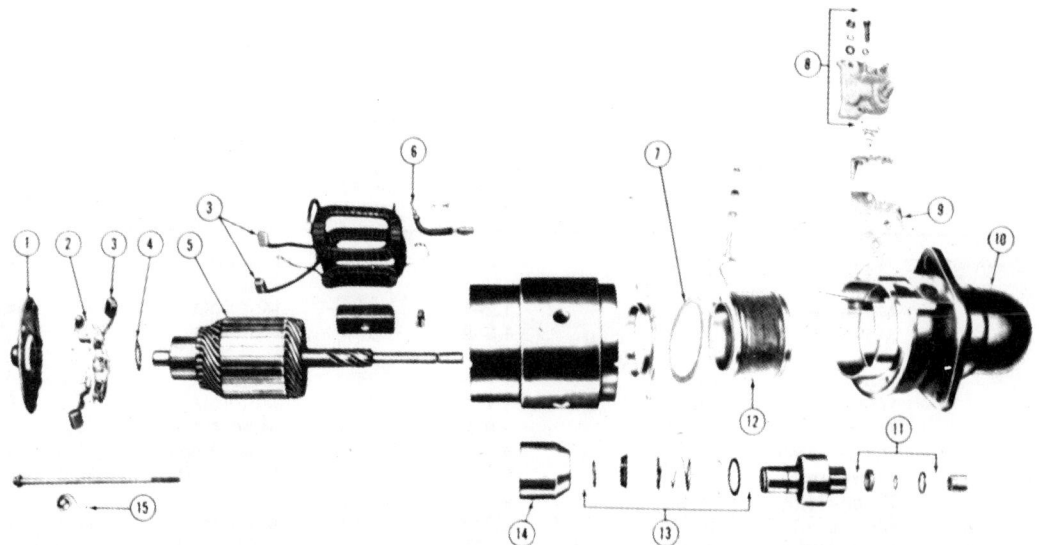

1. Commutator end head
2. Brush plate
3. Brushes
4. Thrust washer
5. Armature
6. Contact lead
7. Gasket
8. Switch cover assembly
9. Pinion gasket
10. Pinion housing
11. Stop and lock-ring
12. Solenoid assembly
13. Spring and washer
14. Solenoid moving core
15. Brush spring

8-327 starter motor

Autolite

To remove the relay, first disconnect the positive battery cable. Remove all of the leads from the relay which is mounted on the right fender well, taking note as to which lead goes to which terminal. Remove the attaching screws which hold the relay to the fender well and remove it from the vehicle. Installation is the reverse of removal.

Delco-Remy

Remove the leads from the solenoid. Remove the drive housing of the starter motor. Remove the shift lever pin and bolt from the shift lever. Remove the attaching bolts that hold the solenoid assembly to the housing of the starter motor. Remove the starter solenoid from the starter housing. To install the solenoid, reverse the above procedure.

Prestolite (F-Head Engine)

Remove the leads to the solenoid assembly. Remove the attaching bolts that hold the solenoid to the starter housing. Remove the bolt from the shift lever. Remove the solenoid assembly from the starter housing. Reverse the procedure for installation.

Prestolite (232 Six and 327 V8)

To remove the solenoid from this starter, it is necessary to disassemble the starter. Remove the two through-bolts and lockwashers that hold the commutator end head to the starter frame and pinion housing. Remove the commutator end head and thrust washer. Remove the two screws and lockwashers holding the intermediate bearing assembly to the pinion housing. Remove the assembled armature, solenoid coil, and over-running clutch drive from the pinion housing. Remove the top washer from the armature shaft. Push down on the retainer to expose the snap-ring; remove the snap-ring. Remove the pinion, washers, return spring, collar, retainer, solenoid winding, and the moving core from the armature shaft. Reassemble the starter in the reverse order of disassembly.

Battery

REMOVAL AND INSTALLATION

Remove the hold-down screws from the battery box. Loosen the nuts that secure the cable ends to the battery terminals. Lift the battery cables from the terminals with a twisting motion.

If there is a battery cable puller available, make use of it. Lift the battery from the vehicle.

Before installing the battery in the vehicle, make sure that the battery terminals are clean and free from corrosion. Use a battery terminal cleaner on the terminals and on the

Alternator and Regulator Specifications

Year	Manufacturer	Alternator Engine No. Cyl (cu in.)	Field Current @ 12V (amps)	Output (amps)	Regulator Manufacturer	Volts @ 75° F
1966–74	Motorola	6-232 ① 8-327	1.2–1.7	35	Motorola	14.2–14.6
	Motorola	8-350	1.2–1.7	35	Motorola	14.2–14.6
	Motorola	6-232, 258 8-304, 360, 401	1.8–2.5	37 ②	Motorola	13.75–14.2
	Motorola	F-Head	1.2–1.7	35	Motorola	14.2–14.6
	Motorola	V6-225	1.2–1.7	35	Motorola	14.2–14.6
1975	Motorola	all V8	1.8–2.5	37 ③	Motorola	12.7–15.3
	Delco-Remy	all 6	4.0–5.0	37 ④	Delco-Remy	12.0–15.5
1976	Motorcraft	all V8	2.5–3.0	40 ⑤	Motorcraft	13.1–14.8
	Delco-Remy	all 6	4.0–5.0	37 ⑥	Delco-Remy	12.0–15.5
1977	Motorcraft	all V8	2.5–3.0	40 ⑤	Motorcraft	13.1–14.8
	Delco-Remy	all 6	4.0–5.0	37 ⑥	Delco-Remy	12.0–15.5
1978–79	Delco-Remy	all	4.0–5.0	37 ⑥	Delco-Remy	12.0–15.5

① —Pre-1971 model.
② —55 amps with air conditioning in 1972.
  —51 amps with air conditioning in 1973.
③ Optional: 51 and 62 amp.
④ Optional: 55 and 63 amp.
⑤ Optional: 60 amp.
⑥ Optional: 63 amp.

Battery and Starter Specifications

		Starters						
		Lock Test			No-Load Test			Brush Spring Tension
Year	Make	Amps	Volts	Torque (ft lbs)	Amps	Volts	RPM	(oz)
1966–74	Prestolite	405	N.A.	9	50	10	5300	32–40
	Delco-Remy	N.A.	N.A.	N.A.	75	10.6	6200	32–40
	Prestolite	405	4	9	60	10	4200	32–40
	Prestolite	405	4	9	60	10	4200	32–40
	Delco-Remy	300–360	3.5	9	65–100	10.6	3600–5100	35
	Autolite	600	3.4	13	65	12	9250	40
1975–77	Motorcraft	600	3.4	13	65	12	9250 max.	40
1978–79	Motorcraft ①	—	—	—	67	12	7380–9356	40
	Motorcraft ②	—	—	—	77	12	8900–9600	40

① 6 cylinder.   ② 8 cylinder.

inside of the battery cable ends. If cleaner is not available, use heavy sandpaper to remove the corrosion. A mixture of baking soda and water poured over the terminals and cable ends will neutralize any acid. Before installing the cables onto the terminals, cut a piece of felt cloth, or something similar, into a circle about 3 in. across. Cut a hole in the middle about the size of the battery terminals at their base. Push the cloth pieces over the terminals so they lay flat on the top of the battery. Soak the pieces of cloth with oil. This will keep the formation of oxidized acid to a minimum. Place the battery in the vehicle. Install the cables onto the terminals. Tighten the nuts on the cable ends. Smear a light coating of grease on the cable ends and the tops of the terminals. This will further prevent buildup of oxidized acid on the terminals and the cable ends. Install and tighten the nuts of the battery box.

ENGINE MECHANICAL

Design

F-Head 4 Cylinder

The F-head, four-cylinder engine is of a combination valve-in-head and valve-in-block construction. The intake valves are mounted in the head and are operated by pushrods through rocker arms. The intake manifold is cast as an integral part of the cylinder head and is completely waterjacketed. This type of construction transfers heat from the cooling system to the intake passages and assists in vaporizing the fuel when the engine is cold. Therefore, there is no heat control valve

(heat riser) needed in the exhaust manifold.

The exhaust valves are mounted in the block with thorough water jacketing to provide effective cooling of the valves.

The engine is pressure-lubricated. An oil pump which is driven by the camshaft forces the lubricant through oil channels and drilled passages in the crankshaft to efficiently lubricate the main and connecting rod bearings. Lubricant is also force-fed to the camshaft bearings, rocker arms, and timing gears. Cylinder walls and piston pins are lubricated from spurt holes in the "follow" side of the connecting rods.

The circulation of the coolant is controlled by a thermostat in the water outlet elbow which is cast as part of the cylinder head.

The engine is equipped with a fully counterbalanced crankshaft that is supported by three main bearings. The counterweights of the crankshaft are independently forged and are permanently attached to the crankshaft with dowels and cap screws that are tack-welded. Crankshaft end-play is adjusted by placing shims between the crankshaft thrust washer and the shoulder on the crankshaft.

The pistons have an extra groove directly above the top ring which acts as a heat dam or insulator.

The engine has a compression ratio ranging from 6.3:1 to 7.8:1; this permits the use of regular octane gas. The displacement of the F-head engine is 134.2 cu in.

V6

The V6 engine has a displacement of 225 cu in. and a compression ratio of 9.0:1 which permits the use of regular octane gas.

It has two banks of three cylinders each

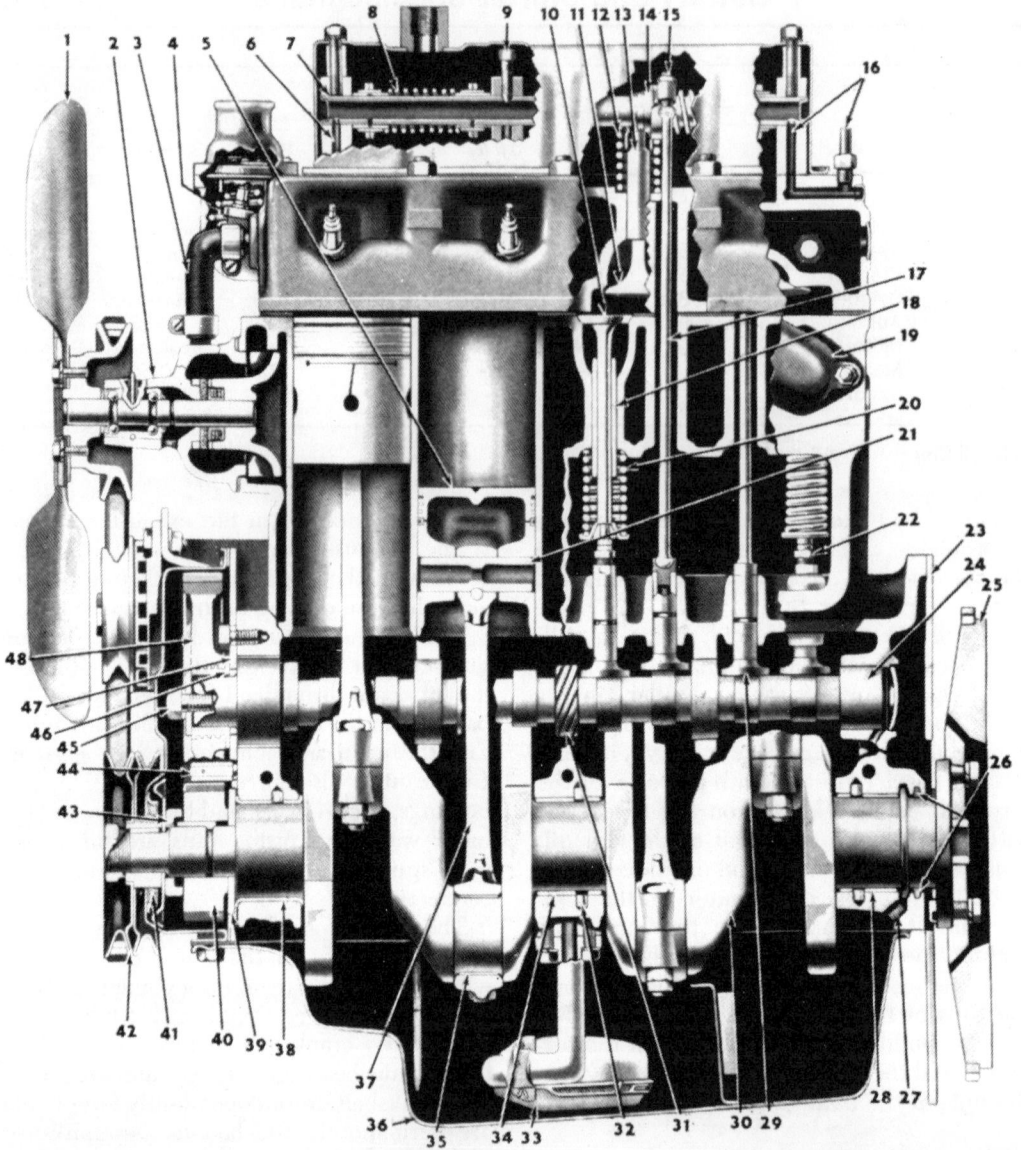

1. Fan assembly
2. Water pump assembly
3. Water by-pass tube
4. Thermostat
5. Piston
6. Oil return tube
7. Rocker arm shaft
8. Rocker arm shaft spring
9. Rocker arm shaft lock screw
10. Exhaust valve
11. Intake valve
12. Intake valve spring
13. Intake valve guide
14. Rocker arm
15. Adjusting screw
16. Oil inlet tube

17. Push rod
18. Exhaust valve guide
19. Exhaust manifold
20. Exhaust valve spring
21. Piston pin
22. Valve tappet adjusting screw
23. Engine rear support plate
24. Camshaft
25. Flywheel
26. Rear bearing oil seal
27. Oil return channel
28. Rear main bearing shell
29. Tappet
30. Crankshaft
31. Oil pump drive gear
32. Main bearing dowel

33. Oil float assembly
34. Center main bearing shell
35. Connecting rod bearing
36. Oil pan
37. Connecting rod
38. Front main bearing shell
39. Front engine plate
40. Crankshaft gear
41. Crankshaft front end seal
42. Fan and generator pully
43. Crankshaft gear spacer
44. Timing gear oil jet
45. Camshaft gear screw
46. Camshaft thrust plate spacer
47. Camshaft thrust plate
48. Camshaft gear

4-134 sectional view

which are opposed to one another at a 90° angle. The left bank of cylinders, as viewed from the driver's seat, is set forward of the right bank so that the connecting rods of opposite pairs of pistons and rods can be attached to the same crankpin.

The crankshaft counterbalance weights are cast as an integral part of the crankshaft. All of the crankshaft bearings are identical in diameter, except for no. 2 bearing which is the thrust bearing; it is larger than the rest.

The cast-iron heads are interchangeable although this is not recommended. They are exactly alike in every way, however.

The camshaft, which is located above the crankshaft—between the two banks, operates hydraulic valve lifters. The rocker arms are not adjustable.

232, 258 Sixes

The American Motors six-cylinder engines are inline sixes with overhead intake and exhaust valves. The valves are operated by separately mounted rocker arms in 1973 models. The rockers are mounted on a common shaft in 1972 models. None of the rocker arms are adjustable.

304 and 360

The 304 and 360 V8s have two banks of cylinders (4 cylinders each) which are opposed to each other at a 90° angle. The camshaft is located above the crankshaft, between the two banks of cylinders. It operates the valves through the use of hydraulic lifters, pushrods and separately mounted rocker arms. The exhaust valves on the 360 V8 rotate. The crankshaft is supported by 5 two-piece main bearings, the thrust being on number 3. A 2-bbl carburetor is used on the 304 and 360. The 360 also has a 4-bbl carburetor as an option.

327 and 350 V8s

The 327 and 350 V8s have two banks of cylinders (4 cylinders each) which are opposed to each other at a 90° angle. The camshaft is located above the crankshaft, between the two banks. It operates the valves through the use of hydraulic lifters, pushrods and rockers mounted in a row on a rocker shaft on top of each cylinder head.

Both engines have offset cylinder heads to facilitate the installation of two connecting rods on the same crankpin. The cylinder heads of the 350 V8 are identical in every

way, but it is recommended that they not be interchanged.

The crankshaft in both engines is supported by 5 two-piece main bearings, the thrust being on number 1 in the 327 V8 and on number 3 in the 350 V8.

ENGINE REMOVAL AND INSTALLATION
232 and 258 Sixes

The engine is removed without the transmission and bellhousing.

1. Mark the location of the hood on the hinges and then remove the hood from the hinges.

2. Drain the cooling system and the oil from the crankcase.

3. Disconnect the battery cables. It might be a good idea to remove the battery from the vehicle; this is left to your discretion.

4. Remove the air cleaner assembly.

5. Disconnect the upper and lower radiator hoses. If the vehicle is equipped with an automatic transmission, disconnect the transmission oil cooler lines from the radiator, plugging the lines to prevent the loss of fluid and the entrance of dirt. If the vehicle is equipped with a fan shroud on the radiator, separate it from the radiator and remove the radiator.

6. Remove the radiator fan and pulley.

7. Disconnect the accelerator linkage at the carburetor.

8. If so equipped, remove the power steering pump and drive belt from the engine and place it aside. Do not disconnect the power steering hoses.

NOTE: *When discharging the A/C system, always wear protective goggles.*

9. If the vehicle is equipped with air conditioning, turn both service valves clockwise to the front seated position.

10. Bleed the compressor refrigerant charge by *SLOWLY* loosening the service valve fittings. Disconnect the condenser and evaporator lines from the compressor. Disconnect the receiver outlet at the disconnect coupling. Remove the condenser and receiver assembly. Cap all openings at once.

11. Disconnect all wires, lines, linkage and hoses which are connected to the engine. This includes fuel lines, vacuum lines, electrical leads, etc.

12. Remove the oil filter.

13. Remove both of the engine front-support bracket-to-frame retaining nuts.

General Engine Specifications

Engine	Type	Horsepower at rpm	Torque (ft lb) at rpm	Bore x Stroke	Compression Ratio	Oil Pressure (psi) at 2000 rpm
4-134 1966–67	1 bbl	75 @ 4000	114 @ 2000	3.125 x 4.375	7.4 : 1	35
1968–69	1 bbl	75 @ 4000	114 @ 2000	3.125 x 4.375	6.7 : 1	35
6-225 1966–69	2 bbl	160 @ 4200	235 @ 2400	3.750 x 3.400	9.0 : 1	33
6-232 1966–69	1 bbl	145 @ 4400	215 @ 1600	3.750 x 3.500	8.5 : 1	50
1970–73	1 bbl	100 @ 3600	185 @ 1800	3.750 x 3.500	8.0 : 1	50
6-258 1971–76	1 bbl	110 @ 3500	195 @ 2000	3.750 x 3.895	8.0 : 1	50
1977–79	2 bbl	114 @ 3600	196 @ 2000	3.750 x 3.895	8.0 : 1	50
8-304 1971–73	2 bbl	150 @ 4200	245 @ 2500	3.750 x 3.7534	8.4 : 1	50
8-327 1966–69	2 bbl	250 @ 4700	340 @ 2600	4.000 x 3.250	8.7 : 1	40
8-350 1966–69	2 bbl	230 @ 4400	350 @ 2400	3.800 x 3.850	9.0 : 1	40
8-360 1972–73	2 bbl	175 @ 4000	285 @ 2400	4.0799 x 4.0831	8.5 : 1	40
1973	4 bbl	195 @ 4400	295 @ 2900	4.0799 x 4.0831	8.5 : 1	40
1974–79	2 bbl	175 @ 4000	285 @ 2900	4.0799 x 4.0831	8.25 : 1	40
1974–79	4 bbl	195 @ 4400	295 @ 2900	4.0799 x 4.0831	8.25 : 1	40
8-401 1974–78	4 bbl	215 @ 4400	320 @ 2800	4.165 x 3.680	8.3 : 1	40

14. Disconnect the exhaust pipe at the support bracket and exhaust manifold.

15. Support the weight of the engine with a lifting device.

16. Remove the front support bracket assemblies from the engine.

17. Remove the transfer case shift lever boot, floor mat (if so equipped) and transmission access cover.

18. Place supports under the transmission.

19. Remove the upper bolts securing the transmission bellhousing to the engine adapter plate on vehicles equipped with automatic transmissions. On vehicles equipped with manual transmissions, remove the upper bolts securing the clutch housing to the engine.

20. Remove the starter motor on vehicles made in 1971 and after.

21. If the vehicle is equipped with an automatic transmission, remove the two engine adapter plate inspection covers (1971 and later vehicles only). Mark the assembled position of the converter and flex plate cap screws. Remove the remaining bolts securing

Valve Specifications

Engine	Seat Angle (deg.)	Face Angle (deg.)	Spring Test Pressure (lbs @ in.)	Spring Installed Height (in.)	Stem-to-Guide Clearance (in.)		Stem Diameter (in.)	
					Intake	Exhaust	Intake	Exhaust
4-134	45	45	153 @ 1.400 ①	1.660 ⑦	.0007–.0022	.0025–.0045	.3733–.3738	.3710–.3720
6-225	45	45	168 @ 1.260	1.640	.0012–.0032	.0015–.0035 ②	.3415–.3427	.3402–.3412 ②
6-232	③	④	195 @ 1.475 ⑥	1.786 ⑧	.0010–.0030	.0010–.0030	.3715–.3725	.3715–.3725
6-258	③	④	195 @ 1.475 ⑥ ⑨	1.786	.0010–.0030	.0010–.0030	.3715–.3725	.3715–.3725
8-304	③	④	218 @ 1.359	1.8125	.0010–.0030	.0010–.0030	.3715–.3725	.3715–.3725
8-327	⑤	④	155 @ 1.475	1.8125	.0010–.0030	.0010–.0030	.3718–.3725	.3718–.3725
8-350	45	45	185 @ 1.340	1.727	.0015–.0035	.0015–.0035 ②	.3720–.3730	.3720–.3780 ②
8-360	③	④	218 @ 1.359	1.8125	.0010–.0030	.0010–.0030	.3715–.3725	.3715–.3725
8-401	③	④	218 @ 1.359	1.8125	.0010–.0030	.0010–.0030	.3715–.3725	.3715–.3725

① Exhaust: 120 @ 1.750.
② Measured at the top.
③ Intake: 30; exhaust: 44.5.
④ Intake: 29; exhaust: 44.
⑤ Intake: 30; exhaust: 45.
⑥ With rotators: 218 @ 1.1875.
⑦ Exhaust: 2.109.
⑧ 1966–70: 1.8125.
⑨ 2 bbl: 204 @ 1.386.

the transmission bellhousing to the engine adapter plate.

22. If the vehicle is equipped with a manual transmission, remove the clutch housing lower cover and the remaining bolts securing the clutch housing to the engine.

23. Remove the engine by lifting it up and moving it forward at the same time. Be careful of any hoses, lines, electrical leads, etc., that might be caught on the engine while it is being lifted out.

24. Install the engine in the reverse order of removal. If the vehicle is equipped with air conditioning, it is recommended that the air conditioning system be charged with refrigerant by a person trained in automotive air conditioning. Don't forget to fill the crankcase with the proper amount and type of oil and the cooling system with the proper mixture of coolant. When the engine is first started, check immediately for leaks of any kind.

NOTE: *The marks made on the converter and flex plate in Step 21 are to be aligned when the rear of the engine is mated to the front of the automatic transmission.*

304, 327, 350, 360 and 401 V8s

The engine is removed without the transmission and bellhousing.

1. Mark the location of the hood on the hinges and then remove the hood from the hinges.

2. Remove the air cleaner assembly.

3. Drain the cooling system at the radiator, remove the upper and lower radiator hoses at the radiator, and if the vehicle is equipped with an automatic transmission, disconnect the transmission oil cooler lines from the bottom of the radiator. Plug the lines to prevent the loss of fluid and the entrance of dirt.

NOTE: *If the vehicle is equipped with a radiator shroud, it is necessary to separate the shroud from the radiator to facilitate*

Crankshaft and Connecting Rod Specifications

Engine	Crankshaft				Connecting Rod		
	Main Bearing Journal Dia.	Main Bearing Oil Clearance	Shaft End Play	Thrust on No.	Journal Dia.	Oil Clearance	Side Clearance
4-134	2.3331–2.3341	.0003–.0029	.004–.006	1	1.9375–1.9383	.0001–.0019	.004–.010
6-225	2.4995	.0005–.0021	.004–.008	2	2.0000	.0020–.0023	.006–.014
6-232	2.4986–2.5001	.0010–.0020	.0015–.0065	3	2.0934–2.0955	.0010–.0020	.005–.014
6-258	2.4986–2.5001	.0010–.0020 ①	.0015–.0065	3	2.0934–2.0955	.0010–.0020 ②	.005–.014
8-304	2.7474–2.7489 ③	.0010–.0020 ④	.003–.008	3	2.0934–2.0955	.0010–.0020	.006–.018
8-327	2.4988–2.4995	.0006–.0030	.003–.007	3	2.2483–2.2490	.0007–.0028	.009–.015
8-350	2.9995	.0004–.0015	.003–.009	3	2.0000	.0020–.0023	.006–.014
8-360	2.7474–2.7489 ③	.0010–.0020 ④	.003–.008	3	2.0934–2.0955	.0010–.0030	.006–.018
8-401	2.7474–2.7489 ③	.0010–.0020 ④	.003–.008	3	2.0934–2.0955	.0010–.0030	.006–.018

① 1974–79: .0010–.0030 (.0025 preferred).
② 1974–76: .0010–.0030 (.0025 preferred).
  1977–79: .0010–.0025 (.0015–.0020 preferred).
③ #5: 2.7464–2.7479.
④ #5: .0020–.0030.

*the removal and installation of the radiator
and cooling fan.*

4. Remove the radiator.

5. Remove the cooling fan, belts and pulley assembly.

6. If the vehicle is equipped with power steering, remove the power steering pump from the engine and lay it aside, out of the way. Do not disconnect the fluid lines from the pump.

7. If the vehicle is equipped with air conditioning, turn both service valves clockwise to the front seated position. Bleed the compressor refrigerant charge by SLOWLY loosening the service valve fittings. Disconnect the condenser and evaporator lines from the compressor. Disconnect the receiver outlet at the disconnect coupling. Remove the condenser and receiver assembly.

8. Disconnect the battery and remove it only if it is required.

9. On 1971 and later Wagoneers, remove the heater core housing and charcoal canister from the firewall.

10. Disconnect all wires, lines, linkage and hoses which are connected to the engine. Be sure to plug all fluid lines.

11. If equipped with an automatic transmission, disconnect the transmission filler tube bracket from the right cylinder head. Do not remove the filler tube from the transmission.

12. Disconnect the exhaust pipes from the exhaust manifolds.

13. Remove both of the retaining nuts from the two front engine mounts.

14. Support the weight of the engine with a lifting device.

15. On Commandos made in 1971 and after, remove the transfer case shift lever boot, floor mat (if so equipped) and transmission access cover. There is a removable cover on pre-1971 Wagoneers, if more access is needed to remove the bolts.

Piston and Ring Specifications

	Ring Gap			Ring Side Clearance			
Engine	#1 Compression	#2 Compression	Oil Control	#1 Compression	#2 Compression	Oil Control	Piston Clearance*
4-134	.007–.017	.007–.017	.007–.017	.002–.004	.0015–.0035	.001–.0025	.0025–.0045
6-225	.010–.020	.010–.020	.015–.035	.002–.0035	.003–.005	.0015–.0085	.0005–.0011
6-232	.010–.020	.010–.020	.015–.055	.0015–.003	.0015–.003	.001–.008	.0009–.0017
6-258	.010–.020	.010–.020	.015–.055	.0015–.003	.0015–.003	.001–.008	.0009–.0017
8-304	.010–.020	.010–.020	.010–.025	.0015–.0035	.0015–.003	.0011–.008	.0010–.0018
8-327	.010–.020	.010–.020	.015–.055	.002–.004	.002–.004	0–.005	.0009–.0025
8-350	.010–.020	.010–.020	.015–.035	.003–.005	.003–.005	.0035–.0095	.0008–.0014
8-360	.010–.020	.010–.020	.015–.045	.0015–.003	.0015–.0035	0–.007	.0012–.0020
8-401	.010–.020	.010–.020	.015–.055	.0015–.003	.0015–.0035	0–.007	.0010–.0018

*Measured at the top of the skirt.

Torque Specifications

Engine	Cyl. Head	Conn. Rod	Main Bearing	Crankshaft Damper	Flywheel	Manifold	
						Intake	Exhaust
4-134	60–70	35–45	65–75	65–75	35–41	—	29–35
6-225	65–85	30–40	80–110	140 min.	50–65	45–55	14–20
6-232	105	28	80	55	105	43	25
6-258	105	33	80	80	105	23	23
8-304	110	28	100	55	105	43	25
8-327	58–62	46–50	80–85 ①	70–80	100–110	20–25	20–25
8-350	②	35	110	140–180	60	50	18
8-360	110	33	100	90	105	43	25
8-401	110	39	100	90	105	43	25

① Rear cap: 50–55.
② With steel gasket: 75; with composition gasket: 80.

16. Remove the upper bolts securing the transmission bellhousing to the engine adapter plate on vehicles equipped with an automatic transmission. If the vehicle is equipped with a manual transmission, remove the upper bolts securing the clutch housing to the engine.

17. On pre-1971 model Wagoneers equipped with manual transmissions, disconnect the clutch linkage from the clutch. Disconnect the clutch cross-shaft support brackets at the flywheel housing and frame.

18. Support the transmission with a floor jack.

19. On 1971 and later model vehicles equipped with automatic transmissions, remove the two engine adapter plate inspection covers and mark the assembled position of the converter and flex plate and remove the converter-to-flex plate attaching screws. Remove the remaining bolts securing the transmission bellhousing to the engine.

On 1971 and later model vehicles equipped with manual transmissions, remove the clutch housing lower cover and remove the remaining bolts securing the clutch housing to the engine.

20. Remove the engine from the vehicle by pulling upward and forward slowly until the engine is clear of the front of the transmission. Care must be taken to avoid damaging the power brake unit while removing the engine.

21. Install the engine in the reverse order of removal. If the vehicle is equipped with air conditioning, it is recommended that the air conditioning system be charged with refrigerant by a person trained in automative air conditioning. Don't forget to fill the crankcase with the proper amount and type of oil and the cooling system with the proper mixture of coolant. When the engine is first started, check immediately for leaks of any kind.

NOTE: *The marks that were made on the converter and the flex plate on 1971 and later models are to be aligned when the rear of the engine is mated to the front of the automatic transmission.*

F-Head

1. Remove the hood.
2. Remove the air cleaner from the fender bracket and disconnect the air bleed hose to the carburetor.
3. Disconnect the battery cables from the battery.

4. Drain the oil from the engine crankcase.
5. Drain the coolant from the radiator and engine.
6. Remove the radiator and radiator grille support rods.
7. Remove the fan from the water pump pulley.
8. Disconnect the sending unit wire from the temperature and oil pressure sending units.
9. Disconnect the engine wiring harness connectors at the front of the cowl.
10. Disconnect the wiring harness from the alternator.
11. Disconnect the electric lead wires and battery cable from the starter solenoid switch. Remove the battery cable.
12. Remove the starter.
13. Disconnect the engine fuel line from the main fuel lines at the right frame rail. Plug both the lines to prevent leakage and the entrance of dirt.
14. Disconnect the throttle and choke cables from the carburetor and the air bleed hose to the air cleaner.
15. Disconnect the exhaust pipe from the exhaust manifold.
16. Remove the bolts securing the front engine mounts to the frame brackets.
17. Support the weight of the engine with a lifting device.
18. Remove the bolts securing the engine to the flywheel housing.
19. Raise the engine slightly and slide the engine forward to remove the transmission mainshaft from the clutch plate spline.
20. When the engine is free of the transmission, raise the engine and remove it from the vehicle.
21. Install the engine in the reverse order of removal. Don't forget to fill the crankcase with the proper amount and type of oil and the cooling system with the proper mixture of coolant. When the engine is started, check immediately for leaks of any kind.

V6

1. Remove the hood.
2. Disconnect the ground cable from the battery and the engine.
3. Remove the air cleaner.
4. Drain the engine oil from the crankcase and the engine coolant from the radiator.
5. Disconnect the headlamp wiring from the block on the left fender.

6. Disconnect the horn wiring harness from the horn.

7. Disconnect the alternator wiring harness from the connector at the regulator.

8. Disconnect the upper and lower radiator hoses from the engine.

9. Disconnect the transmission oil cooler lines from the radiator, if the vehicle is equipped with an automatic transmission.

10. Remove the right and left radiator support bars.

11. Remove the cap screws and washers securing the right and left fenders to the cowl of the body.

12. Remove the bolt securing the radiator grille to the front frame crossmember and remove the front fenders, radiator and radiator grille as a unit from the vehicle.

13. Disconnect the engine wiring harness from the connectors located on the engine firewall.

14. Disconnect the battery cable and wiring from the engine starter assembly.

15. Remove the starter from the engine.

16. Disconnect the fuel hoses from the fuel lines at the right frame rail. Plug the lines to prevent leakage and the entrance of dirt.

17. Disconnect the throttle from the carburetor and cable support bracket mounted on the engine.

18. Disconnect the exhaust pipes from the right and left exhaust manifolds.

19. Place a jack under the transmission and support the weight of the transmission.

20. Remove the bolts securing the engine to the front motor mounts.

21. Attach a lifting device and support the weight of the engine.

22. Remove the bolts securing the engine to the flywheel housing.

23. Raise the engine slightly and slide the engine forward to remove the transmission mainshaft from the clutch plate spline.

NOTE: *The engine and the transmission must be raised slightly to release the mainshaft from the clutch plate, while sliding the engine forward.*

24. When the engine is free of the transmission, raise the engine and remove it from the vehicle.

25. Install the engine in the reverse order of removal. Remember to install the proper amount and type of oil in the crankcase and the proper mixture of coolant in the radiator before starting the engine. As soon as the engine is started, check for leaks of any type immediately.

Rocker Shafts and Rocker Studs

REMOVAL AND INSTALLATION

F-Head

Remove the rocker arm cover attaching bolts and remove the rocker arm cover. Remove the nuts from the rocker arm shaft support studs. Remove the intake valve pushrods

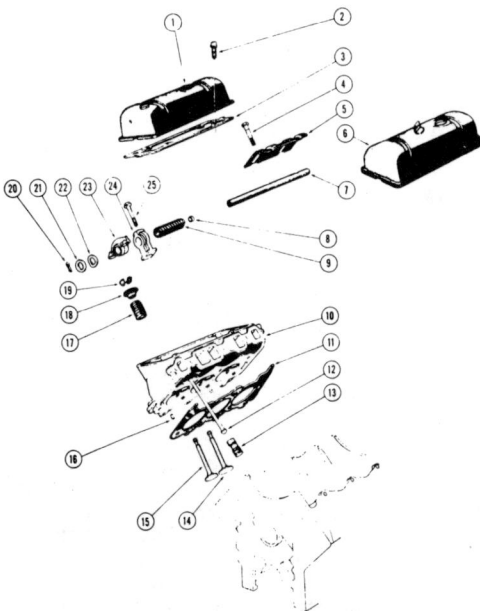

1. Right rocker arm cover
2. Rocker arm cover bolt
3. Gasket
4. Bolt
5. Baffle
6. Left rocker arm cover
7. Rocker arm shaft
8. Plug
9. Rocker arm spring
10. Cylinder head
11. Head gasket
12. Pushrod
13. Valve lifter
14. Intake valve
15. Exhaust valve
16. Dowel pin
17. Valve spring
18. Valve spring cap
19. Valve spring cap key
20. Cotter pin
21. Rocker arm shaft end washer
22. Rocker arm shaft spring
23. Rocker arm
24. Rocker arm shaft bracket
25. Bolt

6-225 valve train

from the engine. Install in the reverse order. Tighten the rocker arm retaining bolts to 30–36 ft lbs.

V6

Remove the crankcase ventilator valve from the right side valve cover. Remove the four attaching bolts from the right and left side valve covers and remove both of the valve covers.

Unscrew, but do not remove, the bolts that attach the rocker arm assemblies to the cylinder heads. Remove the rocker arm assemblies, with the bolts in place, from the cylinder heads. Mark each of the pushrods so that they can be installed in their original positions. Remove the pushrods. Install in the reverse order. Tighten the bolts to 30 ft lbs, a little at a time.

350 V8 and 232, 258 Sixes

For 1973 engines, follow the procedure for the 304 V8. Remove the valve cover by removing the valve cover attaching screws.

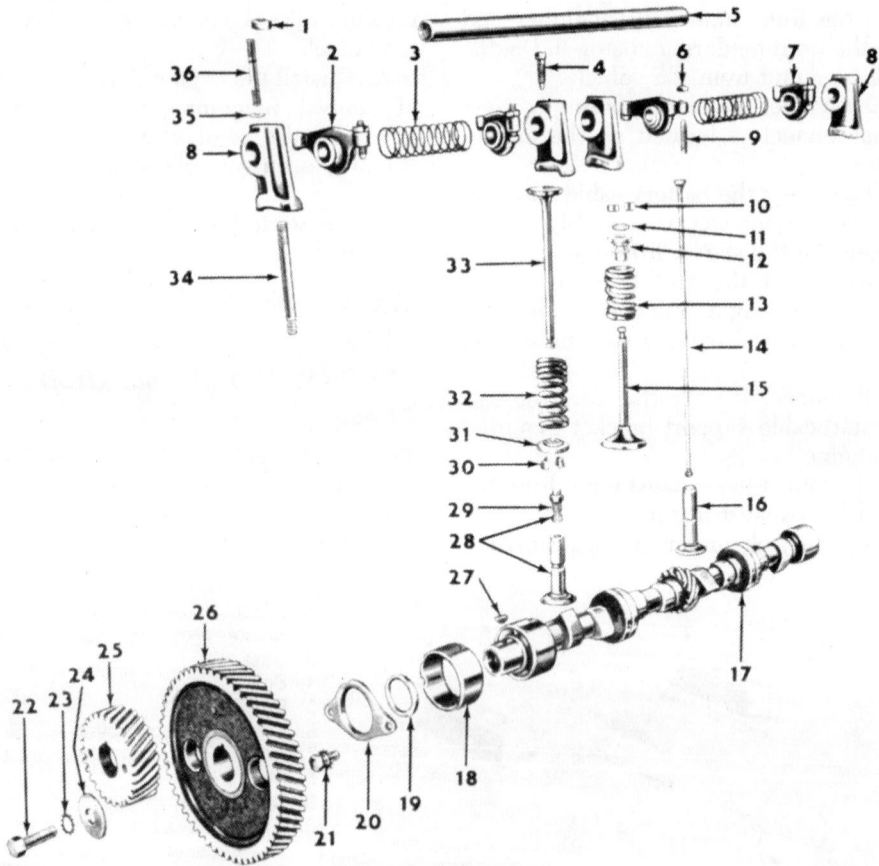

1. Nut	13. Intake valve spring	25. Crankshaft gear
2. Left rocker arm	14. Intake valve push rod	26. Camshaft gear
3. Rocker arm shaft spring	15. Intake valve	27. Woodruff key No. 9
4. Rocker shaft lock screw	16. Intake valve tappet	28. Exhaust valve tappet
5. Rocker shaft	17. Camshaft	29. Tappet adjusting screw
6. Nut	18. Camshaft front bearing	30. Spring retainer lock
7. Right rocker arm	19. Camshaft thrust plate	31. Roto cap assembly
8. Rocker arm shaft bracket	spacer	32. Exhaust valve spring
9. Intake valve tappet adjusting	20. Camshaft thrust plate	33. Exhaust valve
screw	21. Bolt and lock washer	34. Rocker shaft support
10. Intake valve upper retainer lock	22. Bolt	stud
11. Oil seal	23. Lockwasher	35. Washer
12. Intake valve spring upper retainer	24. Camshaft gear washer	36. Rocker arm cover stud

4-134 valve train

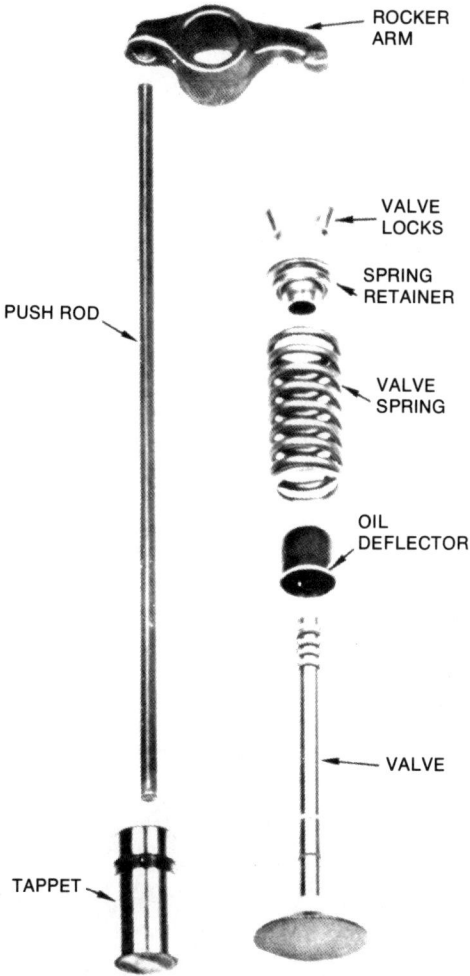

ROCKER ARM

PUSH ROD

VALVE LOCKS

SPRING RETAINER

VALVE SPRING

OIL DEFLECTOR

VALVE

TAPPET

6-232, 258 valve train

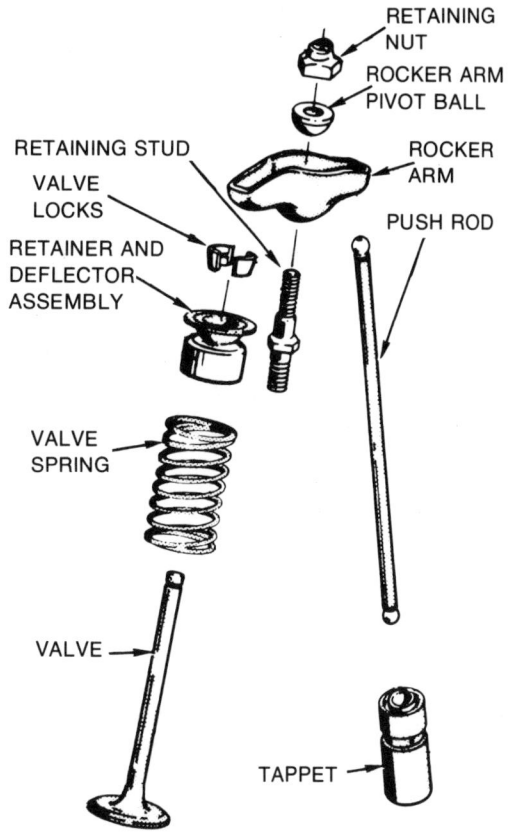

RETAINING NUT

ROCKER ARM PIVOT BALL

RETAINING STUD

VALVE LOCKS

RETAINER AND DEFLECTOR ASSEMBLY

ROCKER ARM

PUSH ROD

VALVE SPRING

VALVE

TAPPET

8-304, 360, 401 valve train

Loosen evenly, but do not remove, the bolts that attach the rocker arm assembly to the cylinder head. Lift the whole rocker arm assembly off the head with the bolts in place. Identify each of the pushrods so that they can be replaced in their original positions. Remove the pushrods. Install in reverse order. Tighten the mounting bolts, working evenly, from the center outward. Tighten to 22 ft lbs.

304, 360, and 401 V8s

The 304, 360 and 401 V8s have each rocker arm individually mounted on a separate stud. Each rocker assembly consists of the following: a rocker arm retaining stud, a rocker arm pivot ball, a rocker arm, and a retaining nut which screws onto the rocker arm retaining stud. Each assembly is removed and installed separately. To remove, unscrew the rocker retaining nut from the stud and lift off the rocker arm and its pivot ball. To remove the

stud from the block use a wrench. Label the pushrods so that they can be installed in their original positions and remove them from the block.

When installing the rocker arm retaining studs, use caution not to cross thread them. They are designed to cause an interference fit. Lubricate the studs with high pressure grease before installing them in the head. Install the rocker arm assemblies in the reverse order of removal. Tighten the rocker arm retaining nuts to 23 ft lbs.

327 V8

The 327 V8 has the rocker arms mounted on a rocker arm shaft. Remove the rocker arm cover to gain access to the rocker shafts by removing the wing nuts, flat washers and grommets holding the covers to the cylinder heads and lift the covers off together with the gaskets. Loosen the rocker arm hold-down bolts evenly to release the strain on the shaft and binding on the bolts. Remove the bolts and rocker arms. If the rocker arm shaft dowels stay with the cylinder head, remove them and install them in the supports.

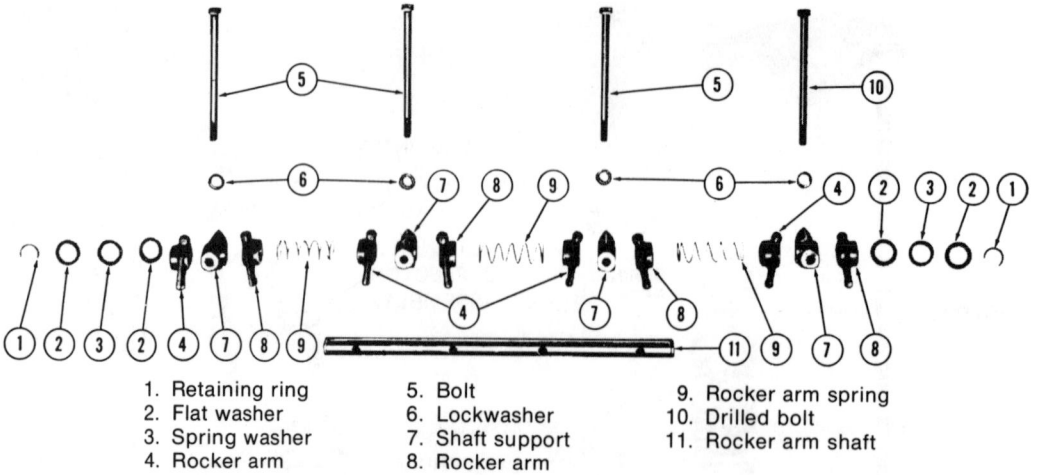

1. Retaining ring
2. Flat washer
3. Spring washer
4. Rocker arm
5. Bolt
6. Lockwasher
7. Shaft support
8. Rocker arm
9. Rocker arm spring
10. Drilled bolt
11. Rocker arm shaft

8-327 Rocker arm shaft

NOTE: *When removing the rocker arms, while the engine is installed in the vehicle, loosen the radiator cap to relieve the pressure in the cooling system before loosening the bolts to prevent coolant seepage.*

If the rocker arm assembly is disassembled, mark the rockers themselves so that they can be reassembled onto the shaft in the same order from which they were removed.

If the pushrods are removed, they should also be identified so that they can be reinstalled in the same positions from which they were removed.

Cylinder Head

REMOVAL AND INSTALLATION

NOTE: *It is important to note that each engine has its own head bolt tightening sequence and torque. Incorrect tightening procedure may cause head warpage and compression loss. Correct sequence and torque for each engine model is shown in this chapter.*

F-Head

1. Drain the coolant.
2. Remove the upper radiator hose.
3. Remove the carburetor.
4. On early engines remove the by-pass hose on the front of the cylinder head.
5. Remove the rocker arm cover.
6. Remove the rocker arm attaching stud nuts and rocker arm shaft assembly.
7. Remove the cylinder head bolts. One of the bolts is located below the carburetor mounting, inside the intake manifold.
8. Lift off the cylinder head.

9. Remove the pushrods and the valve lifters if further disassembly is to be carried out.

10. Reverse the above procedure to install the cylinder head. Tighten the head bolts first to 40 ft lbs then to the specified torque in the correct sequence.

4-134 cylinder head tightening sequence

V6

1. Remove the intake manifold.
2. Remove the rocker cover.
3. Remove the exhaust pipes at the flanges.

6-225 cylinder head tightening sequence

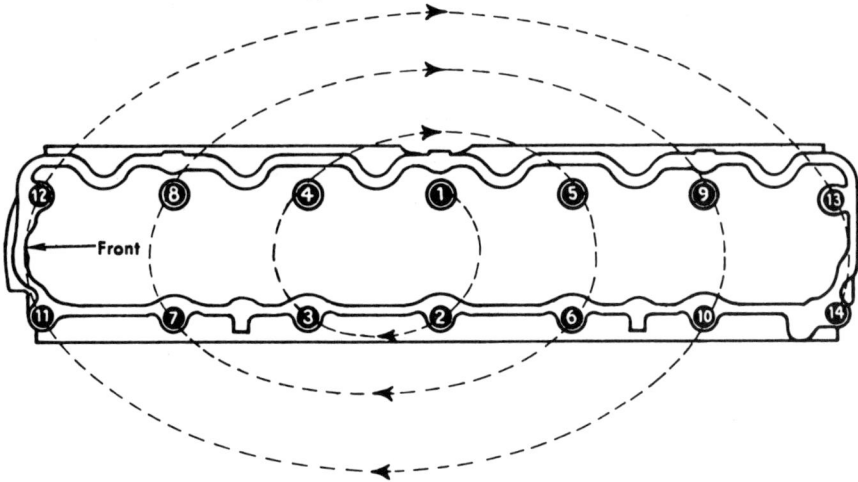

6-232, 258 cylinder head tightening sequence

4. Remove the alternator in order to remove the right head.

5. Remove the dipstick and power steering pump, if so equipped, in order to remove the left head.

6. Remove the valve cover and the rocker assemblies. Mark these parts so that they can be re-installed in exactly the same positions.

7. Unbolt the head bolts and lift off the cylinder head(s). It is very important that the inside of the engine be protected from dirt. The hydraulic lifters are particularly susceptible to being damaged by dirt.

8. To install, use the reverse procedure.

232, 258 Sixes

1. Drain the cooling system and disconnect the hoses at the thermostat housing.

2. Remove the cylinder head cover (valve cover), the gasket, the rocker arm assembly, and the pushrods.

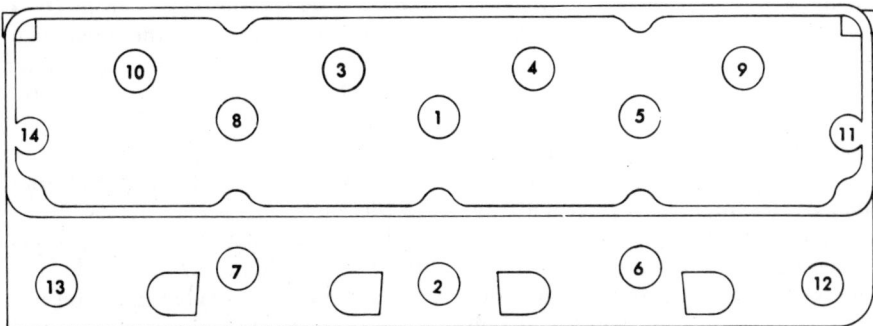

NOTE: *The pushrods must be replaced in their original positions.*

3. Remove the intake and exhaust manifold from the cylinder head.

4. Disconnect the spark plug wires and the spark plugs to avoid damaging them.

5. Disconnect the temperature sending unit wire, ignition coil and bracket assembly from the engine.

6. Remove the cylinder head bolts, the cylinder head and gasket from the block.

7. To install reverse the above procedure. Tighten the headbolts to the specified torque, using the proper sequence.

NOTE: *On engines through 1969, the front left (Number 11) bolt, must be coated with waterproof, non-hardening sealer to prevent coolant leakage.*

304, 360 and 401 V8s

1. Drain the cooling system.

2. When removing the right cylinder head, remove the heater core housing from the firewall.

3. Remove the valve cover(s) and gasket(s).

4. Remove the rocker arm assemblies and the pushrods.

8-304, 360, 401 cylinder head tightening sequence

NOTE: *The pushrods must be replaced in their original positions.*

5. Remove the spark plugs to avoid damaging them.

6. Remove the intake manifold with the carburetor still attached.

7. Remove the exhaust pipes at the flange of the exhaust manifold. When replacing the exhaust pipes it is advisable to install new gaskets at the flange.

8. Loosen all of the drive belts.

9. Disconnect the battery ground cable from the right cylinder head.

10. Disconnect the air pump bracket from the left cylinder head.

11. Remove the cylinder head bolts and lift the head(s) from the cylinder block.

12. Remove the cylinder head gasket from the head or the block.

13. To install, reverse the above procedure.

NOTE: *The second head bolt from the front, on the bottom row, must have its threads coated with sealer when installed in order to prevent coolant leakage. First, tighten all bolts to 80 ft lbs then tighten to the specified torque. Follow the correct sequence.*

327, 350 V8s

1. Drain the cooling system at the radiator and at the engine block.

2. Disconnect the battery cables from the battery and remove the alternator and alternator bracket assembly from the engine.

3. Remove the air cleaner assembly.

4. Remove the rocker covers and gaskets and remove the rocker assemblies together with the pushrods.

NOTE: *The pushrods must be reinstalled in the same positions from which they were removed. Identify each pushrod so that this can be accomplished.*

5. Disconnect the exhaust pipes from the exhaust manifolds by removing the nuts from the attaching bolts of the clamp. Remove the donut gasket.

6. Remove the distributor.

7. Remove the cap screws, lockwashers, nuts, and flat washers that secure the intake manifold to the cylinder heads.

8. Remove the intake manifold with the carburetor attached from the engine. Remove the gaskets and discard them.

9. Remove the bolts holding the cylinder head to the block. Removing one head bolt on the left bank frees the oil dipstick and

8-350 cylinder head tightening sequence

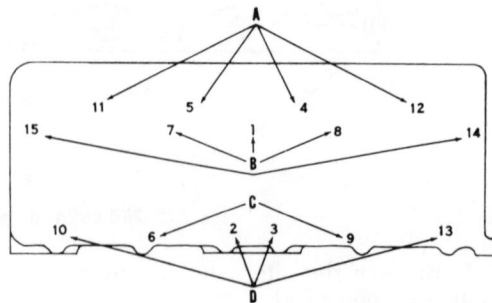

| A. 6⅜ in. bolts | C. 3 in. bolts |
| B. 4¼ in. bolts | D. 3 15/16 in. bolts |

8-327 cylinder head tightening sequence

tube which may now be lifted off. There is a rubber bushing on the lower end of the tube; take care not to lose the rubber bushing.

10. Lift the heads off the block and remove the discard the cylinder head gaskets. Remove all traces of gasket material from all mating surfaces on both the heads and the intake manifold.

11. Install the cylinder heads and assemble the engine in the reverse order of removal and disassembly. Torque the head bolts and intake manifold bolts in the correct sequence and to the proper specifications.

OVERHAUL

V6, 232 and 258 Sixes, 304, 327, 350, 360 and 401 V8s
See the "Engine Rebuilding" Section

F-Head

After removing the head from the cylinder block, remove the valves from the cylinder head. These valves are the intake valves, and are removed in the following manner:

1. Depress the valve spring with a valve spring tool.

2. Remove the valve spring retainer locks.

3. Remove the rubber O-ring from the top of the valve.

4. Remove the valve spring retainer and the valve spring from the head.

5. Remove the valve from the head.

6. Replace the valves by reversing the above procedure.

NOTE: *All of the valves must be replaced in their original positions. Even though the exhaust valves are located in the block of the engine and not the head the procedure for removing them will be given here.*

1. Remove the attaching bolts from the side valve spring cover. Remove the side valve spring cover and gasket.

2. Use rags to block off the three holes in the exhaust chamber to prevent the valve retaining locks from falling into the crankcase, should they be accidentally dropped.

3. Using a valve spring compressor, compress the valve springs only on those valves which are in the closed position (valve seated against the head). Remove the exhaust valve spring retainer locks, the retainer, and the exhaust valve spring. Close the other valves by rotating the camshaft and repeat the above operation for the remaining valves.

4. Lift all of the valves from the cylinder block. If the valve cannot be removed from the block, pull the valve upward as far as possible and remove the spring. Lower the valve and remove any carbon deposits from the valve stem. This will permit removal of the valve.

5. Install by reversing the above procedure.

NOTE: *All of the valves must be replaced in their original positions. Do not get the exhaust and intake valve springs mixed. They are not interchangeable.*

Refer to the engine rebuilding section for further servicing of the cylinder head and its components.

Intake Manifold

REMOVAL AND INSTALLATION

F-Head

On the F-head engine the intake manifold is cast as an integral part of the head.

V6

1. Drain the cooling system.

2. Disconnect the crankcase vent hose, distributor vacuum hose, and the fuel line from the carburetor.

3. Disconnect the two distributor leads from the coil.

4. Disconnect the wire from the temperature sending unit.

5. Remove the ten cap bolts that hold the intake manifold to the cylinder head. They *must* be replaced in their original location.

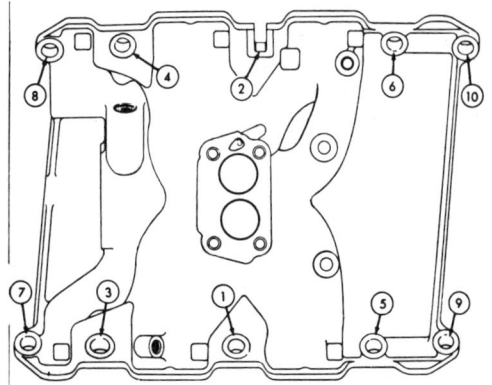

6-225 intake manifold tightening sequence

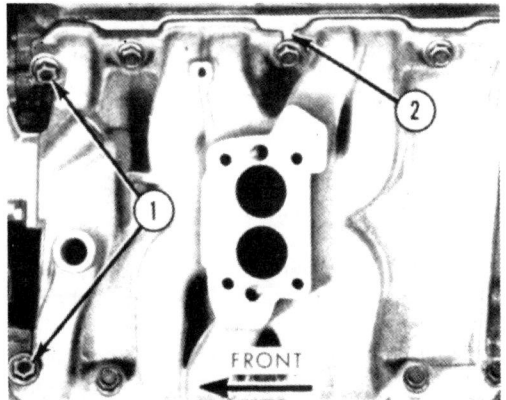

Replace all of the bolts in their original locations on the 6-225

6. Remove the intake manifold assembly and gasket from the engine.

7. Reverse the above procedure for installation. Tighten the bolts to the correct torque, and in the proper sequence.

232, 258 Sixes

The intake manifold and exhaust manifold are mounted externally on the left side of the engine and are attached to the cylinder head. The intake and exhaust manifolds are removed as a unit. On some engines, an exhaust gas recirculation valve is mounted on the side of the intake manifold.

1. Remove the air cleaner and carburetor.

2. Disconnect the accelerator cable from the accelerator bellcrank.

3. Disconnect the PCV vacuum hose from the intake manifold.

4. Disconnect the distributor vacuum hose and electrical wires at the TCS solenoid vacuum valve.

5. Remove the TCS solenoid vacuum valve and bracket from the intake manifold.

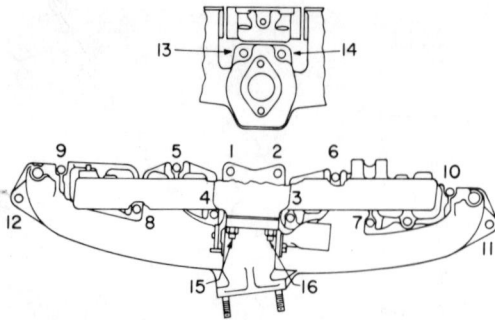

6-232, 258 intake manifold tightening sequence

In some cases it might not be necessary to remove the TCS unit.

6. If so equipped, disconnect the EGR valve vacuum hoses.

7. Disconnect the exhaust pipe from the manifold flange.

8. Remove the manifold attaching bolts, nuts and clamps.

9. Separate the intake manifold and exhaust manifold from the engine as an assembly. Discard the gasket.

10. If either manifold is to be replaced, they should be separated at the heat riser area.

11. Clean the mating surfaces of the manifolds and the cylinder head before replacing the manifolds. Replace them in reverse order of the above procedure. Tighten the bolts to the specified torque.

304, 360 and 401 V8s

1. Drain the coolant from the radiator.

2. Remove the air cleaner assembly.

3. Disconnect the spark plug wires. Remove the spark plug wire bracket from the valve covers, and the bypass valve bracket.

4. Disconnect the upper radiator hose and the by-pass hose from the intake manifold.

5. Disconnect the ignition coil bracket and lay the coil aside.

6. Disconnect the TCS solenoid vacuum valve from the right side valve cover.

7. Disconnect all lines, hoses, linkages and wires from the carburetor and intake manifold and TCS components as required.

8. Disconnect the air delivery hoses at the air distribution manifolds.

9. Disconnect the by-pass valve bracket and lay the valve and the bracket assembly, including the hoses, forward of the engine.

10. Remove the intake manifold after removing the cap bolts that hold it in place.

Remove and discard the side gaskets and the end seals.

11. Clean the mating surfaces of the intake manifold and the cylinder head before replacing the intake manifold. Use new gaskets and tighten the cap bolts to the correct torque. Install in reverse order of the above procedure.

NOTE: *There is no specified tightening sequence for the 8-304, 360 and 401 intake manifold. Begin at the center and work outward.*

327 V8

1. Drain the cooling system.

2. Remove the air cleaner assembly.

3. Remove the distributor.

4. Disconnect the throttle linkage at the carburetor and the throttle cable bracket attached to the manifold.

5. Disconnect the fuel line at the carburetor inlet. Also remove the vacuum lines at the thermostatic vacuum switch, if so equipped.

6. Disconnect the vacuum lines, the coil and ignition primary leads.

7. Remove the cap screws, lockwashers, nuts, and flat washers that secure the intake manifold to the cylinder heads.

8. Lift the intake manifold and carburetor assembly from the engine and remove all traces of gasket material from the intake manifold, cylinder block and cylinder heads mating surfaces.

9. Install the intake manifold assembly on the engine in the reverse order of removal. Torque the intake manifold mounted bolts to the proper torque value.

NOTE: *There is no specific tightening sequence for the intake manifold of the 8-327. Begin at the center and work outward.*

350 V8

1. Drain the cooling system.

2. Remove the carburetor air cleaner. Disconnect all tubes and hoses from the carburetor. Disconnect and remove the ignition coil.

3. Disconnect the temperature indicator wire from the sending unit.

4. Disconnect the accelerator and transmission linkage at the carburetor.

5. Slide the front thermostat by-pass hose clamp back on the hose. Disconnect the upper radiator hose at the outlet.

8-350 intake manifold tightening sequence

6. Disconnect the heater hose at the temperature control valve inlet. Force the end of the hose down to allow the coolant to drain from the intake manifold.

7. Remove the intake manifold-to-head attaching bolts.

8. Remove the intake manifold and carburetor as an assembly, by sliding the assembly rearward to disengage the thermostat by-pass hose from the water pump. Remove all traces of gasket material from the mating surfaces of the intake manifold and the cylinder heads and block.

9. Install the intake manifold in the reverse order of removal, tightening the attaching bolts in the correct sequence and to the proper torque.

Exhaust Manifold

REMOVAL AND INSTALLATION

F-Head

1. Remove the air delivery hose from the air injection tube assembly if the engine is so equipped. If not proceed to step two.

2. Remove the five nuts from the manifold studs.

3. Pull the manifold from the mounting studs. Be careful not to damage the air injection tubes if the engine is equipped with the emission control air pump.

4. Remove the gaskets from the cylinder block.

5. If the exhaust manifold is to be replaced it will be necessary to remove the air injec-

tion tubes from the exhaust manifold. The application of heat may be necessary to aid removal.

6. Use new gaskets when replacing the exhaust manifold. Make sure that the mating surfaces of both the exhaust manifold and the cylinder head are clean. Tighten the attaching nuts to the correct torque specification. Replace in reverse order of the above procedure.

V6

1. Remove the five attaching screws, one nut, and exhaust manifold(s) from the side of the cylinder head(s).

2. Use a new gasket when replacing the exhaust manifolds. Make sure that the mating surfaces of the manifold and the cylinder head are clean. Tighten the manifold nuts and bolts to the correct torque.

6-225 exhaust manifold bolts

232, 258 Sixes

The intake and exhaust manifolds of the 232 and 258 cu in. Sixes must be removed together. See the procedure for removing and installing the intake manifold.

304, 360 and 401 V8s

1. Disconnect the spark plug wires.

2. Disconnect the air delivery hose at the distribution manifold.

3. Remove the air distribution manifold and the injection tubes.

4. Disconnect the exhaust pipe at the manifold.

5. Remove the exhaust manifold attaching bolts and washers along with the spark plug shields.

6. Separate the exhaust manifold from the cylinder head.

7. Install in reverse order of the above procedure. Clean the mating surfaces and

1. Heat stove
2. Exhaust manifold

8-350 left exhaust manifold

tighten the attaching bolts to the correct torque.

327, 350 V8s

1. Disconnect the exhaust pipe at the exhaust manifold/exhaust pipe joining flange by removing the retaining nuts.

2. Remove the donut gasket and save it, if it is in good condition. If not, discard and replace it.

3. Remove the bolts attaching the exhaust manifold to the cylinder head, leaving one upper bolt toward the center until last.

4. Remove all traces of gasket material from the exhaust manifold and cylinder head mounting surfaces.

NOTE: *The sheet metal heat stove is removed with the left exhaust manifold on the 350 V8.*

5. Install the exhaust manifold in the reverse order of removal, tightening the attaching bolts in the correct pattern and to the proper torque.

Timing Gear Cover

TIMING GEAR COVER OIL SEAL REPLACEMENT

F-Head

1. Remove the drive belts and crankshaft pulley.

2. Remove the attaching bolts, nuts and lock washers that hold the timing gear cover to the engine.

3. Remove the timing gear cover.

4. Remove the timing pointer.

5. Remove the timing gear cover gasket.

6. Remove and discard the crankshaft oil seal from the timing gear cover.

7. Replace in reverse order of the above procedure. Replace the crankshaft oil seal. Use a new timing gear cover gasket.

V6

1. Remove the water pump and crankshaft pulley.

2. Remove the two bolts that attach the oil pan to the timing chain cover.

3. Remove the five bolts that attach the timing chain cover to the engine block.

4. Remove the cover and gasket.

5. Remove the crankshaft front oil seal.

6. From the rear of the timing chain cover, coil new packing around the crankshaft hole in the cover so that the ends of the packing are at the top. Drive in the new packing with a punch. It will be necessary to ream out the hole to obtain clearance for the crankshaft vibration damper hub.

Water pump and timing chain cover bolts location: 6-225

232, 258 Sixes

1. Remove the drive belts, engine fan and hub assembly, the accessory pulley and vibration damper.

2. Remove the oil pan to timing chain cover screws and the screws that attach the cover to the block.

3. Raise the timing chain cover just high enough to detach the retaining nibs of the oil pan neoprene seal from the bottom side of the cover. This must be done to prevent pulling the seal end tabs away from the tongues of the oil pan gaskets which would cause a leak.

4. Remove the timing chain cover and gasket from the engine.

5. Use a razor blade to cut off the oil pan

6-232, 258 oil pan front seal trimming

seal end tabs flush with the front face of the cylinder block and remove the seal. Clean the timing chain cover, oil pan, and cylinder block surfaces.

6. Remove the crankshaft oil seal from the timing chain cover.

7. Install in reverse order of the above procedure. It will be necessary to cut the same amount from the end tabs of a new oil pan seal as was cut from the original seal, before installing the new gasket.

304, 360 and 401 V8s

1. Remove the negative battery cable.

2. Drain the cooling system and disconnect the radiator hoses and by-pass hose.

3. Remove all of the drive belts.

4. Remove the alternator and the front portion of the alternator bracket as an assembly.

5. Disconnect the heater hose.

6. Remove the power steering pump, and/or the air pump, and the mounting

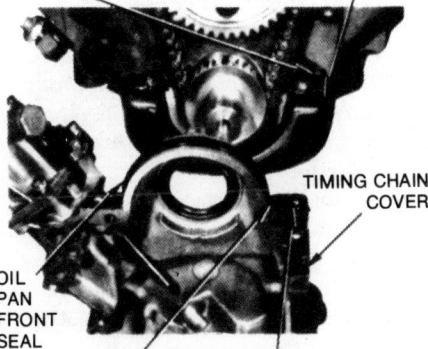

REMOVE DOWEL BEFORE INSTALLING COVER. INSTALL NEW DOWEL AFTER COVER IS INSTALLED

CUT OFF ORIGINAL GASKET FLUSH WITH END OF BLOCK. APPLY GASKET CEMENT (BOTH SIDES)

TIMING CHAIN COVER

OIL PAN FRONT SEAL

APPLY GASKET CEMENT. INDEX GASKET TAB UNDER SEAL

CEMENT PIECE OF NEW GASKET CUT OFF FLUSH WITH EDGE OF COVER (BOTH SIDES)

Timing chain cover and oil pan seal installation: 304, 360, 401

bracket as an assembly. Do not disconnect the power steering hoses.

7. Remove the distributor cap and note the position of the rotor. Remove the distributor. (See the Engine Electrical Section.)

8. Remove the fuel pump.

9. Remove the vibration damper and pulley.

10. Remove the two front oil pan bolts and the bolts which secure the timing chain cover to the engine block.

11. Remove the cover by pulling forward until it is free of the locating dowel pins.

12. Clean the gasket surface of the cover and the engine block.

13. Pry out the original seal from inside the timing chain cover and clean the seal bore.

14. Drive the new seal into place with a punch until it contacts the outer flange of the cover.

15. Apply a light film of motor oil to the lips of the new seal.

16. Install the timing chain cover in reverse order of removal. Refer to the "Engine Electrical" section for the distributor installation procedure.

327 V8

1. Drain the cooling system.

2. Loosen all drive belts and remove them.

3. Remove the radiator.

4. Remove the bolt, lockwasher, fly washer and the cork washer attaching the vibration damper to the crankshaft. Use a puller to remove the vibration damper from the crankshaft. A standard gear puller will do.

5. After the vibration damper is removed, the timing chain cover oil seal can be removed and a new one installed without removing the timing chain cover, by using a timing chain cover oil seal removal and installation tool (Jeep Tool Number J-9233). To remove the seal using the tool, turn the heavy threads into the oil seal until the seal is tightly engaged, then turn down the puller screw against the crankshaft to pull the seal out of the cover.

6. Install the seal with a timing chain cover oil seal installer (Jeep Tool Number J-9163). First, apply sealer lightly to the outer diameter of the seal and lubricate the inner diameter with clean engine oil.

If the special tools are not available, it is possible that the seal can be removed by prying it out with the tip of a screwdriver or another similar tool. The new seal can be installed with sealer applied to the outside and lubricant to the inside, by gently tapping it evenly in place with a hammer and a block of wood.

The only other way to replace the oil seal is to remove the entire timing chain cover from the engine and replace the seal with the cover on a work bench. Follow the procedure given below for removing the timing chain cover.

1. Drain the cooling system and remove the radiator, vibration damper, water pump and the fuel pump. Remove the alternator and power steering pump, if necessary.

2. Remove the water outlet manifold by removing the 4 capscrews and lockwashers holding the manifold to the engine. Remove the manifold, manifold gaskets, and the timing chain cover O-ring. Discard the O-ring.

3. Remove the two nuts and lockwashers from the studs at the bottom of the timing case cover and the nut and lockwasher from the stud below the water intake.

4. Remove the remaining capscrews of the timing case cover. Insert a screwdriver between the block and the timing case cover and lightly pry the cover from the block gasket.

5. Once the timing case cover is removed, the removal and installation of the oil seal is obvious. Replace the timing case cover in the reverse order of removal.

NOTE: *Do not tighten the timing case cover attaching bolts until the vibration damper is installed. This will assure that the cover and the damper are properly aligned.*

350 V8

1. Drain the cooling system and remove the radiator, shroud, fan, pulleys and drive belts.

2. Remove the crankshaft pulley, fuel pump and distributor.

3. Remove the alternator and the power steering pump, if necessary.

4. Loosen and slide rearward the front clamp on the thermostat by-pass hose.

5. Remove the vibration damper.

6. Remove the bolts attaching the timing chain cover to the cylinder block and oil pan. Remove the timing chain cover assembly and gasket.

7. Use a punch to drive the oil seal out of

8-350 timing chain cover attaching bolts

the cover. Drive the seal from the front toward the rear of the timing chain cover.

8. Tap the new seal into place from the rear of the timing case cover. If a rope seal is used, position the joint of the two ends at the top, then drive in a new shedder with a suitable punch. Stake the shedder in at least 3 places to secure it in position. Size the packing by rotating a hammer handle or similar smooth tool around the inside of the seal to obtain clearance for the crankshaft hub.

9. If the oil pump has not been removed from the timing chain cover, remove the attaching screws, oil pump cover and gasket from the timing chain cover.

10. Completely pack the space around the oil pump gears with petroleum jelly. There must be no air space left inside the pump. Secure the oil pump cover and a new gasket to the timing chain cover with the 5 slotted attaching screws. Torque the screws alternately and evenly to 8–12 ft lbs.

NOTE: *Unless the oil pump gears are packed with petroleum jelly, the pump may not prime itself when the engine is started, thus oil will not be pumped throughout the engine for lubrication.*

11. The gasket surfaces of the cylinder block and the timing chain cover must be smooth and clean before positioning the gasket in place on the engine. Use sealer on the engine and timing chain cover mating surfaces.

12. Position the timing chain cover to the engine block. Be sure that the dowel pins on the engine engage the holes in the cover before installing the bolts.

13. Install the attaching bolts and tighten them evenly to 30 ft lbs.

Timing Chain and Tensioner or Gears

REMOVAL AND INSTALLATION

F-Head

1. Remove the timing gear cover.
2. Use a puller to remove both the crankshaft and the camshaft gear from the engine after removing all attaching nuts and bolts.
3. Remove the Woodruff keys.

Installation is as follows:

1. Install the Woodruff key in the longer of the two keyways on the front end of the crankshaft.
2. Install the crankshaft timing gear on the front end of the crankshaft with the timing mark facing away from the cylinder block.
3. Align the keyway in the gear with the Woodruff key and then drive or press the gear onto the crankshaft firmly against the thrust washer.
4. Turn the camshaft or the crankshaft as necessary so that the timing marks on the two gears will be together after the camshaft gear is installed.
5. Install the Woodruff key in the keyway on the front of the camshaft.
6. Start the large timing gear on the camshaft with the timing mark facing out.

NOTE: *Do not drive the gear onto the camshaft as the camshaft may drive the plug out of the rear of the engine and cause an oil leak.*

7. Install the camshaft retaining screw and torque it to 30–40 ft lbs. This will draw the gear onto the camshaft as the screw is tightened. Standard running tolerance between the timing gears is 0.000 to 0.002 in.

V6

1. Remove the timing chain cover.
2. Make sure that the timing marks on the crankshaft and the camshaft sprockets are aligned. This will make installing the parts easier.

NOTE: *It is not necessary to remove the timing chain dampers (tensioners) unless they are worn or damaged and require replacement.*

3. Remove the front crankshaft oil slinger.
4. Remove the bolt and the special washer that hold the camshaft distributor drive gear and fuel pump eccentric at the forward end of the camshaft. Remove the eccentric and the gear from the camshaft.
5. Alternately pry forward the camshaft sprocket and then the crankshaft sprocket until the camshaft sprocket is pried from the camshaft.

1. Puller
2. Camshaft gear

4-134 timing gear removal

4-134 timing gear alignment

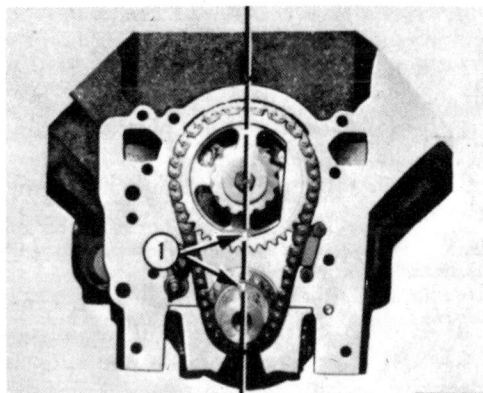

6-225 timing gear alignment

6. Remove the camshaft sprocket, sprocket key, and timing chain from the engine.

7. Pry the crankshaft sprocket from the crankshaft.

Install as follows:

1. If the engine has not been disturbed proceed to step Number 4 for installation procedures.

2. If the engine has been disturbed turn the crankshaft so that number one piston is at top dead center of the compression stroke.

3. Temporarily install the sprocket key and the camshaft sprocket on the camshaft. Turn the camshaft so that the index mark of the sprocket is downward. Remove the key and sprocket from the camshaft.

4. Assemble the timing chain and sprockets. Install the keys, sprockets, and chain assembly on the camshaft and crankshaft so that the index marks of both the sprockets are aligned.

NOTE: *It will be necessary to hold the spring loaded timing chain damper out of the way while installing the timing chain and sprocket assembly.*

5. Install the front oil slinger on the crankshaft with the inside diameter against the sprocket (concave side toward the front of the engine).

6. Install the fuel pump eccentric on the camshaft and the key, with the oil groove of the eccentric forward.

7. Install the distributor drive gear on the camshaft. Secure the gear and eccentric to the camshaft with the retaining washer and bolt.

8. Torque the bolt to 40–55 ft lbs.

232, 258 Sixes

1. Remove the drive belts, engine fan and hub assembly, accessory pulley, vibration damper and timing chain cover.

2. Remove the oil seal from the timing chain cover.

3. Remove the camshaft sprocket retaining bolt and washer.

4. Rotate the crankshaft until the timing mark on the crankshaft sprocket is closest to and in a center line with the timing pointer of the camshaft sprocket.

5. Remove the crankshaft sprocket, camshaft sprocket and timing chain as an assembly. Disassemble the chain and sprockets.

Installation is as follows:

1. Assemble the timing chain, crankshaft

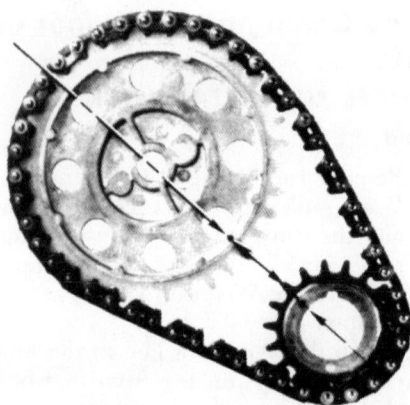

6-232, 258 timing mark alignment

sprocket and camshaft sprocket with the timing marks aligned.

2. Install the assembly to the crankshaft and the camshaft.

3. Install the camshaft sprocket retaining bolt and washer and tighten to 45–55 ft lbs.

4. Install the timing chain cover and a new oil seal.

5. Install the vibration damper, accessory pulley, engine fan and hub assembly and drive belts. Tighten the belts to the proper tension.

304, 360 and 401 V8s

1. Remove the timing chain cover and gasket.

2. Remove the crankshaft oil slinger.

8-304, 360, 401 timing mark alignment

3. Remove the camshaft sprocket retaining bolt and washer, distributor drive gear and fuel pump eccentric.

4. Rotate the crankshaft until the timing mark on the crankshaft sprocket is adjacent to, and on a center line with, the timing mark on the camshaft sprocket.

5. Remove the crankshaft sprocket, camshaft sprocket and timing chain as an assembly.

6. Clean all of the gasket surfaces.

Installation is as follows:

1. Assemble the timing chain, crankshaft sprocket and camshaft sprocket with the timing marks on both sprockets aligned.

2. Install the assembly to the crankshaft and the camshaft.

3. Install the fuel pump eccentric, distributor drive gear, washer and retaining bolt. Tighten the bolt to 25–35 ft lbs.

4. Install the crankshaft oil slinger.

5. Install the timing chain cover using a new gasket and oil seal.

327, 350 V8s

1. Remove the timing chain cover, and rotate the crankshaft until the timing marks on both the crankshaft and camshaft sprockets are aligned.

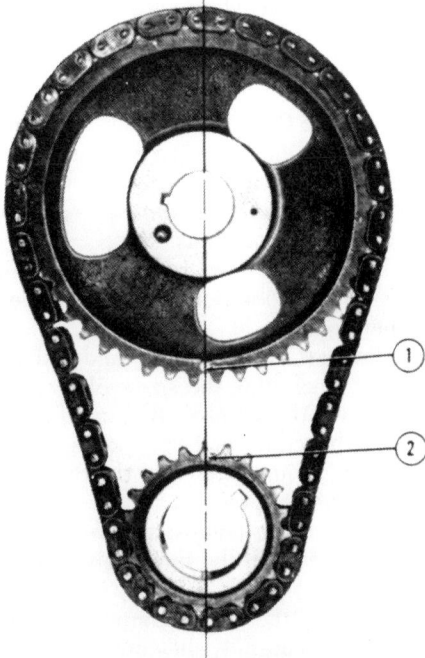

1. Camshaft sprocket   3. Timing chain
2. Crankshaft sprocket   4. Timing marks

8-350 timing mark alignment

2. Remove the crankshaft oil slinger from the crankshaft.

3. Remove the fuel pump eccentric from the face of the camshaft sprocket by removing the bolt, lockwasher and flat washer. On the 350 V8, remove the distributor drive gear.

4. Pry the camshaft sprocket and the crankshaft sprocket forward a little at a time until the camshaft sprocket is free from the camshaft. Remove the timing chain from both sprockets.

5. Continue prying the crankshaft sprocket forward until it is free of the crankshaft.

To install the timing chain and sprockets:

6. If the engine has been disturbed while the timing chain was removed from the engine or the timing marks on both the sprockets were not aligned prior to being removed, then it is necessary to establish the correct valve timing.

7. Index the "0" marks on the camshaft and crankshaft sprockets on an imaginary line drawn vertically through the centerline of each sprocket. See Step 1.

8. Adjust the positioning of the crankshaft and camshaft so that the sprockets can

1. Camshaft sprocket timing mark
2. Crankshaft sprocket timing mark

8-327 timing mark alignment

be installed in the position described in Step 7.

9. Without disturbing the positioning of the indexed sprockets and chain assembly, lift up the indexed assembly and slide the sprockets onto their shafts. Make sure the longer hubs of both sprockets face the block.

10. To check the assembly, rotate the crankshaft until the timing mark on the camshaft sprocket is on a horizontal line at either the 3 or 9 o'clock position. Count the number of links or pins on the timing chain between the two timing marks. There should be 10 links or 20 pins between the timing marks.

11. On the 327 V8, install the fuel pump eccentric, putting the dowel in the eccentric in the depression in the sprocket. On the 350 V8, install the fuel pump eccentric and the distributor drive gear. Install the bolt and washer and torque to 50–55 ft lbs.

12. Install the oil slinger on the crankshaft with the flange facing outward. Install any timing chain cover studs that may have been removed and install the timing chain cover.

Camshaft

REMOVAL AND INSTALLATION

F-Head

1. Remove the engine.
2. Remove the exhaust manifold.
3. Remove the oil pump and the distributor.
4. Remove the crankshaft pulley.
5. Remove the cylinder head.
6. Remove the exhaust valves.
7. Remove the timing gear and cover and the crankshaft and camshaft timing gears.
8. Remove the front end plate.
9. Push the intake and exhaust valve lifters into the cylinder block as far as possible so that the ends of the lifters are not in contact with the camshaft.
10. Secure each tappet in the raised position by installing a clip-type clothes pin on the shank of each tappet or tie them up in the valve chamber.
11. Remove the camshaft thrust plate attaching screws. Remove the camshaft thrust plate and spacer.
12. Pull the camshaft forward out of the cylinder block being careful to prevent damage to the camshaft bearing surfaces.
13. Install in the reverse order of the above procedure.

V6

1. Remove the engine.
2. Remove the intake manifold and carburetor assembly.
3. Remove the distributor.
4. Remove the fuel pump.
5. Remove the alternator, drive belts, cooling fan, fan pulley, and water pump.
6. Remove the crankshaft pulley and the vibration damper.
7. Remove the oil pump.
8. Remove the timing chain cover.
9. Remove the timing chain and the camshaft sprocket, along with the distributor drive gear and the fuel pump eccentric.
10. Remove the rocker arm assemblies.
NOTE: *The pushrods need not be removed. But if they are, be sure that they are replaced in their original positions.*
11. Lift the tappets up so that they are not in contact with the camshaft. Use wire clips or clip-type clothes pins to hold the tappets up.
12. Carefully guide the camshaft forward out of the engine. Avoid marring the bearing surfaces.
13. Install in reverse order of the above procedure.

232, 258 Sixes

1. Drain the cooling system and remove the radiator. If equipped with air-conditioning, discharge the system and remove the condenser.
2. Remove the valve cover and gasket, the rocker assemblies, pushrods, cylinder head and gasket and the lifters.
NOTE: *The pushrods must be replaced in their original locations.*
3. Remove the drive belts, cooling fan, fan hub assembly, vibration damper and the timing chain cover.
4. Remove the fuel pump and distributor assembly, including the spark plug wires.
5. Rotate the crankshaft until the timing mark of the crankshaft sprocket is adjacent to, and on a center line with, the timing mark of the camshaft sprocket.
6. Remove the crankshaft sprocket, camshaft sprocket, and the timing chain as an assembly.
7. Remove the front bumper or grille as required and carefully slide out the camshaft.
8. Install in reverse order of the above procedure.

304, 360 and 401 V8s

1. Disconnect the battery cables.
2. Drain the radiator and both banks of the block. Remove the lower hose at the radiator, the by-pass hose at the pump, the thermostat housing and the radiator. If the vehicle is equipped with air conditioning, remove the condenser and receiver assembly as a charged unit.
3. Remove the distributor, all wires, and the coil from the manifold.
4. Remove the intake manifold as an assembly.
5. Remove the valve covers, rocker arms and pushrods.
6. Remove the lifters.
7. Remove the cooling fan and hub assembly, fuel pump, and heater hose at the water pump.
8. Remove the alternator and bracket as an assembly. Just move it aside, do not disconnect the wiring.
9. Remove the crankshaft pulley and the damper. Remove the lower radiator hose at the water pump.
10. Remove the timing chain cover.
11. Remove the distributor-oil pump drive gear, fuel pump eccentric, sprockets and the timing chain.
12. Remove the hood catch support on the Commando.
13. Remove the camshaft carefully by sliding it forward out of the engine.
14. Install by reversing the above procedure.

327 V8

1. Disconnect the battery cables.
2. Drain the radiator and both banks of the cylinder block. Remove the lower hose at the radiator, the upper hose at the water pump outlet manifold and the radiator.
3. Remove the distributor, all wires and the ignition coil from the manifold.
4. Remove the intake manifold and carburetor as an assembly.
5. Remove the valve covers, rocker arms and pushrods. Identify the pushrods so that they can be installed in the same positions from which they were removed.
6. Remove the valve lifter cover and oil baffle.
7. Remove the valve lifters. Identify them so that they can be installed in the same positions from which they were removed.

8. Remove the alternator and bracket as an assembly. Do not disconnect the wiring, just move it aside.
9. Remove the water pump outlet manifold.
10. Remove the power steering pump bracket from the engine and lay the pump aside, out of the way. Do not disconnect the fluid lines.
11. Remove the fan.
12. Remove the vibration damper and the fuel pump.
13. Remove the timing chain cover.
14. Remove the crankshaft oil slinger, fuel pump eccentric and the timing chain and sprockets.
15. Remove the 3 screws and lockwashers that attach the camshaft thrust plate to the engine block and remove the thrust plate.
16. Remove the camshaft by carefully sliding the camshaft forward and out of the engine.
17. Before installing the camshaft, liberally lubricate it with clean engine oil.
18. Install the camshaft and assemble the engine in the reverse order of removal and disassembly.

350 V8

1. Disconnect the battery and drain the cooling system at both the radiator and the engine block. Disconnect the upper and lower hoses and the engine side of the hoses and the transmission oil cooling lines at the radiator, if the vehicle is equipped with an automatic transmission. Plug the transmission oil lines and remove the radiator.
2. If the vehicle is equipped with air conditioning, the system must be discharged and the condenser removed. It is best to consult an air conditioning mechanic as to the proper procedure. Better yet, allow him to do it for you.
3. Remove the intake manifold and carburetor assembly.
4. Remove the distributor.
5. Remove the fuel pump.
6. Remove the alternator bracket and alternator as an assembly from the engine and lay it aside. Do not disconnect any of the leads from the alternator.
7. Remove the cooling fan and water pump.
8. Remove the crankshaft pulley and the crankshaft vibration damper.

9. Remove the power steering pump bracket and remove the power steering pump and bracket as an assembly from the engine. Lay it aside, out of the way. Do not disconnect the fluid lines.

10. Remove the timing chain cover.

11. Remove the timing chain and sprockets.

12. Remove the rocker arm covers and remove the rocker assemblies, pushrods and the lifters. Identify the pushrods and lifters so that they can be installed in the same positions from which they were removed.

13. Slide the camshaft forward out of its bearing bores carefully, to avoid marring the bearing surfaces and remove the camshaft from the cylinder block.

14. Coat the camshaft with clean engine oil before installing the camshaft in the engine. Install the camshaft and assemble the engine in the reverse order of removal and disassembly.

Pistons and Connecting Rods

REMOVAL AND INSTALLATION
F-Head

The pistons and connecting rods can be removed from the engine with the engine in the vehicle. Refer to the engine rebuilding section for the proper procedure. The connecting rod identifying number must be toward the camshaft side of the block. Mark all pistons and rods so that they can be replaced in their original positions.

V6 and 350 V8

Use connecting rod bolt guides on the bolts to hold the upper half of the bearing shell in place when removing the bearing caps and bearings from the lower half of the connecting rods toward the camshaft.

Place the notch on the pistons forward and the oil spurt hole in the bottom of the connecting rod facing up.

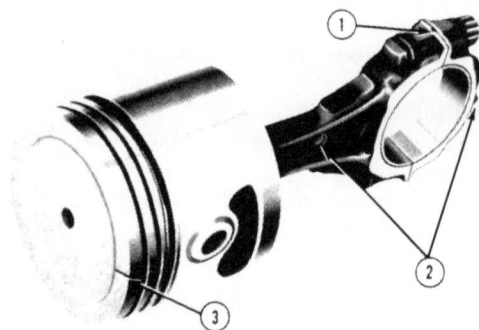

1. Oil spurt hole pointing up
2. Boss on rod and cap pointing forward
3. Notch on piston pointing forward

6-225 right bank piston and rod assembly

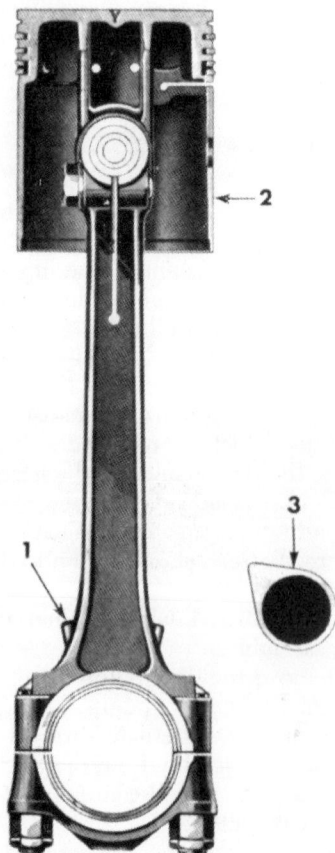

1. Oil spray hole
2. Piston skirt T-slot
3. Relative position of camshaft

4-134 piston and rod assembly

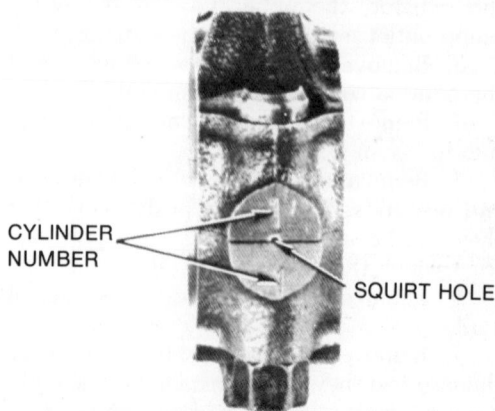

CYLINDER NUMBER

SQUIRT HOLE

6-232, 258 connecting rod and cap mating

232, 258 Sixes and 304, 327, 360 and 401 V8s

The connecting rods and caps are stamped with the number of the cylinder to which they belong. Replace them in their original positions.

The numbered sides and squirt hole on the connecting rods must face toward the camshaft when assembled in the sixes. The numbered sides must face out on the V8 engines.

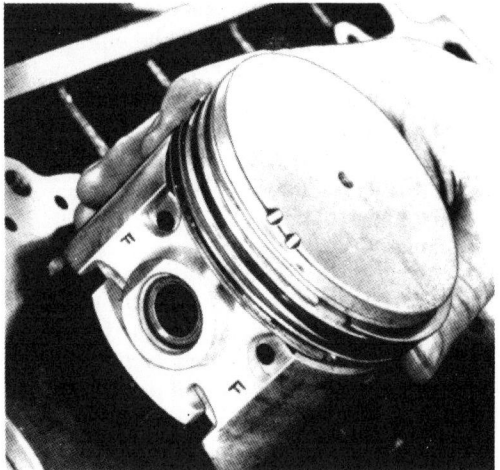

ROD AND CYLINDER NUMBERS TO OUTSIDE

SQUIRT HOLE TO INSIDE

8-304, 360, 401 connecting rod and cap mating

The pistons of all the engines listed above have notches in their tops. In addition, the 327 V8 pistons have the letter "F" stamped on their sides, near the wrist pin holes. Both of these markings must face toward the front of the engine when the pistons are installed in the cylinder block.

ENGINE LUBRICATION

Oil Pan

REMOVAL AND INSTALLATION

F-Head and V6

To remove the oil pan on the F-Head and V6 engines installed in the Commando, simply remove the oil pan attaching bolts and remove the oil pan. Clean all the mating surfaces and install a new gasket.

232 and 258 Sixes

To remove the oil pan from the 232 and 258 Sixes installed in any of the vehicles, follow the procedure given below.

1. Raise the vehicle and drain the engine oil.
2. Remove the starter motor.
3. On the Commando, place a jack under the transmission bellhousing. Disconnect the engine right support cushion bracket from the engine block and raise the engine to

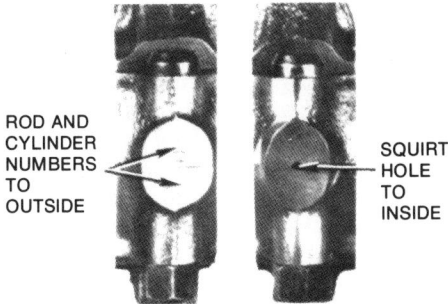

8-327 piston installation

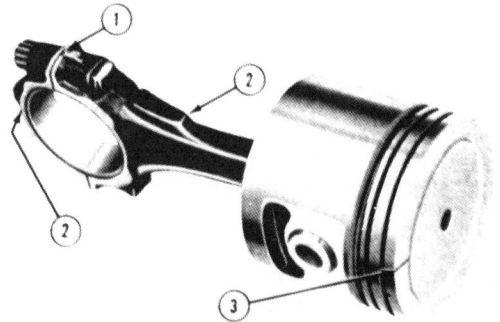

1. Oil spurt hole pointing up
2. Boss on rod and cap facing rearward
3. Notch on piston facing forward

8-350 piston positioning for installation

allow sufficient clearance for the removal of the oil pan.

4. Remove all of the oil pan attaching bolts and remove the oil pan.

5. Remove the oil pan front and rear oil seals and side gaskets. Thoroughly clean the gasket surfaces of the oil pan and engine block. Remove all sludge and dirt from the oil pan sump.

6. When installing the front oil pan seal to the timing chain cover, apply a generous amount of Permatex® No. 2 to the end tabs. Also, cement the oil pan side gaskets to the mating surface on the bottom of the engine block. Coat the inside curved surface of the new oil pan rear seal with soap and apply a generous amount of Permatex® No. 2 to the gasket contacting surface of the seal end tabs.

7. Install the seal in the recess of the rear main bearing cap, making certain that it is fully seated.

8. Apply engine oil to the oil pan contact-

ing surface of the front and rear oil pan seals.

9. Install the oil pan, lower the engine and connect the engine mount. Install the starter motor. Fill the crankcase with oil.

All V8 Engines

1. Drain the engine oil.
2. Remove the starter.
3. Remove the oil pan.
4. Install in the reverse order.

NOTE: *On the 304, 360 and 401 V8s, follow the procedure given for the Sixes pertaining to the removal and installation of the rubber front and rear oil pan end seals. The 327 and 350 V8s have no rubber end seals for the oil pan. Thoroughly clean all mating surfaces and remove all sludge and dirt from the oil pan.*

Rear Main Oil Seal

REPLACEMENT

All Engines

1. Remove the engine from the vehicle.
2. Remove the timing chain cover and the crankshaft timing gear.
3. Remove the oil pan, oil float support and the oil float.
4. Slide the crankshaft thrust washer and all of the end-play adjusting shims off the front end of the crankshaft.
5. Move the two pieces of the rear main bearing cap packing away from the side of the bearing cap and the cylinder block.
6. You will now be able to see the marks on the bearing caps and the cylinder block for bearing number and position.
7. Remove the screws and lockwashers that attach the main bearing caps to the cylinder block. Use a lifting bar beneath the ends of each bearing cap. Be careful not to exert too much pressure which could damage the cap or the dowels they fit onto. Lift each cap evenly on both sides until free of the dowels. If there is any reason to believe that any of the dowels have become bent during bearing cap removal, remove them and install new dowels. Remove the connecting rod caps and bearings.
8. Remove the crankshaft.
9. Remove the upper half of the rear main bearing oil seal from the cylinder block and the lower half from the oil seal groove in the rear main bearing cap.
10. Install the main bearing caps and bearings on the cylinder block in their original positions.

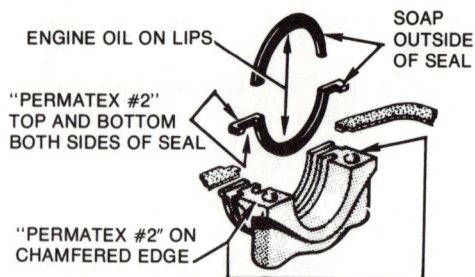

American motors engines rear main seal and cap assembly

11. Reassemble the engine in reverse order of the above procedure.

NOTE: *It is possible to replace the rear main rubber seal without removing the crankshaft. The procedure is as follows:*

1. Loosen the crankshaft cap bolts and lower the crankshaft not more than $1/32$ in. Remove the rear main bearing cap. Do not turn the crankshaft while it is loosened.
2. Push the seal around the groove until it can be gripped and removed with a pair of pliers.

1. Upper seal
2. Cylinder block
3. Rear main bearing shell
4. Lower seal
5. Side seal groove
6. Rear bearing cap
7. Bearing cap
8. Main bearing shell

8-327 rear main seal and cap

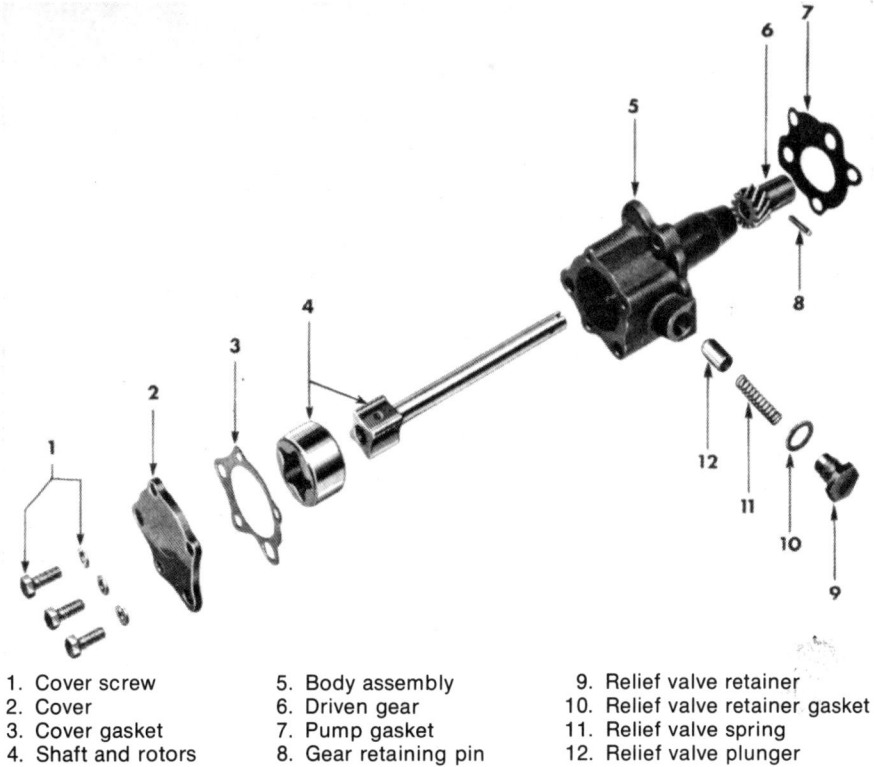

1. Cover screw
2. Cover
3. Cover gasket
4. Shaft and rotors
5. Body assembly
6. Driven gear
7. Pump gasket
8. Gear retaining pin
9. Relief valve retainer
10. Relief valve retainer gasket
11. Relief valve spring
12. Relief valve plunger

4-134 oil pump

3. Coat the new seal with clean motor oil and slide it into the groove until ⅜ in. protrudes from the groove. On American Motors engines push the seal around until its ends are flush with the block.

4. Install the other half of the seal in the bearing cap with ⅜ in. protruding from the opposite side. This is so that the juncture of the two halves of the seal is not at the same point as the juncture of the bearing cap and the cylinder block. On American Motors engines install the seal so that the ends are flush.

5. Install the rear main bearing cap and tighten all the caps to the proper torque. Be sure that the caps have not fallen out of place and that they are tightened in a straight manner.

Oil Pump

REMOVAL AND INSTALLATION

F-Head

1. Set number one piston at TDC in order to install the oil pump without disturbing the ignition timing.

2. Remove the distributor cover and note the position of the rotor. Keep the rotor in that position when the oil pump is installed.

3. Remove the cap screws and lockwashers that attach the oil pump to the cylinder block. Carefully slide the oil pump and its drive shaft out of the cylinder block.

The oil pump is driven by the camshaft by means of a spiral gear. The distributor in turn is driven by the oil pump by means of a tongue on the end of the distributor shaft which engages a slot in the end of the oil pump shaft. Because the tongue and the slot are both machined off center, the two shafts can be meshed in only one position. Since the position of the distributor shaft determines the timing of the engine, and is controlled by the oil pump shaft, the position of the oil pump shaft with respect to the camshaft is important.

If only the oil pump has been removed, install it so that the slot in the end of the shaft lines up with the tip of the distributor shaft and allows that shaft to slip into it without disturbing the original position of the distributor. If the engine has been disturbed or both the distributor and the oil pump have been removed, follow the procedure given below.

1. Turn the crankshaft to align the timing

marks on the crankshaft and camshaft timing gears.

2. Install the oil pump gasket on the pump.

3. With the wider side of the slot on top, start the oil pump drive shaft into the opening in the cylinder keeping the mounting holes in the body of the pump in alignment with the holes in the cylinder block.

4. Insert a long blade screwdriver into the distributor shaft opening in the side of the cylinder block and engage the slot in the oil pump shaft. Turn the shaft so that the slot is positioned at what would be roughly the nine-thirty position on a clock face.

5. Remove the screwdriver and observe the position of the slot in the end of the oil pump shaft to make certain it is properly positioned.

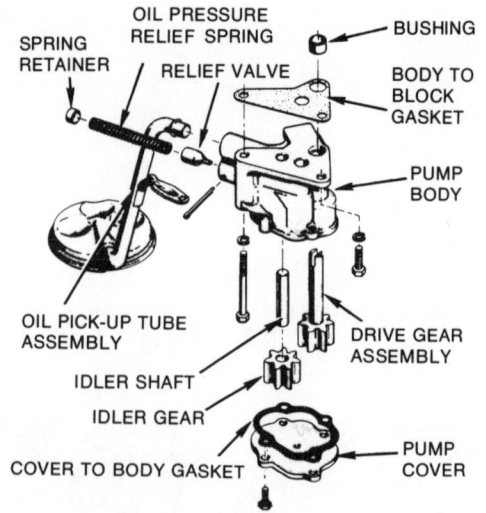

6-232, 258 oil pump

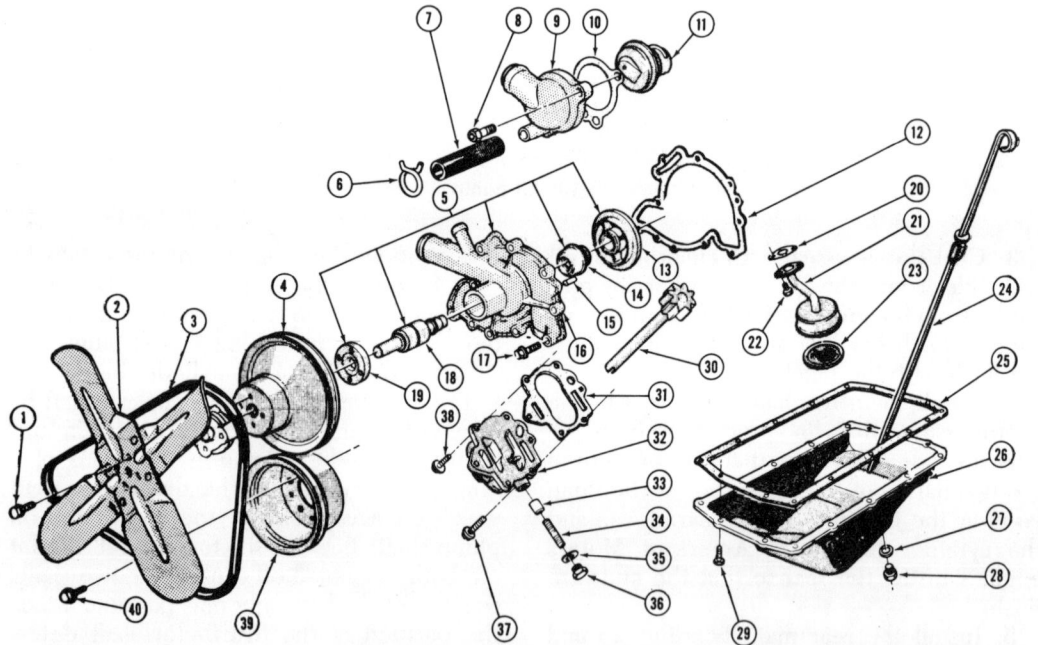

1. Bolt and lock washer
2. Fan assembly
3. Fan and alternator belt
4. Fan driven pulley
5. Water pump assembly
6. Hose clamp
7. Thermostat by-pass hose
8. Hex head bolt
9. Water outlet elbow
10. Water outlet elbow gasket
11. Thermostat
12. Water pump gasket
13. Impeller and insert, water pump
14. Water pump seal
15. Dowel pin
16. Water pump cover
17. Bolt
18. Water pump shaft and bearing
19. Fan hub
20. Oil suction pipe gasket
21. Oil suction housing, pipe and flange
22. Bolt
23. Oil pump screen
24. Oil dipstick
25. Oil pan gasket
26. Oil pan assembly
27. Drain plug gasket
28. Drain plug
29. Screw and lockwasher
30. Oil pump shaft and gear
31. Oil pump cover gasket
32. Valve by-pass and cover assembly
33. Oil pressure valve
34. Valve by-pass spring
35. Oil pressure valve cap gasket
36. Oil pressure valve cap
37. Screw
38. Screw
39. Fan driving pulley
40. Hex head bolt

6-225 lubrication system

6. Replace the screwdriver and, while turning the screwdriver clockwise to guide the oil pump drive shaft gear into engagement with the camshaft gear, press against the oil pump to force it into position.

7. Remove the screwdriver and again observe the position of the slot. If the installation was properly made, the slot will be in a position roughly equivalent to the eleven o'clock position on the face of a clock, with the wider side of the slot still on the top. If the slot is improperly positioned, remove the oil pump and repeat the operation.

8. Coat the threads of the capscrews with gasket cement and secure the oil pump in place.

V6

1. Remove the oil filter.
2. Disconnect the wire from the oil pressure indicator switch in the filter by-pass valve cap.
3. Remove the screws that attach the oil pump cover assembly to the timing chain cover.
4. Remove the cover assembly and slide out the oil pump.
5. Install in reverse order of the above procedure.

232, 258 Sixes

1. Drain the oil and remove the oil pan.
2. Remove the oil pump retaining screws and separate the oil pump and gasket from the engine block.

NOTE: *Do not disturb the position of the oil pick-up tube and screen assembly in the pump body. If the tube is moved within the pump body, a new assembly must be installed to assure an airtight seal.*

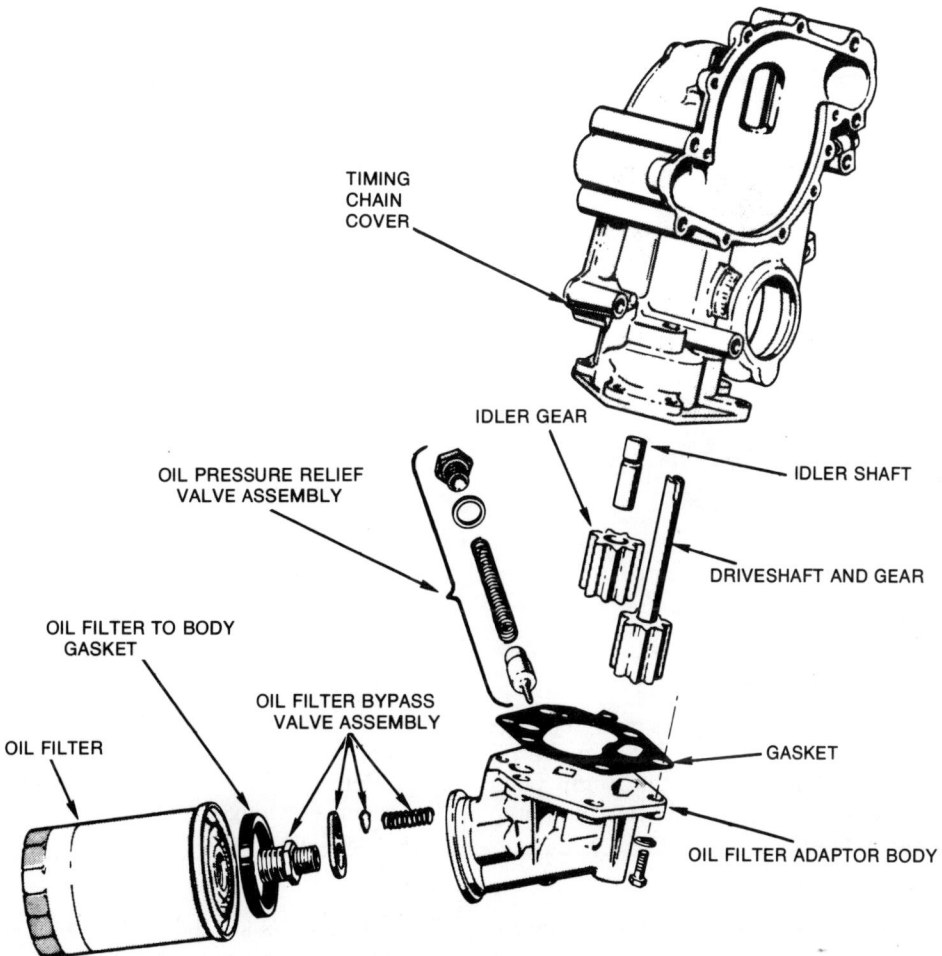

8-304, 360, 401 oil pump

1. Stud
2. Gasket
3. Filter mounting bracket
4. Flat washer
5. Lockwasher
6. Nut
7. Filter cartridge
8. Relief valve spring
9. Idler shaft
10. Idler gear
11. Oil pump cover
12. Discharge tube
13. Capscrew
14. Lockwasher
15. Capscrew
16. Lockwasher
17. Inlet tube and screen
    assembly
18. Drive gear
19. Woodruff key
20. Driveshaft
21. Relief valve plunger
22. Gasket
23. Pump body
24. Pump gasket

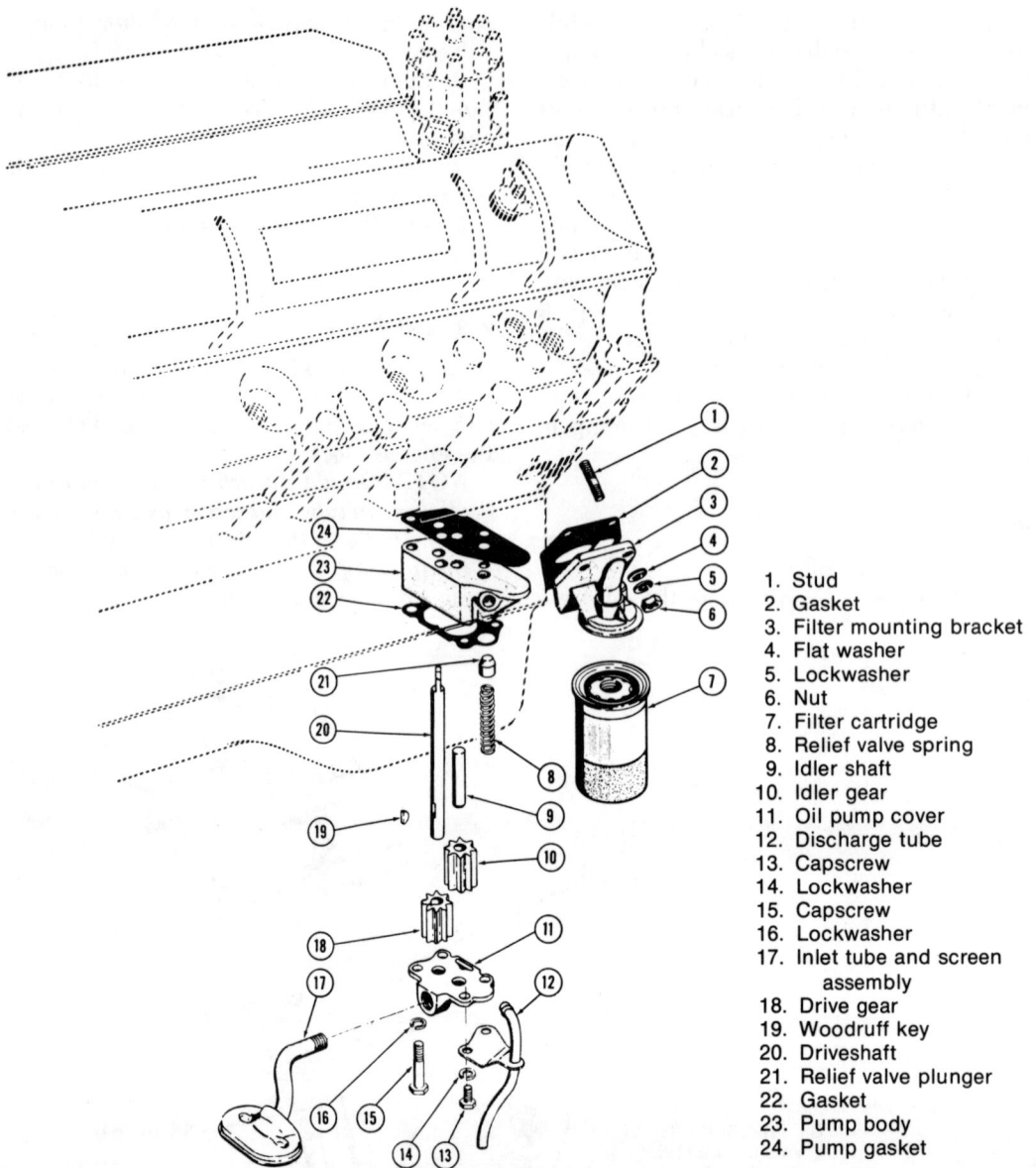

8-327 oil pump and filter

3. Install in reverse order of the above procedure.

304, 360 and 401 V8s

Remove the retaining screws and separate the oil pump cover and gasket and the oil filter as an assembly from the pump body (timing chain cover). Install the oil pump with a new filter.

327 V8

The oil pump is driven by the distributor driveshaft. Oil pump replacement does not, however, affect the timing of the distributor because the drive gear remains in mesh with the camshaft gear.

1. Drain the oil and remove the oil pan.
2. Remove the oil pump attaching screws. Remove the pump and gasket from the engine.
3. Remove the pump cover by removing the cover attaching screws.
4. Installation is the reverse of removal.

350 V8

The oil pump is located on the right side of the timing chain cover. The housing of the oil pump is actually part of the timing chain

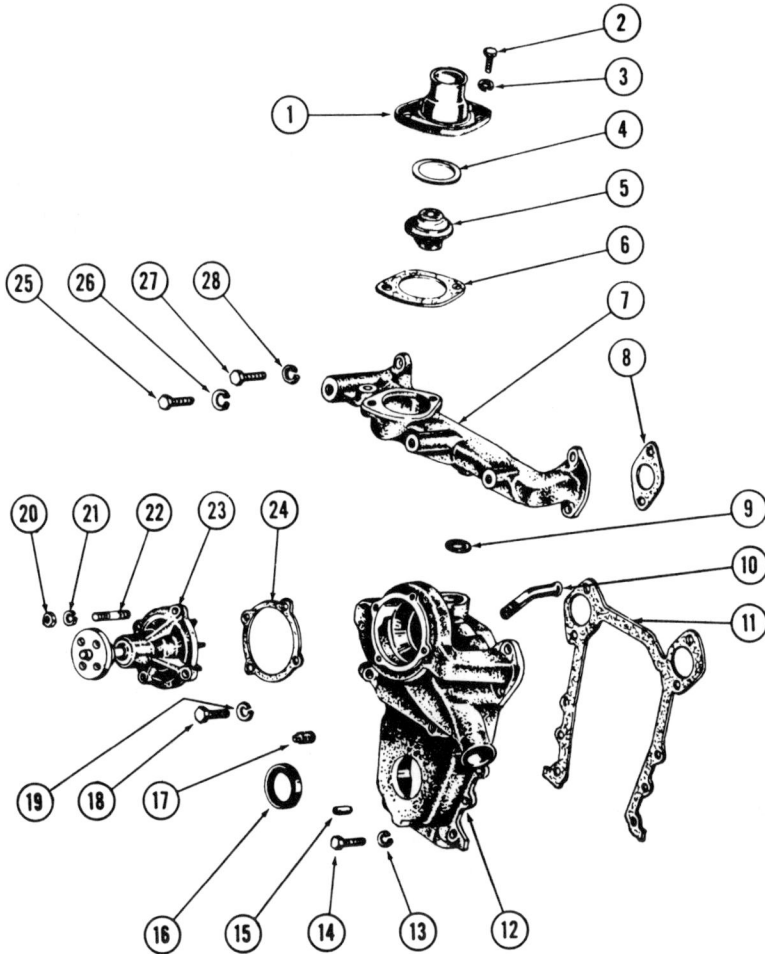

1. Thermostat housing	8. Gasket	15. Dowel	22. Stud
2. Capscrew	9. O-ring	16. Seal	23. Water pump
3. Lockwasher	10. Hose adapter	17. Plug	24. Gasket
4. Gasket	11. Gear cover gasket	18. Capscrew	25. Capscrew
5. Thermostat	12. Gear cover	19. Lockwasher	26. Lockwasher
6. Gasket	13. Lockwasher	20. Nut	27. Capscrew
7. Water outlet manifold	14. Capscrew	21. Lockwasher	28. Lockwasher

8-327 water pump assembly

cover. To remove the oil pump, remove the cover attaching screws, the cover and slide out the oil pump gears. When installing the oil pump, pack the spaces around the oil pump gears completely with petroleum jelly. Leave no air space between the gears. This is done so that the oil pump will start pumping oil immediately when the engine is started. Installation is the reverse of removal.

ENGINE COOLING

The satisfactory performance of any water cooled engine is determined to the greatest extent by the proper operation of the cooling system. The engine block is fully water jacketed to prevent distortion of the cylinder walls. Directed cooling and water holes in the cylinder head causes water to flow past the valve seats, which are one of the hottest parts of any engine, and carry the heat away from the valves and seats.

The minimum temperature of the coolant is controlled by a thermostat mounted in the outlet passage of the engine. When the coolant temperature is below the temperature rating of the thermostat, the thermostat remains closed and the coolant is directed through the radiator-bypass hose to the water pump. If the coolant temperature is too high, the thermostat opens and coolant flow is directed to the top of the radiator. The radiator dissipates the excess engine heat before the coolant is recirculated through the engine.

The cooling system is pressurized and the operating pressure is regulated by the rating of the radiator cap which contains a relief valve.

Radiator

REMOVAL AND INSTALLATION

1. Drain the radiator by opening the drain cock and removing the radiator pressure cap.

2. Remove the upper and lower hose clamps and hoses at the radiator. If the vehicle is equipped with an automatic transmission, disconnect the transmission oil lines at the radiator. Plug the lines to prevent loss of fluid and the entrance of dirt.

3. Remove all attaching screws that secure the radiator to the radiator body support.

4. Remove the radiator.

5. Replace in reverse order of the above procedure.

Thermostat

REMOVAL AND INSTALLATION

To remove the thermostats from all of these engines, first drain the cooling system. It is

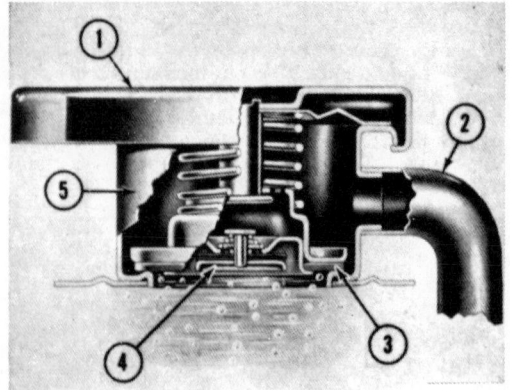

1. Pressure cap
2. Overflow tube
3. Pressure seal
4. Vacuum release valve
5. Radiator neck

A typical radiator cap

not necessary to disconnect or remove any of the hoses. Remove the two attaching screws and lift the housing from the engine. Remove the thermostat and the gasket. To install, place the thermostat in the housing with the spring inside the engine. Install a new gasket with a small amount of sealing compound applied to both sides. Install the water outlet and tighten the attaching bolts to 30 ft lbs. Refill the cooling system.

Engine Rebuilding

This section describes, in detail, the procedures involved in rebuilding a typical engine. The procedures specifically refer to an inline engine, however, they are basically identical to those used in rebuilding engines of nearly all design and configurations. Procedures for servicing atypical engines (i.e., horizontally opposed) are described in the appropriate section, although in most cases, cylinder head reconditioning procedures described in this chapter will apply.

The section is divided into two sections. The first, Cylinder Head Reconditioning, assumes that the cylinder head is removed from the engine, all manifolds are removed, and the cylinder head is on a workbench. The camshaft should be removed from overhead cam cylinder heads. The second section, Cyl-

Torque (ft. lbs.)*

U.S./Bolt Grade (SAE)							Metric/Bolt Grade				
Bolt Diameter (inches)	⬡ 1 and 2	⬡ 5	⬡ 6	⬡ 8	Wrench Size (inches) Bolt	Nut	Bolt Diameter (mm)	5D 5D	8G 8G	10K 10K	12K 12K	Wrench Size (mm) Bolt and Nut
1/4	5	7	10	10.5	3/8	7/16	6	5	6	8	10	10
5/16	9	14	19	22	1/2	9/16	8	10	16	22	27	14
3/8	15	25	34	37	9/16	5/8	10	19	31	40	49	17
7/16	24	40	55	60	5/8	3/4	12	34	54	70	86	19
1/2	37	60	85	92	3/4	13/16	14	55	89	117	137	22
9/16	53	88	120	132	7/8	7/8	16	83	132	175	208	24
5/8	74	120	167	180	15/16	1	18	111	182	236	283	27
3/4	120	200	280	296	1 1/8	1 1/8	22	182	284	394	464	32
7/8	190	302	440	473	1 5/16	1 5/16	24	261	419	570	689	36
1	282	466	660	714	1 1/2	1 1/2						

*—Torque values are for lightly oiled bolts. CAUTION: Bolts threaded into aluminum require much less torque.

inder Block Reconditioning, covers the block, pistons, connecting rods and crankshaft. It is assumed that the engine is mounted on a work stand, and the cylinder head and all accessories are removed.

Procedures are identified as follows:

Unmarked—Basic procedures that must be performed in order to successfully complete the rebuilding process.

Starred (*)—Procedures that should be performed to ensure maximum performance and engine life.

Double starred (**)—Procedures that may be performed to increase engine performance and reliability. These procedures are usually reserved for extremely heavy-duty or competition usage.

In many cases, a choice of methods is also provided. Methods are identified in the same manner as procedures. The choice of method for a procedure is at the discretion of the user.

The tools required for the basic rebuilding procedure should, with minor exceptions, be those included in a mechanic's tool kit. An accurate torque wrench, and a dial indicator (reading in thousandths) mounted on a universal base should be available. Bolts and nuts with no torque specification should be tightened according to size (see chart). Special tools, where required, all are readily available from the major tool suppliers (i.e., Craftsman, Snap-On, K-D). The services of a competent automotive machine shop must also be readily available.

When assembling the engine, any parts that will be in frictional contact must be pre-lubricated, to provide protection on initial start-up. Vortex Pre-Lube, STP, or any product specifically formulated for this purpose may be used. NOTE: *Do not use engine oil.* Where semi-permanent (locked but removable) installation of bolts or nuts is desired, threads should be cleaned and coated with Loctite. Studs may be permanently installed using Loctite Stud and Bearing Mount.

Aluminum has become increasingly popular for use in engines, due to its low weight and excellent heat transfer characteristics. The following precautions must be observed when handling aluminum engine parts:

—Never hot-tank aluminum parts.

—Remove all aluminum parts (identification tags, etc.) from engine parts before hot-tanking (otherwise they will be removed during the process).

—Always coat threads lightly with engine oil or anti-seize compounds before installation, to prevent seizure.

—Never over-torque bolts or spark plugs in aluminum threads. Should stripping occur, threads can be restored according to the following procedure, using Heli-Coil thread inserts:

Tap drill the hole with the stripped threads to the specified size (see chart). Using the specified tap (NOTE: *Heli-Coil tap sizes refer to the size thread being replaced, rather than the actual tap size*), tap the hole for the Heli-

STANDARD SCREW FITS IN

HELI-COIL INSERT IN HELI-COIL TAPPED HOLE

Heli-Coil installation (© Chrysler Corp.)

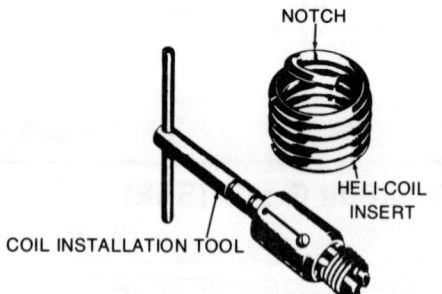

NOTCH

HELI-COIL INSERT

COIL INSTALLATION TOOL

Heli-Coil and installation tool

Heli-Coil Specifications

Heli-Coil Insert			Drill	Tap	Insert. Tool	Extracting Tool
Thread Size	Part No.	Insert Length (In.)	Size	Part No.	Part No.	Part No.
1/2 -20	1185-4	3/8	17/64(.266)	4 CPB	528-4N	1227-6
5/16-18	1185-5	15/32	Q(.332)	5 CPB	528-5N	1227-6
3/8 -16	1185-6	9/16	X(.397)	6 CPB	528-6N	1227-6
7/16-14	1185-7	21/32	29/64(.453)	7 CPB	528-7N	1227-16
1/2 -13	1185-8	3/4	33/64(.516)	8 CPB	528-8N	1227-16

Coil. place the insert on the proper installation tool (see chart). Apply pressure on the insert while winding it clockwise into the hole, until the top of the insert is one turn below the surface. Remove the installation tool, and break the installation tang from the bottom of the insert by moving it up and down. If the Heli-Coil must be removed, tap the removal tool firmly into the hole, so that it engages the top thread, and turn the tool counterclockwise to extract that insert.

Snapped bolts or studs may be removed, using a stud extractor (unthreaded) or Vise-Grip pliers (threaded). Penetrating oil (e.g., Liquid Wrench) will often aid in breaking frozen threads. In cases where the stud or bolt is flush with, or below the surface, proceed as follows:

Drill a hole in the broken stud or bolt, approximately 1/2 its diameter. Select a screw extractor (e.g., Easy-Out) of the proper size, and tap it into the stud or bolt. Turn the extractor counter-clockwise to remove the stud or bolt.

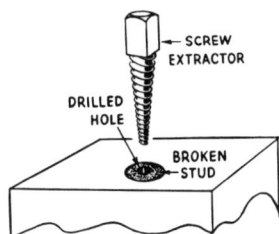

Screw extractor

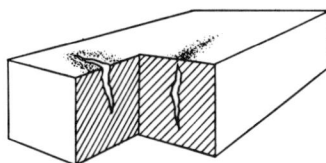

Magnaflux indication of cracks

Magnaflux and Zyglo are inspection techniques used to locate material flaws, such as stress cracks. Magnafluxing coats the part with fine magnetic particles, and subjects the part to a magnetic field. Cracks cause breaks in the magnetic field, which are outlined by the particles. Since Magnaflux is a magnetic process, it is applicable only to ferrous materials. The Zyglo process coats the material with a fluorescent dye penetrant, and then subjects it to blacklight inspection, under which cracks glow brightly. Parts made of any material may be tested using Zyglo. While Magnaflux and Zyglo are excellent for general inspection, and locating hidden defects, specific checks of suspected cracks may be made at lower cost and more readily using spot check dye. The dye is sprayed onto the suspected area, wiped off, and the area is then sprayed with a developer. Cracks then will show up brightly. Spot check dyes will only indicate surface cracks; therefore, structural cracks below the surface may escape detection. When questionable, the part should be tested using Magnaflux or Zyglo.

Cylinder Head Reconditioning

NOTE: *This engine rebuilding section is a guide to accepted engine rebuilding procedures. Every effort is made to illustrate the engine(s) used by this manufacturer; but, occasionally, typical examples of standard engine rebuilding practice are illustrated.*

Procedure	Method
Identify the valves: 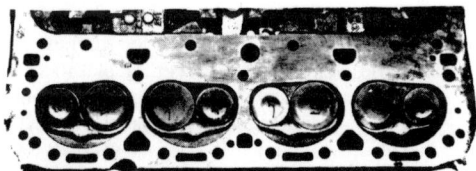 Valve identification	Invert the cylinder head, and number the valve faces front to rear, using a permanent felt-tip marker.
Remove the rocker arms:	Remove the rocker arms with shaft(s) or balls and nuts. Wire the sets of rockers, balls and nuts together, and identify according to the corresponding valve.

Procedure	Method
Remove the valves and springs:	Using an appropriate valve spring compressor (depending on the configuration of the cylinder head), compress the valve springs. Lift out the keepers with needlenose pliers, release the compressor, and remove the valve, spring, and spring retainer.
Check the valve stem-to-guide clearance:  Checking the valve stem-to-guide clearance	Clean the valve stem with lacquer thinner or a similar solvent to remove all gum and varnish. Clean the valve guides using solvent and an expanding wire-type valve guide cleaner. Mount a dial indicator so that the stem is at 90° to the valve stem, as close to the valve guide as possible. Move the valve off its seat, and measure the valve guide-to-stem clearance by moving the stem back and forth to actuate the dial indicator. Measure the valve stems using a micrometer, and compare to specifications, to determine whether stem or guide wear is responsible for excessive clearance.
De-carbon the cylinder head and valves: 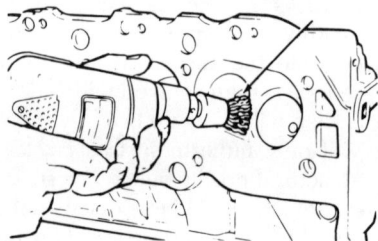 Removing carbon from the cylinder head	Chip carbon away from the valve heads, combustion chambers, and ports, using a chisel made of hardwood. Remove the remaining deposits with a stiff wire brush. **NOTE: *Ensure that the deposits are actually removed, rather than burnished.***
Hot-tank the cylinder head:	Have the cylinder head hot-tanked to remove grease, corrosion, and scale from the water passages. **NOTE: *In the case of overhead cam cylinder heads, consult the operator to determine whether the camshaft bearings will be damaged by the caustic solution.***
Degrease the remaining cylinder head parts:	Using solvent (e.g., Gunk), clean the rockers, rocker shaft(s) (where applicable), rocker balls and nuts, springs, spring retainers, and keepers. Do not remove the protective coating from the springs.
Check the cylinder head for warpage: 1 & 3 CHECK DIAGONALLY 2 CHECK ACROSS CENTER 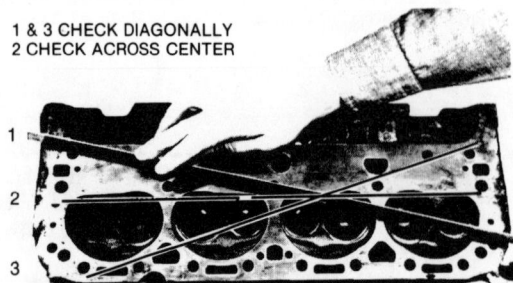 Checking the cylinder head for warpage	Place a straight-edge across the gasket surface of the cylinder head. Using feeler gauges, determine the clearance at the center of the straight-edge. Measure across both diagonals, along the longitudinal centerline, and across the cylinder head at several points. If warpage exceeds .003″ in a 6″ span, or .006″ over the total length, the cylinder head must be resurfaced. **NOTE: *If warpage exceeds the manufacturers maximum tolerance for material removal, the cylinder head must be replaced.*** When milling the cylinder heads of V-type engines, the intake manifold mounting position is altered, and must be corrected by milling the manifold flange a proportionate amount.

Procedure	Method
** Porting and gasket matching:	** Coat the manifold flanges of the cylinder head with Prussian blue dye. Glue intake and exhaust gaskets to the cylinder head in their installed position using rubber cement and scribe the outline of the ports on the manifold flanges. Remove the gaskets. Using a small cutter in a hand-held power tool (i.e., Dremel Moto-Tool), gradually taper the walls of the port out to the scribed outline of the gasket. Further enlargement of the ports should include the removal of sharp edges and radiusing of sharp corners. Do not alter the valve guides. **NOTE:** *The most efficient port configuration is determined only by extensive testing. Therefore, it is best to consult someone experienced with the head in question to determine the optimum alterations.*

Marking the cylinder head for gasket matching

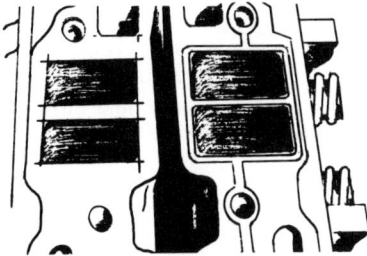

Port configuration before and after gasket matching

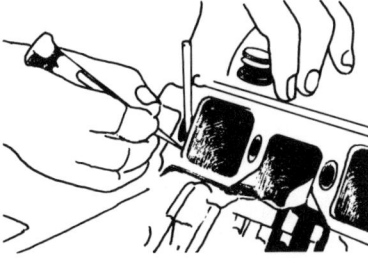

| ** Polish the ports: | ** Using a grinding stone with the above mentioned tool, polish the walls of the intake and exhaust ports, and combustion chamber. Use progressively finer stones until all surface imperfections are removed. **NOTE:** *Through testing, it has been determined that a smooth surface is more effective than a mirror polished surface in intake ports, and vice-versa in exhaust ports.* |

Relieved and polished ports

| * Knurling the valve guides: | * Valve guides which are not excessively worn or distorted may, in some cases, be knurled rather than replaced. Knurling is a process in which metal is displaced and raised, thereby reducing clearance. Knurling also provides excellent oil control. The possibility of knurling rather than replacing valve guides should be discussed with a machinist. |

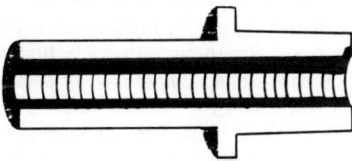

Cut-away view of a knurled valve guide

| Replacing the valve guides: **NOTE:** *Valve guides should only be replaced if damaged or if an oversize valve stem is not available.* | Depending on the type of cylinder head, valve guides may be pressed, hammered, or shrunk in. In cases where the guides are shrunk into the head, replacement should be left to an equipped machine shop. In other cases, the guides are replaced as follows: Press or tap the valve guides out of the head using a stepped drift (see illustration). Determine the height above the boss that the guide must extend, and obtain a stack of washers, their I.D. similar to the guide's O.D., of that height. Place the stack of washers on the guide, and insert the guide into the boss. **NOTE:** *Valve* |

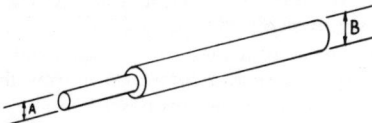

A-VALVE GUIDE I.D.
B-SLIGHTLY SMALLER THAN VALVE GUIDE O.D.

Valve guide removal tool

Procedure	Method

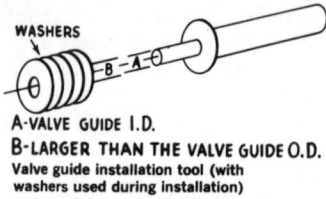

WASHERS

B — A

A-VALVE GUIDE I.D.
B-LARGER THAN THE VALVE GUIDE O.D.
Valve guide installation tool (with
washers used during installation)

guides are often tapered or beveled for installation. Using the stepped installation tool (see illustration), press or tap the guides into position. Ream the guides according to the size of the valve stem.

Replacing valve seat inserts:

Replacement of valve seat inserts which are worn beyond resurfacing or broken, if feasible, must be done by a machine shop.

Resurfacing (grinding) the valve face:

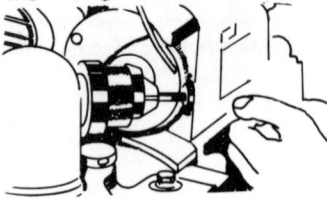

Grinding a valve

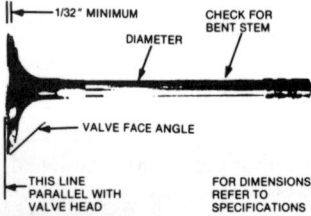

1/32" MINIMUM   CHECK FOR BENT STEM

DIAMETER

VALVE FACE ANGLE

THIS LINE   FOR DIMENSIONS,
PARALLEL WITH   REFER TO
VALVE HEAD   SPECIFICATIONS
Critical valve dimensions

Using a valve grinder, resurface the valves according to specifications. **CAUTION:** *Valve face angle is not always identical to valve seat angle.* A minimum margin of $1/32''$ should remain after grinding the valve. The valve stem tip should also be squared and resurfaced, by placing the stem in the V-block of the grinder, and turning it while pressing lightly against the grinding wheel.

Resurfacing the valve seats using reamers:

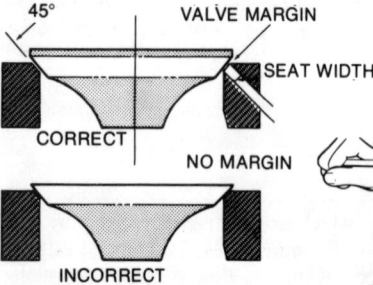

45°   VALVE MARGIN

SEAT WIDTH

CORRECT

NO MARGIN

INCORRECT
Valve seat width and centering

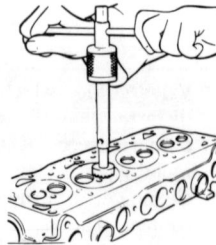

Reaming the valve seat

Select a reamer of the correct seat angle, slightly larger than the diameter of the valve seat, and assemble it with a pilot of the correct size. Install the pilot into the valve guide, and using steady pressure, turn the reamer clockwise. **CAUTION:** *Do not turn the reamer counter-clockwise.* Remove only as much material as necessary to clean the seat. Check the concentricity of the seat (see below). If the dye method is not used, coat the valve face with Prussian blue dye, install and rotate it on the valve seat. Using the dye marked area as a centering guide, center and narrow the valve seat to specifications with correction cutters. **NOTE:** *When no specifications are available, minimum seat width for exhaust valves should be* $5/64''$, *intake valves* $1/16''$. After making correction cuts, check the position of the valve seat on the valve face using Prussian blue dye.

* Resurfacing the valve seats using a grinder:

Grinding a valve seat

Select a pilot of the correct size, and a coarse stone of the correct seat angle. Lubricate the pilot if necessary, and install the tool in the valve guide. Move the stone on and off the seat at approximately two cycles per second, until all flaws are removed from the seat. Install a fine stone, and finish the seat. Center and narrow the seat using correction stones, as described above.

Procedure	Method

Checking the valve seat concentricity:

Checking the valve seat concentricity using a dial gauge

Coat the valve face with Prussian blue dye, install the valve, and rotate it on the valve seat. If the entire seat becomes coated, and the valve is known to be concentric, the seat is concentric.

* Install the dial gauge pilot into the guide, and rest the arm on the valve seat. Zero the gauge, and rotate the arm around the seat. Run-out should not exceed .002".

---

* **Lapping the valves: NOTE:** *Valve lapping is done to ensure efficient sealing of resurfaced valves and seats. Valve lapping alone is not recommended for use as a resurfacing procedure.*

Hand lapping the valves

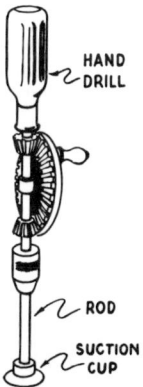

Home made mechanical valve lapping tool

* Invert the cylinder head, lightly lubricate the valve stems, and install the valves in the head as numbered. Coat valve seats with fine grinding compound, and attach the lapping tool suction cup to a valve head (**NOTE:** *Moisten the suction cup*). Rotate the tool between the palms, changing position and lifting the tool often to prevent grooving. Lap the valve until a smooth, polished seat is evident. Remove the valve and tool, and rinse away all traces of grinding compound.

** Fasten a suction cup to a piece of drill rod, and mount the rod in a hand drill. Proceed as above, using the hand drill as a lapping tool. **CAUTION:** *Due to the higher speeds involved when using the hand drill, care must be exercised to avoid grooving the seat.* Lift the tool and change direction of rotation often.

Check the valve springs:

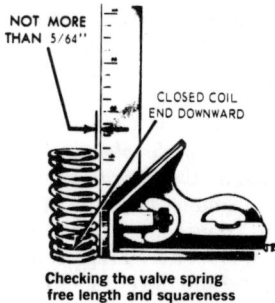

Checking the valve spring free length and squareness

Checking the valve spring tension

Place the spring on a flat surface next to a square. Measure the height of the spring, and rotate it against the edge of the square to measure distortion. If spring height varies (by comparison) by more than $1/16''$ or if distortion exceeds $1/16''$, replace the spring.

** In addition to evaluating the spring as above, test the spring pressure at the installed and compressed (installed height minus valve lift) height using a valve spring tester. Springs used on small displacement engines (up to 3 liters) should be ± 1 lb of all other springs in either position. A tolerance of ± 5 lbs is permissible on larger engines.

Procedure	Method

* Install valve stem seals:

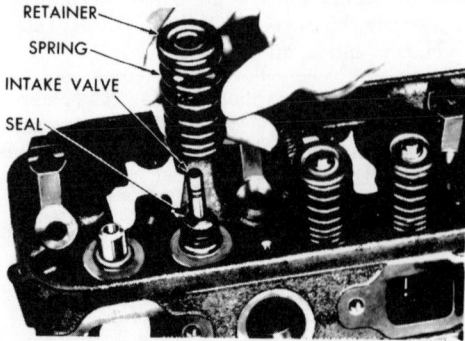

RETAINER

SPRING

INTAKE VALVE

SEAL

Valve stem seal installation

* Due to the pressure differential that exists at the ends of the intake valve guides (atmospheric pressure above, manifold vacuum below), oil is drawn through the valve guides into the intake port. This has been alleviated somewhat since the addition of positive crankcase ventilation, which lowers the pressure above the guides. Several types of valve stem seals are available to reduce blow-by. Certain seals simply slip over the stem and guide boss, while others require that the boss be machined. Recently, Teflon guide seals have become popular. Consult a parts supplier or machinist concerning availability and suggested usages. **NOTE:** *When installing seals, ensure that a small amount of oil is able to pass the seal to lubricate the valve guides; otherwise, excessive wear may result.*

Install the valves:

Lubricate the valve stems, and install the valves in the cylinder head as numbered. Lubricate and position the seals (if used, see above) and the valve springs. Install the spring retainers, compress the springs, and insert the keys using needlenose pliers or a tool designed for this purpose. **NOTE:** *Retain the keys with wheel bearing grease during installation.*

Checking valve spring installed height:

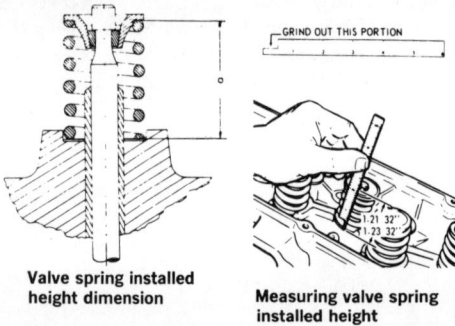

GRIND OUT THIS PORTION

Valve spring installed
height dimension

Measuring valve spring
installed height

Measure the distance between the spring pad and the lower edge of the spring retainer, and compare to specifications. If the installed height is incorrect, add shim washers between the spring pad and the spring. **CAUTION:** *Use only washers designed for this purpose.*

** CC'ing the combustion chambers:

** Invert the cylinder head and place a bead of sealer around a combustion chamber. Install an apparatus designed for this purpose (burette mounted on a clear plate; see illustration) over the combustion chamber, and fill with the specified fluid to an even mark on the burette. Record the burette reading, and fill the combustion chamber with fluid. (**NOTE:** *A hole drilled in the plate will permit air to escape.*) Subtract the burette reading, with the combustion chamber filled, from the previous reading, to determine combustion chamber volume in cc's. Duplicate this procedure in all combustion chambers on the cylinder head,

Procedure	Method

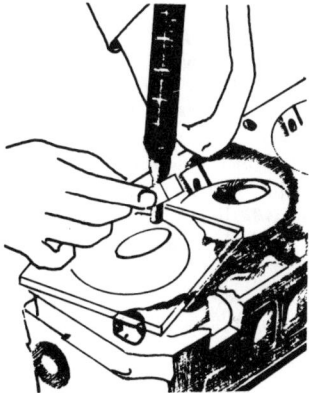

CC'ing the combustion chamber

and compare the readings. The volume of all combustion chambers should be made equal to that of the largest. Combustion chamber volume may be increased in two ways. When only a small change is required (usually), a small cutter or coarse stone may be used to remove material from the combustion chamber. NOTE: *Check volume frequently.* Remove material over a wide area, so as not to change the configuration of the combustion chamber. When a larger change is required, the valve seat may be sunk (lowered into the head). NOTE: *When altering valve seat, remember to compensate for the change in spring installed height.*

Inspect the rocker arms, balls, studs, and nuts (where applicable):

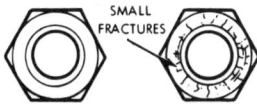

Stress cracks in rocker nuts

Visually inspect the rocker arms, balls, studs, and nuts for cracks, galling, burning, scoring, or wear. If all parts are intact, liberally lubricate the rocker arms and balls, and install them on the cylinder head. If wear is noted on a rocker arm at the point of valve contact, grind it smooth and square, removing as little material as possible. Replace the rocker arm if excessively worn. If a rocker stud shows signs of wear, it must be replaced (see below). If a rocker nut shows stress cracks, replace it. If an exhaust ball is galled or burned, substitute the intake ball from the same cylinder (if it is intact), and install a new intake ball. NOTE: *Avoid using new rocker balls on exhaust valves.*

Replacing rocker studs:

Reaming the stud bore for oversize rocker studs

FLAT WASHERS

AS STUD BEGINS TO PULL UP, IT WILL BE NECESSARY TO REMOVE THE NUT AND ADD MORE WASHERS.

Extracting a pressed in rocker stud

In order to remove a threaded stud, lock two nuts on the stud, and unscrew the stud using the lower nut. Coat the lower threads of the new stud with Loctite, and install.

Two alternative methods are available for replacing pressed in studs. Remove the damaged stud using a stack of washers and a nut (see illustration). In the first, the boss is reamed .005–.006″ oversize, and an oversize stud pressed in. Control the stud extension over the boss using washers, in the same manner as valve guides. Before installing the stud, coat it with white lead and grease. To retain the stud more positively, drill a hole through the stud and boss, and install a roll pin. In the second method, the boss is tapped, and a threaded stud installed. Retain the stud using Loctite Stud and Bearing Mount.

Procedure	Method

Inspect the rocker shaft(s) and rocker arms (where applicable):

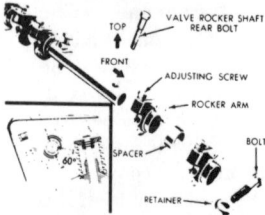

VALVE ROCKER SHAFT
REAR BOLT
TOP
FRONT
ADJUSTING SCREW
ROCKER ARM
SPACER
BOLT
RETAINER

Disassembled rocker shaft parts arranged for inspection

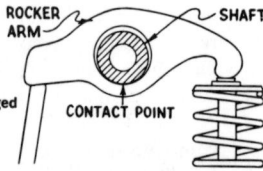

ROCKER ARM — SHAFT
CONTACT POINT

Rocker arm to rocker shaft contact.

Remove rocker arms, springs and washers from rocker shaft. **NOTE:** *Lay out parts in the order they are removed.* Inspect rocker arms for pitting or wear on the valve contact point, or excessive bushing wear. Bushings need only be replaced if wear is excessive, because the rocker arm normally contacts the shaft at one point only. Grind the valve contact point of rocker arm smooth if necessary, removing as little material as possible. If excessive material must be removed to smooth and square the arm, it should be replaced. Clean out all oil holes and passages in rocker shaft. If shaft is grooved or worn, replace it. Lubricate and assemble the rocker shaft.

Inspect the camshaft bushings and the camshaft (overhead cam engines):

See next section.

Inspect the pushrods:

Remove the pushrods, and, if hollow, clean out the oil passages using fine wire. Roll each pushrod over a piece of clean glass. If a distinct clicking sound is heard as the pushrod rolls, the rod is bent, and must be replaced.

* The length of all pushrods must be equal. Measure the length of the pushrods, compare to specifications, and replace as necessary.

Inspect the valve lifters:

CHECK FOR CONCAVE WEAR ON FACE OF TAPPET USING TAPPET FOR STRAIGHT EDGE

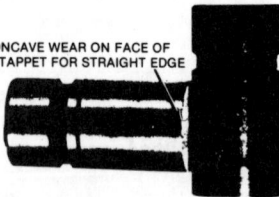

Checking the lifter face

Remove lifters from their bores, and remove gum and varnish, using solvent. Clean walls of lifter bores. Check lifters for concave wear as illustrated. If face is worn concave, replace lifter, and carefully inspect the camshaft. Lightly lubricate lifter and insert it into its bore. If play is excessive, an oversize lifter must be installed (where possible). Consult a machinist concerning feasibility. If play is satisfactory, remove, lubricate, and reinstall the lifter.

* Testing hydraulic lifter leak down:

TAPPET BODY
VALVE RETAINER
PUSH ROD SOCKET   PLUNGER CAP
VALVE SEAT
LOCK RING
VALVE SPRING   PLUNGER
PLUNGER RETURN SPRING   VALVE   METERING DISC

Exploded view of a typical hydraulic lifter

Submerge lifter in a container of kerosene. Chuck a used pushrod or its equivalent into a drill press. Position container of kerosene so pushrod acts on the lifter plunger. Pump lifter with the drill press, until resistance increases. Pump several more times to bleed any air out of lifter. Apply very firm, constant pressure to the lifter, and observe rate at which fluid bleeds out of lifter. If the fluid bleeds very quickly (less than 15 seconds), lifter is defective. If the time exceeds 60 seconds, lifter is sticking. In either case, recondition or replace lifter. If lifter is operating properly (leak down time 15–60 seconds), lubricate and install it.

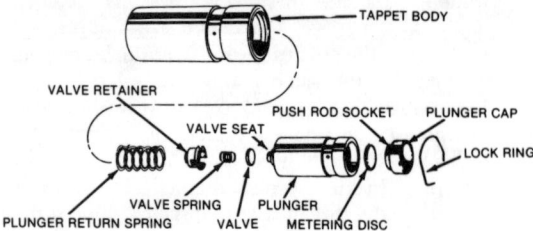

Cylinder Block Reconditioning

Procedure	Method

Checking the main bearing clearance:

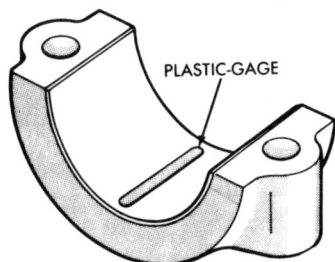

Installing Plastigage on lower bearing shell

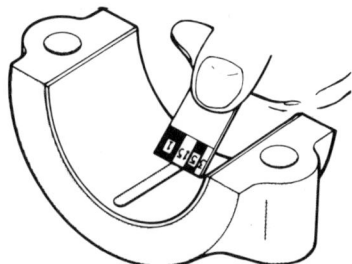

Measuring Plastigage to determine bearing clearance

Invert engine, and remove cap from the bearing to be checked. Using a clean, dry rag, thoroughly clean all oil from crankshaft journal and bearing insert. **NOTE:** *Plastigage is soluble in oil; therefore, oil on the journal or bearing could result in erroneous readings.* Place a piece of Plastigage along the full length of journal, reinstall cap, and torque to specifications. Remove bearing cap, and determine bearing clearance by comparing width of Plastigage to the scale on Plastigage envelope. Journal taper is determined by comparing width of the Plastigage strip near its ends. Rotate crankshaft 90° and retest, to determine journal eccentricity. **NOTE:** *Do not rotate crankshaft with Plastigage installed.* If bearing insert and journal appear intact, and are within tolerances, no further main bearing service is required. If bearing or journal appear defective, cause of failure should be determined before replacement.

* Remove crankshaft from block (see below). Measure the main bearing journals at each end twice (90° apart) using a micrometer, to determine diameter, journal taper and eccentricity. If journals are within tolerances, reinstall bearing caps at their specified torque. Using a telescope gauge and micrometer, measure bearing I.D. parallel to piston axis and at 30° on each side of piston axis. Subtract journal O.D. from bearing I.D. to determine oil clearance. If crankshaft journals appear defective, or do not meet tolerances, there is no need to measure bearings; for the crankshaft will require grinding and/or undersize bearings will be required. If bearing appears defective, cause for failure should be determined prior to replacement.

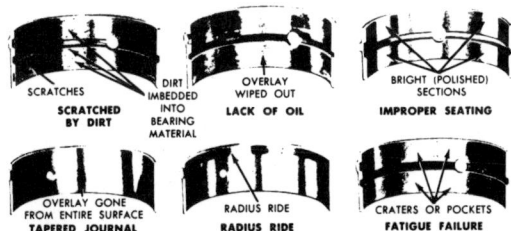

SCRATCHES SCRATCHED BY DIRT	DIRT IMBEDDED INTO BEARING MATERIAL	OVERLAY WIPED OUT LACK OF OIL	BRIGHT (POLISHED) SECTIONS IMPROPER SEATING
OVERLAY GONE FROM ENTIRE SURFACE TAPERED JOURNAL		RADIUS RIDE RADIUS RIDE	CRATERS OR POCKETS FATIGUE FAILURE

Causes of bearing failure

Checking the connecting rod bearing clearance:

Connecting rod bearing clearance is checked in the same manner as main bearing clearance, using Plastigage. Before removing the crankshaft, connecting rod side clearance also should be measured and recorded.

* Checking connecting rod bearing clearance, using a micrometer, is identical to checking main bearing clearance. If no other service is required, the piston and rod assemblies need not be removed.

Procedure	Method
Removing the crankshaft:  Connecting rod matching marks	Using a punch, mark the corresponding main bearing caps and saddles according to position (i.e., one punch on the front main cap and saddle, two on the second, three on the third, etc.). Using number stamps, identify the corresponding connecting rods and caps, according to cylinder (if no numbers are present). Remove the main and connecting rod caps, and place sleeves of plastic tubing over the connecting rod bolts, to protect the journals as the crankshaft is removed. Lift the crankshaft out of the block.
Remove the ridge from the top of the cylinder: 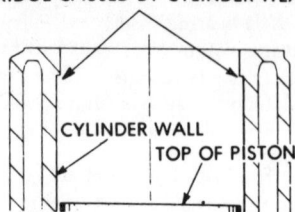 Cylinder bore ridge	In order to facilitate removal of the piston and connecting rod, the ridge at the top of the cylinder (unknown area; see illustration) must be removed. Place the piston at the bottom of the bore, and cover it with a rag. Cut the ridge away using a ridge reamer, exercising extreme care to avoid cutting too deeply. Remove the rag, and remove cuttings that remain on the piston. **CAUTION:** *If the ridge is not removed, and new rings are installed, damage to rings will result.*
Removing the piston and connecting rod: 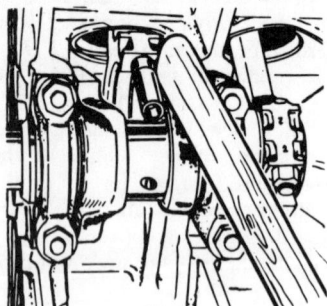 Removing the piston	Invert the engine, and push the pistons and connecting rods out of the cylinders. If necessary, tap the connecting rod boss with a wooden hammer handle, to force the piston out. **CAUTION:** *Do not attempt to force the piston past the cylinder ridge* (see above).
Service the crankshaft:	Ensure that all oil holes and passages in the crankshaft are open and free of sludge. If necessary, have the crankshaft ground to the largest possible undersize.
	** Have the crankshaft Magnafluxed, to locate stress cracks. Consult a machinist concerning additional service procedures, such as surface hardening (e.g., nitriding, Tuftriding) to improve wear characteristics, cross drilling and chamfering the oil holes to improve lubrication, and balancing.
Removing freeze plugs:	Drill a small hole in the center of the freeze plugs. Thread a large sheet metal screw into the hole and remove the plug with a slide hammer.
Remove the oil gallery plugs:	Threaded plugs should be removed using an appropriate (usually square) wrench. To remove soft, pressed in plugs, drill a hole in the plug, and thread in a sheet metal screw. Pull the plug out by the screw using a slide hammer.

Procedure	Method
Hot-tank the block:	Have the block hot-tanked to remove grease, corrosion, and scale from the water jackets. **NOTE:** *Consult the operator to determine whether the camshaft bearings will be damaged during the hot-tank process.*

Check the block for cracks:	Visually inspect the block for cracks or chips. The most common locations are as follows:
	    Adjacent to freeze plugs.
	    Between the cylinders and water jackets.
	    Adjacent to the main bearing saddles.
	    At the extreme bottom of the cylinders.
	Check only suspected cracks using spot check dye (see introduction). If a crack is located, consult a machinist concerning possible repairs.

	** Magnaflux the block to locate hidden cracks. If cracks are located, consult a machinist about feasibility of repair.
Install the oil gallery plugs and freeze plugs:	Coat freeze plugs with sealer and tap into position using a piece of pipe, slightly smaller than the plug, as a driver. To ensure retention, stake the edges of the plugs. Coat threaded oil gallery plugs with sealer and install. Drive replacement soft plugs into block using a large drift as a driver.
	* Rather than reinstalling lead plugs, drill and tap the holes, and install threaded plugs.
Check the bore diameter and surface: 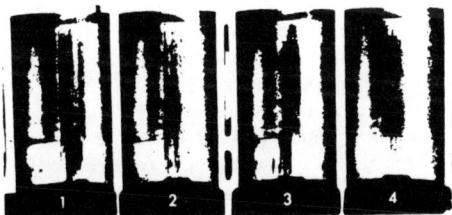 1, 2, 3 Piston skirt seizure resulted in this pattern. Engine must be rebored 4. Piston skirt and oil ring seizure caused this damag Engine must be rebored  5, 6 Score marks caused by a split piston skirt. Damage is not serious enough to warrant reboring 7. Ring seized longitudinally, causing a score mark 1 3/16" wide, on the land side of the piston groove. The honing pattern is destroyed and the cylinder must be rebored	Visually inspect the cylinder bores for roughness, scoring, or scuffing. If evident, the cylinder bore must be bored or honed oversize to eliminate imperfections, and the smallest possible oversize piston used. The new pistons should be given to the machinist with the block, so that the cylinders can be bored or honed exactly to the piston size (plus clearance). If no flaws are evident, measure the bore diameter using a telescope gauge and micrometer, or dial gauge, parallel and perpendicular to the engine centerline, at the top (below the ridge) and bottom of the bore. Subtract the bottom measurements from the top to determine taper, and the parallel to the centerline measurements from the perpendicular measurements to determine eccentricity. If the measurements are not within specifications, the cylinder must be bored or honed, and an oversize piston installed. If the measurements are within specifications the cylinder may be used as is, with only finish honing (see below). **NOTE:** *Prior to submitting the block for boring, perform the following operation(s).*

Procedure	Method

8. Result of oil ring seizure. Engine must be rebored

Cylinder wall damage (© Daimler-Benz A.G.)

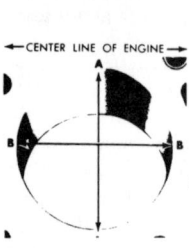

9. Oil ring seizure here was not serious enough to warrant reboring. The honing marks are still visible

Cylinder bore measuring positions

Measuring the cylinder bore with a telescope gauge

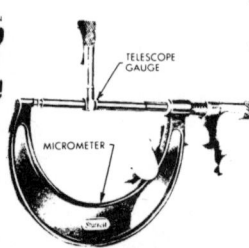

Determining the cylinder bore by measuring the telescope gauge with a micrometer

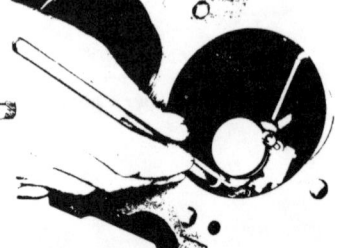

Measuring the cylinder bore with a dial gauge

Procedure	Method
Check the block deck for warpage:	Using a straightedge and feeler gauges, check the block deck for warpage in the same manner that the cylinder head is checked (see Cylinder Head Reconditioning). If warpage exceeds specifications, have the deck resurfaced. **NOTE:** *In certain cases a specification for total material removal (Cylinder head and block deck) is provided. This specification must not be exceeded.*
* Check the deck height:	The deck height is the distance from the crankshaft centerline to the block deck. To measure, invert the engine, and install the crankshaft, retaining it with the center main cap. Measure the distance from the crankshaft journal to the block deck, parallel to the cylinder centerline. Measure the diameter of the end (front and rear) main journals, parallel to the centerline of the cylinders, divide the diameter in half, and subtract it from the previous measurement. The results of the front and rear measurements should be identical. If the difference exceeds .005″, the deck height should be corrected. **NOTE:** *Block deck height and warpage should be corrected concurrently.*

Procedure	Method

Check the cylinder block bearing alignment:

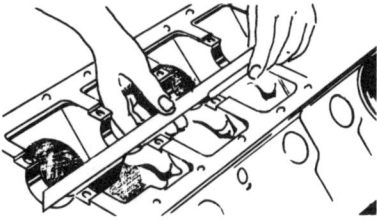

Checking main bearing saddle alignment

Remove the upper bearing inserts. Place a straightedge in the bearing saddles along the centerline of the crankshaft. If clearance exists between the straightedge and the center saddle, the block must be align-bored.

Clean and inspect the pistons and connecting rods:

RING EXPANDER

Removing the piston rings

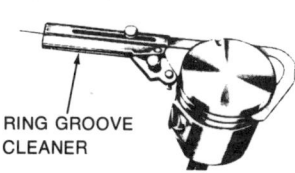

RING GROOVE CLEANER

Cleaning the piston ring grooves

Connecting rod length checking dimension

Using a ring expander, remove the rings from the piston. Remove the retaining rings (if so equipped) and remove piston pin. **NOTE:** *If the piston pin must be pressed out, determine the proper method and use the proper tools; otherwise the piston will distort.* Clean the ring grooves using an appropriate tool, exercising care to avoid cutting too deeply. Thoroughly clean all carbon and varnish from the piston with solvent. **CAUTION:** *Do not use a wire brush or caustic solvent on pistons.* Inspect the pistons for scuffing, scoring, cracks, pitting, or excessive ring groove wear. If wear is evident, the piston must be replaced. Check the connecting rod length by measuring the rod from the inside of the large end to the inside of the small end using calipers (see illustration). All connecting rods should be equal length. Replace any rod that differs from the others in the engine.

* Have the connecting rod alignment checked in an alignment fixture by a machinist. Replace any twisted or bent rods.

* Magnaflux the connecting rods to locate stress cracks. If cracks are found, replace the connecting rod.

Fit the pistons to the cylinders:

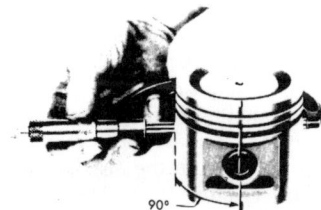

90°

Measuring the piston for fitting (© Buick Div.)

Using a telescope gauge and micrometer, or a dial gauge, measure the cylinder bore diameter perpendicular to the piston pin, 2½″ below the deck. Measure the piston perpendicular to its pin on the skirt. The difference between the two measurements is the piston clearance. If the clearance is within specifications or slightly below (after boring or honing), finish honing is all that is required. If the clearance is excessive, try to obtain a slightly larger piston to bring clearance within specifications. Where this is not possible, obtain the first oversize piston, and hone (or if necessary, bore) the cylinder to size.

Procedure	Method

Assemble the pistons and connecting rods:

Installing piston pin lock rings

Inspect piston pin, connecting rod small end bushing, and piston bore for galling, scoring, or excessive wear. If evident, replace defective part(s). Measure the I.D. of the piston boss and connecting rod small end, and the O.D. of the piston pin. If within specifications, assemble piston pin and rod. **CAUTION:** *If piston pin must be pressed in, determine the proper method and use the proper tools; otherwise the piston will distort.* Install the lock rings; ensure that they seat properly. If the parts are not within specifications, determine the service method for the type of engine. In some cases, piston and pin are serviced as an assembly when either is defective. Others specify reaming the piston and connecting rods for an oversize pin. If the connecting rod bushing is worn, it may in many cases be replaced. Reaming the piston and replacing the rod bushing are machine shop operations.

Clean and inspect the camshaft:

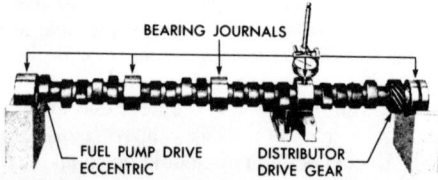

BEARING JOURNALS

FUEL PUMP DRIVE ECCENTRIC   DISTRIBUTOR DRIVE GEAR

Checking the camshaft for straightness
(© Chevrolet Motor Div. G.M. Corp.)

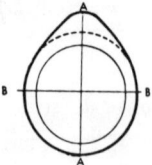

Camshaft lobe measurement

Degrease the camshaft, using solvent, and clean out all oil holes. Visually inspect cam lobes and bearing journals for excessive wear. If a lobe is questionable, check all lobes as indicated below. If a journal or lobe is worn, the camshaft must be reground or replaced. **NOTE:** *If a journal is worn, there is a good chance that the bushings are worn.* If lobes and journals appear intact, place the front and rear journals in V-blocks, and rest a dial indicator on the center journal. Rotate the camshaft to check straightness. If deviation exceeds .001", replace the camshaft.

* Check the camshaft lobes with a micrometer, by measuring the lobes from the nose to base and again at 90° (see illustration). The lift is determined by subtracting the second measurement from the first. If all exhaust lobes and all intake lobes are not identical, the camshaft must be reground or replaced.

Replace the camshaft bearings:

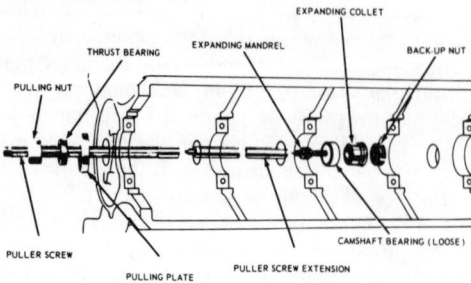

EXPANDING COLLET
THRUST BEARING   EXPANDING MANDREL   BACK-UP NUT
PULLING NUT
PULLER SCREW   CAMSHAFT BEARING (LOOSE)
PULLING PLATE   PULLER SCREW EXTENSION

Camshaft removal and installation tool (typical)

If excessive wear is indicated, or if the engine is being completely rebuilt, camshaft bearings should be replaced as follows: Drive the camshaft rear plug from the block. Assemble the removal puller with its shoulder on the bearing to be removed. Gradually tighten the puller nut until bearing is removed. Remove remaining bearings, leaving the front and rear for last. To remove front and rear bearings, reverse position of the tool, so as to pull the bearings in toward the center of the block. Leave the tool in this position, pilot the new front and rear bearings on the installer, and pull them into position. Return the tool to its original position and pull remaining bearings into position. **NOTE:** *Ensure that oil holes align when installing bearings.* Replace camshaft rear plug, and stake it into position to aid retention.

Procedure	Method

Finish hone the cylinders:

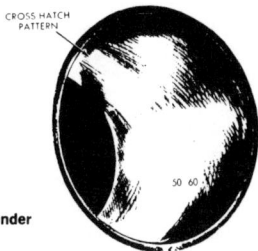

CROSS-HATCH PATTERN

50 60

Finish honed cylinder

Chuck a flexible drive hone into a power drill, and insert it into the cylinder. Start the hone, and move it up and down in the cylinder at a rate which will produce approximately a 60° cross-hatch pattern (see illustration). **NOTE:** *Do not extend the hone below the cylinder bore.* After developing the pattern, remove the hone and recheck piston fit. Wash the cylinders with a detergent and water solution to remove abrasive dust, dry, and wipe several times with a rag soaked in engine oil.

Check piston ring end-gap:

Checking ring end-gap

Compress the piston rings to be used in a cylinder, one at a time, into that cylinder, and press them approximately 1″ below the deck with an inverted piston. Using feeler gauges, measure the ring end-gap, and compare to specifications. Pull the ring out of the cylinder and file the ends with a fine file to obtain proper clearance. **CAUTION:** *If inadequate ring end-gap is utilized, ring breakage will result.*

Install the piston rings:

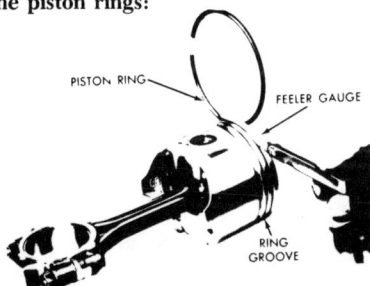

PISTON RING

FEELER GAUGE

RING GROOVE

Checking ring side clearance

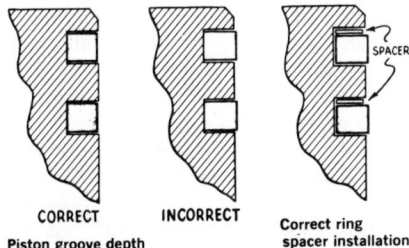

SPACER

CORRECT   INCORRECT

Piston groove depth

Correct ring spacer installation

Inspect the ring grooves in the piston for excessive wear or taper. If necessary, recut the groove(s) for use with an overwidth ring or a standard ring and spacer. If the groove is worn uniformly, overwidth rings, or standard rings and spacers may be installed without recutting. Roll the outside of the ring around the groove to check for burrs or deposits. If any are found, remove with a fine file. Hold the ring in the groove, and measure side clearance. If necessary, correct as indicated above. **NOTE:** *Always install any additional spacers above the piston ring.* The ring groove must be deep enough to allow the ring to seat below the lands (see illustration). In many cases, a "go-no-go" depth gauge will be provided with the piston rings. Shallow grooves may be corrected by recutting, while deep grooves require some type of filler or expander behind the piston. Consult the piston ring supplier concerning the suggested method. Install the rings on the piston, lowest ring first, using a ring expander. **NOTE:** *Position the ring markings as specified by the manufacturer (see car section).*

Install the camshaft:

Liberally lubricate the camshaft lobes and journals, and slide the camshaft into the block. **CAUTION:** *Exercise extreme care to avoid damaging the bearings when inserting the camshaft.* Install and tighten the camshaft thrust plate retaining bolts.

Procedure	Method

Check camshaft end-play:

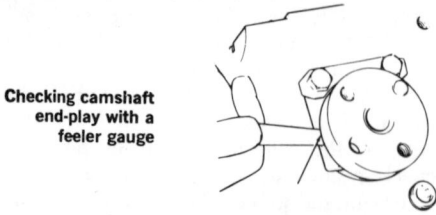

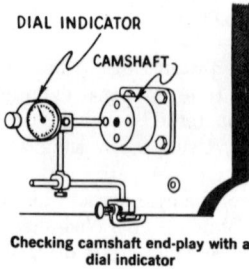

Checking camshaft
end-play with a
feeler gauge

DIAL INDICATOR

CAMSHAFT

Checking camshaft end-play with a
dial indicator

Using feeler gauges, determine whether the clearance between the camshaft boss (or gear) and backing plate is within specifications. Install shims behind the thrust plate, or reposition the camshaft gear and retest end-play.

* Mount a dial indicator stand so that the stem of the dial indicator rests on the nose of the camshaft, parallel to the camshaft axis. Push the camshaft as far in as possible and zero the gauge. Move the camshaft outward to determine the amount of camshaft end-play. If the end-play is not within tolerance, install shims behind the thrust plate, or reposition the camshaft gear and retest.

Install the rear main seal (where applicable):

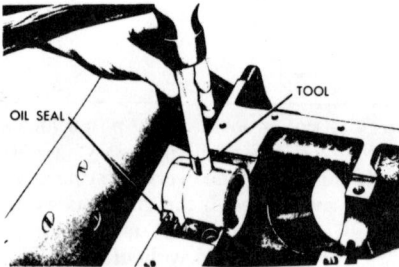

TOOL

OIL SEAL

Seating the rear
main seal

Position the block with the bearing saddles facing upward. Lay the rear main seal in its groove and press it lightly into its seat. Place a piece of pipe the same diameter as the crankshaft journal into the saddle, and firmly seat the seal. Hold the pipe in position, and trim the ends of the seal flush if required.

Install the crankshaft:

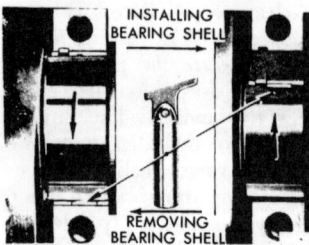

INSTALLING
BEARING SHELL

REMOVING
BEARING SHELL

Removal and installation of upper bearing insert using a
roll-out pin (© Buick Div. G.M. Corp.)

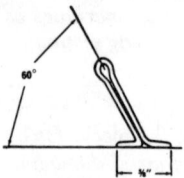

60°

Home made bearing roll-out pin

Thoroughly clean the main bearing saddles and caps. Place the upper halves of the bearing inserts on the saddles and press into position. **NOTE: *Ensure that the oil holes align.*** Press the corresponding bearing inserts into the main bearing caps. Lubricate the upper main bearings, and lay the crankshaft in position. Place a strip of Plastigage on each of the crankshaft journals, install the main caps, and torque to specifications. Remove the main caps, and compare the Plastigage to the scale on the Plastigage envelope. If clearances are within tolerances, remove the Plastigage, turn the crankshaft 90°, wipe off all oil and retest. If all clearances are correct, remove all Plastigage, thoroughly lubricate the main caps and bearing journals, and install the main caps. If clearances are not within tolerance, the upper bearing inserts may be removed, without removing the crankshaft, using a bearing roll out pin (see illustration). Roll in a bearing that will provide proper clearance, and retest. Torque all main caps, excluding the thrust bearing cap, to specifications. Tighten the thrust bearing cap finger tight. To properly align the thrust bearing, pry the crankshaft the extent of its axial travel several

Procedure	Method

times, the last movement held toward the front of the engine, and torque the thrust bearing cap to specifications. Determine the crankshaft end-play (see below), and bring within tolerance with thrust washers.

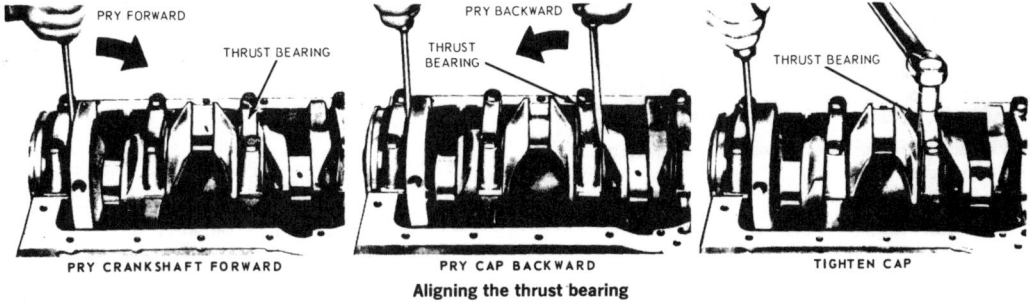

Aligning the thrust bearing

Measure crankshaft end-play:

Checking crankshaft end-play with a dial indicator

Mount a dial indicator stand on the front of the block, with the dial indicator stem resting on the nose of the crankshaft, parallel to the crankshaft axis. Pry the crankshaft the extent of its travel rearward, and zero the indicator. Pry the crankshaft forward and record crankshaft end-play. **NOTE:** *Crankshaft end-play also may be measured at the thrust bearing, using feeler gauges (see illustration).*

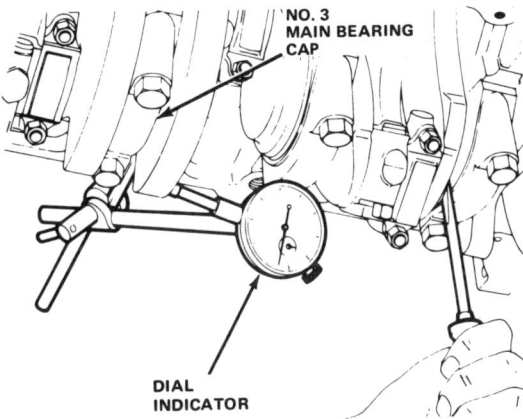

Checking crankshaft end-play with a feeler gauge

Install the pistons:

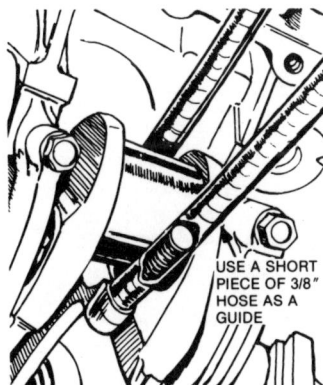

Tubing used as guide when installing a piston

Press the upper connecting rod bearing halves into the connecting rods, and the lower halves into the connecting rod caps. Position the piston ring gaps according to specifications (see car section), and lubricate the pistons. Install a ring compresser on a piston, and press two long (8″) pieces of plastic tubing over the rod bolts. Using the plastic tubes as a guide, press the pistons into the bores and onto the crankshaft with a wooden hammer handle. After seating the rod on the crankshaft journal, remove the tubes and install the cap finger tight. Install the remaining pistons in the same manner. Invert the engine and check the bearing clearance at two points (90° apart) on each journal with Plastigage. **NOTE:** *Do not turn the crankshaft with Plastigage installed.* If clearance is within tolerances, remove *all* Plastigage, thoroughly lubricate the journals, and torque the

Procedure	*Method*

Installing a piston

rod caps to specifications. If clearance is not within specifications, install different thickness bearing inserts and recheck. **CAUTION:** *Never shim or file the connecting rods or caps.* Always install plastic tube sleeves over the rod bolts when the caps are not installed, to protect the crankshaft journals.

Check connecting rod side clearance:

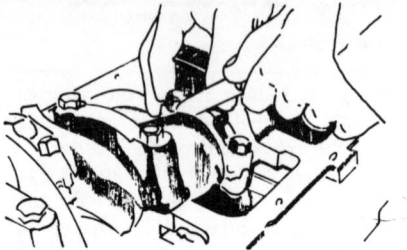

Checking connecting rod side clearance

Determine the clearance between the sides of the connecting rods and the crankshaft, using feeler gauges. If clearance is below the minimum tolerance, the rod may be machined to provide adequate clearance. If clearance is excessive, substitute an unworn rod, and recheck. If clearance is still outside specifications, the crankshaft must be welded and reground, or replaced.

Inspect the timing chain:

Visually inspect the timing chain for broken or loose links, and replace the chain if any are found. If the chain will flex sideways, it must be replaced. Install the timing chain as specified. **NOTE:** *If the original timing chain is to be reused, install it in its original position.*

Check timing gear backlash and runout:

Checking camshaft gear backlash

Checking camshaft gear runout

Mount a dial indicator with its stem resting on a tooth of the camshaft gear (as illustrated). Rotate the gear until all slack is removed, and zero the indicator. Rotate the gear in the opposite direction until slack is removed, and record gear backlash. Mount the indicator with its stem resting on the edge of the camshaft gear, parallel to the axis of the camshaft. Zero the indicator, and turn the camshaft gear one full turn, recording the runout. If either backlash or runout exceed specifications, replace the worn gear(s).

Completing the Rebuilding Process

Following the above procedures, complete the rebuilding process as follows:

Fill the oil pump with oil, to prevent cavitating (sucking air) on initial engine start up. Install the oil pump and the pickup tube on the engine. Coat the oil pan gasket as necessary, and install the gasket and the oil pan. Mount the flywheel and the crankshaft vibrational damper or pulley on the crankshaft. NOTE: *Always use new bolts when installing the flywheel.* Inspect the clutch shaft pilot bushing in the crankshaft. If the bushing is excessively worn, remove it with an expanding puller and a slide hammer, and tap a new bushing into place.

Position the engine, cylinder head side up. Lubricate the lifters, and install them into their bores. Install the cylinder head, and torque it as specified in the car section. Insert the pushrods (where applicable), and install the rocker shaft(s) (if so equipped) or position the rocker arms on the pushrods. If solid lifters are utilized, adjust the valves to the "cold" specifications.

Mount the intake and exhaust manifolds, the carburetor(s), the distributor and spark plugs. Adjust the point gap and the static ignition timing. Mount all accessories and install the engine in the car. Fill the radiator with coolant, and the crankcase with high quality engine oil.

Break-in Procedure

Start the engine, and allow it to run at low speed for a few minutes, while checking for leaks. Stop the engine, check the oil level, and fill as necessary. Restart the engine, and fill the cooling system to capacity. Check the point dwell angle and adjust the ignition timing and the valves. Run the engine at low to medium speed (800–2500 rpm) for approximately ½ hour, and retorque the cylinder head bolts. Road test the car, and check again for leaks.

Follow the manufacturer's recommended engine break-in procedure and maintenance schedule for new engines.

4

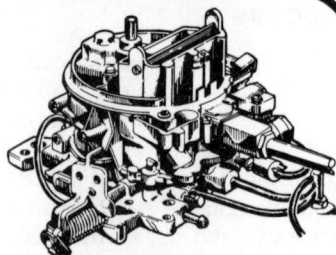

Emission Controls and Fuel System

EMISSION CONTROLS

There are three types of automotive pollutants; crankcase fumes, exhaust gases and gasoline evaporation. The equipment that is used to limit these pollutants is commonly called emission control equipment.

Crankcase Emission Controls

The crankcase emission control equipment consists of a positive crankcase ventilation valve (PCV), a closed or open oil filler cap and hoses to connect this equipment.

When the engine is running, a small portion of the gases which are formed in the combustion chamber during combustion leak by the piston rings and enter the crankcase. Since these gases are under pressure they tend to escape from the crankcase and enter into the atmosphere. If these gases were allowed to remain in the crankcase for any length of time, they would contaminate the engine oil and cause sludge to build up. If the gases are allowed to escape into the atmosphere, they would pollute the air, as they contain unburned hydrocarbons. The crankcase emission control equipment recycles these gases back into the engine combustion chamber where they are burned.

Crankcase gases are recycled in the following manner: while the engine is running, clean filtered air is drawn into the crankcase either directly through the oil filler cap, or through the carburetor air filter and then through a hose leading to the oil filler cap. As the air passes through the crankcase it picks up the combustion gases and carries them out of the crankcase, up through the PCV valve and into the intake manifold. After they enter the intake manifold they are drawn into the combustion chamber and burned.

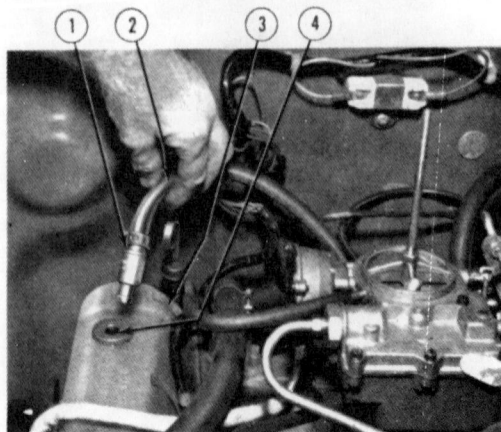

1. Ventilation valve
2. Hose to carburetor inlet
3. Right rocker cover
4. Grommet

Removing the PCV valve from a 6-225

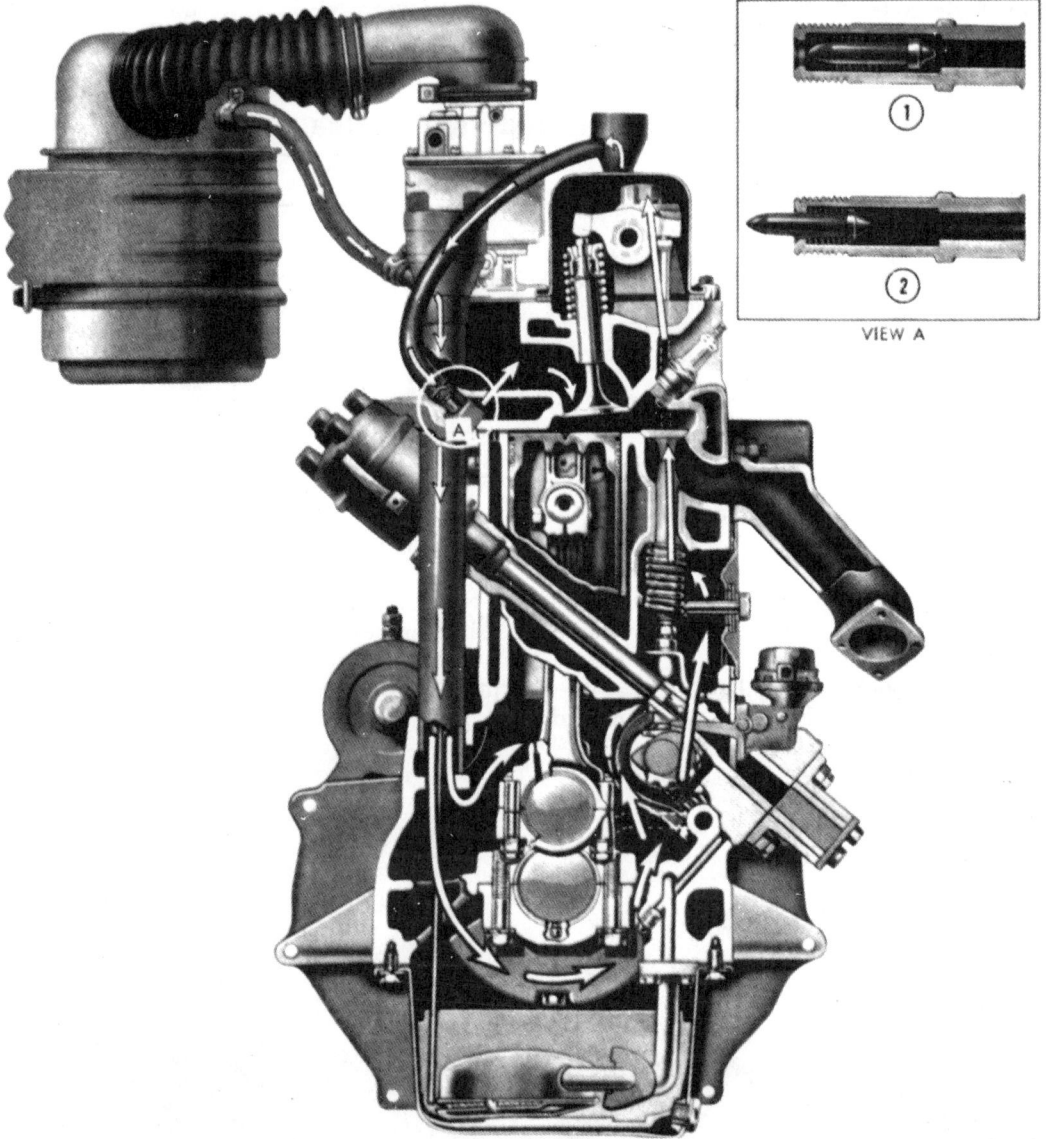

PCV operation: 4-134

The most critical component in the system is the PCV valve. This vacuum controlled valve regulates the amount of gases which are recycled into the combustion chamber. At low engine speeds the valve is partially closed, limiting the flow of gases into the intake manifold. As engine speed increases, the valve opens to admit greater quantities of the gases into the intake manifold. If the valve should become blocked or plugged, the gases will be prevented from escaping from the crankcases by the normal route. Since these gases are under pressure, they will find their own way out of the crankcase. This alternate route is usually a weak oil seal or

gasket in the engine. As the gas escapes by the gasket, it also creates an oil leak. Besides causing oil leaks, a clogged PCV valve also allows these gases to remain in the crankcase for an extended period of time, promoting the formation of sludge in the engine.

The above explanation and the troubleshooting procedure which follows applies to all engines with PCV systems.

TROUBLESHOOTING

With the engine running, pull the PCV valve and hose from the engine. Block off the end of the valve with your finger. The engine speed should drop at least 50 rpm when the end of the valve is blocked. If the engine speed does not drop at least 50 rpm, then the valve is defective and should be replaced.

2. Connect the PCV valve to its hose.
3. Install the PCV valve on the engine.

Emission Control Systems Usage

System	Application
Positive Crankcase Ventilation	All engines
Air Guard air injection reactor	All engines except 8-350
Thermostatically controlled air cleaner (TAC)	All 1971 and later engines
Transmission controlled spark (TCS)	All AMC engines
Exhaust gas recirculation (EGR)	All 1971 and later engines
Evaporative emission control canister	All 1971 and later engines
Catalytic converter	All 1979
Vacuum throttle modulation	All 1975 and later

REMOVAL AND INSTALLATION

1. Pull the PCV valve and hose from the engine.

2. Remove the PCV valve from the hose. Inspect the inside of the PCV valve from the hose. If it is dirty, disconnect it from the intake manifold and clean it.

To install, proceed as follows:

1. If the PCV valve hose was removed, connect it to the intake manifold.

Exhaust Emission Controls

All of the engines used in Jeeps, except the 350 V8, at one time incorporated the air injection system for controlling the emission of exhaust gases into the atmosphere. Since this type of emission control system is common to most of the engines, it will be explained here.

The exhaust emission air injection system

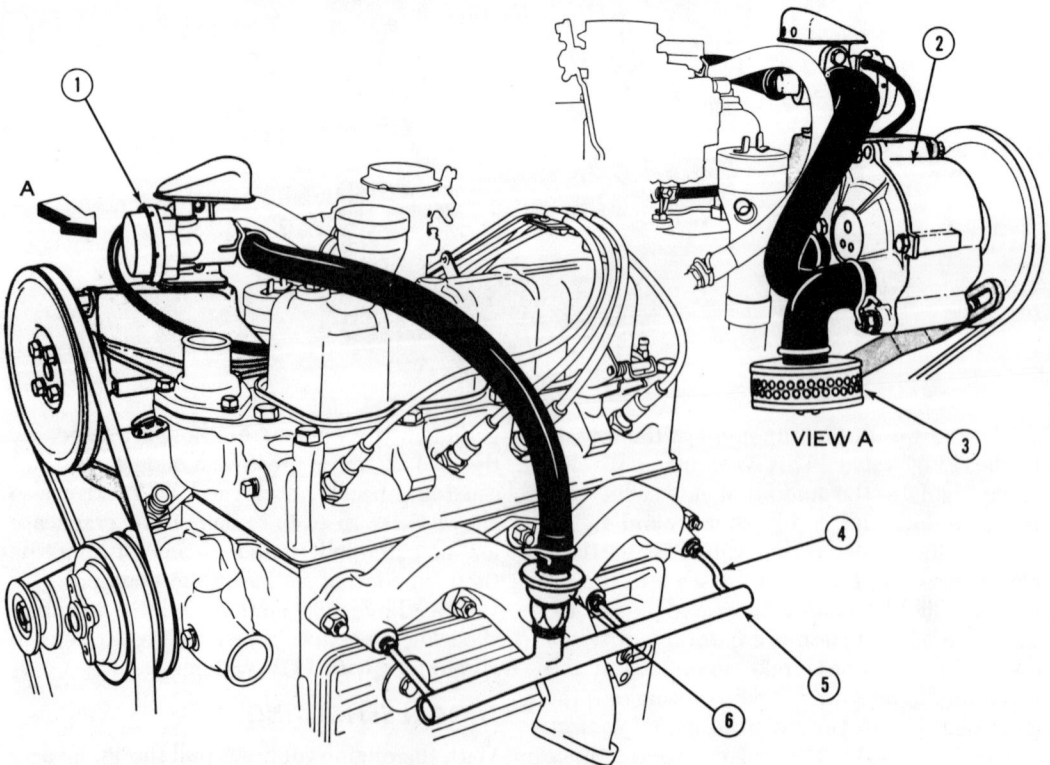

VIEW A

1. Anti-backfire diverter valve
2. Air pump
3. Pump air filter
4. Air injection tubes
5. Air delivery manifold
6. Check valve

4-134 air injection

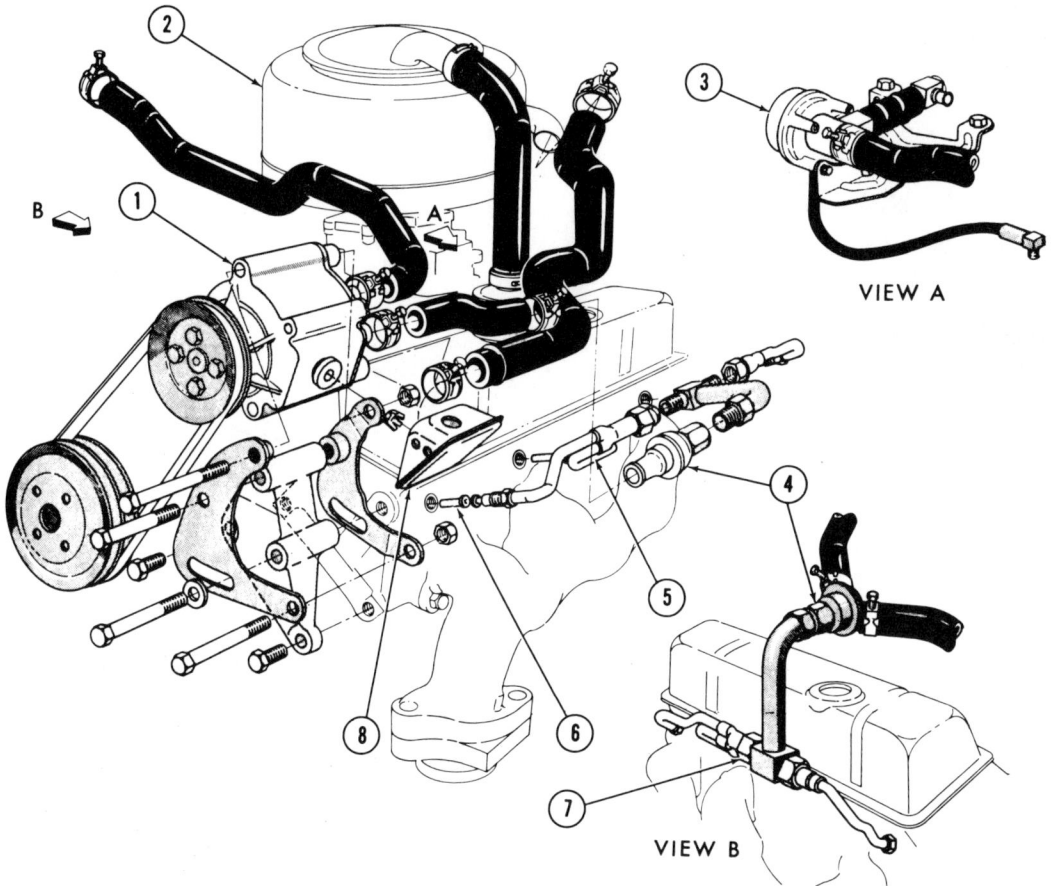

A. Top rear of engine
B. Right side of engine
1. Air pump
2. Air filter
3. Anti-backfire valve
4. Check valve
5. Distribution manifold assembly (left side)
6. Injection nozzle
7. Distribution manifold assembly (right side)
8. Relief valve muffler

6-225 air injection

consists of a belt driven air pump which directs compresed air through connecting hoses to a steel distribution manifold into stainless steel injection tubes in the exhaust port adjacent to each exhaust valve. The air, with its normal oxygen content, reacts with the hot, but incompletely burned exhaust gases and permits further combustion in the exhaust port or manifold.

AIR PUMP

The air injection pump is a positive displacement vane type which is permanently lubricated and requires little periodic maintenance. The only serviceable parts on the air pump are the filter, exhaust tube, and relief valve. The relief valve relieves the air flow when the pump pressure reaches a preset level. This occurs at high engine rpm. This serves to prevent damage to the pump and to limit maximum exhaust manifold temperatures.

Pump Air Filter

The air filter attached to the pump is a replaceable element type. The filter should be replaced every 12,000 miles under normal conditions and sooner under off-road use. Some models draw their air supply through the carburetor air filter.

Air Delivery Manifold

The air delivery manifold distributes the air from the pump to each of the air delivery tubes in a uniform manner. A check valve is integral with the air delivery manifold. Its function is to prevent the reverse flow of exhaust gases to the pump should the pump fail. This reverse flow would damage the air pump and connecting hose.

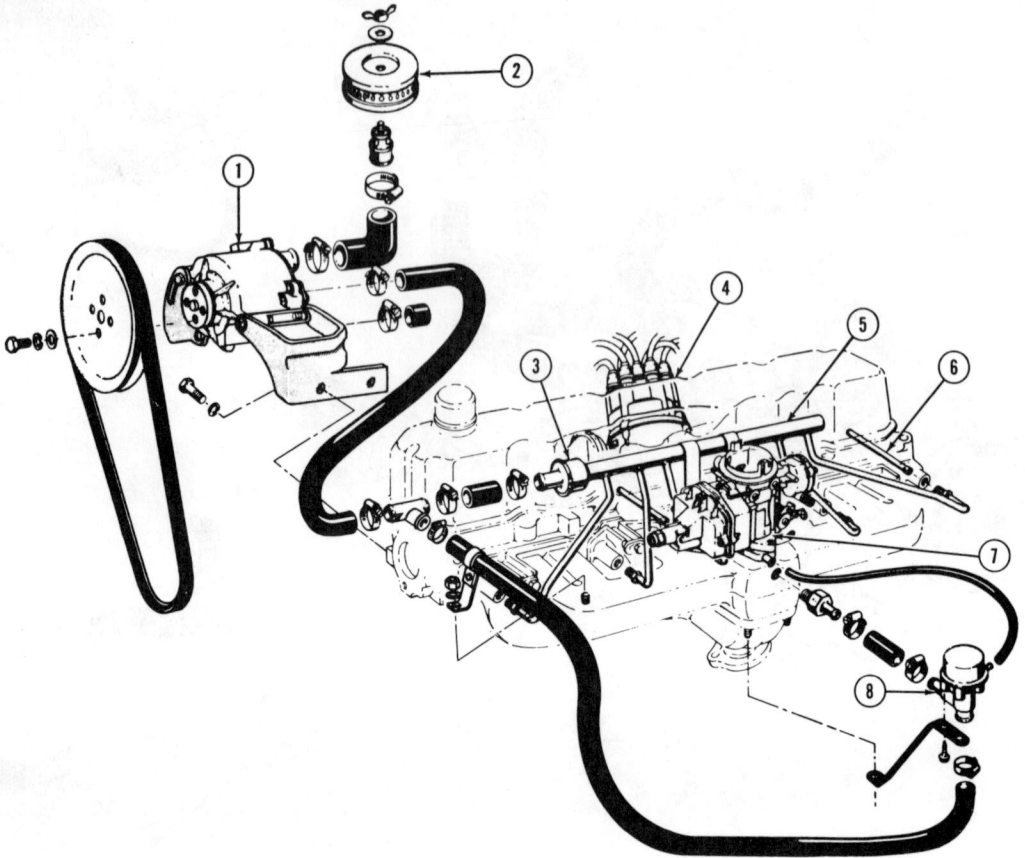

1. Air pump
2. Filter
3. Check valve
4. Distributor (special calibration)
5. Air delivery distribution manifold
6. Air injection tube(s)
7. Carburetor (special calibration)
8. Anti-backfire (gulp) valve

Air injection system: 1966–70 6-232

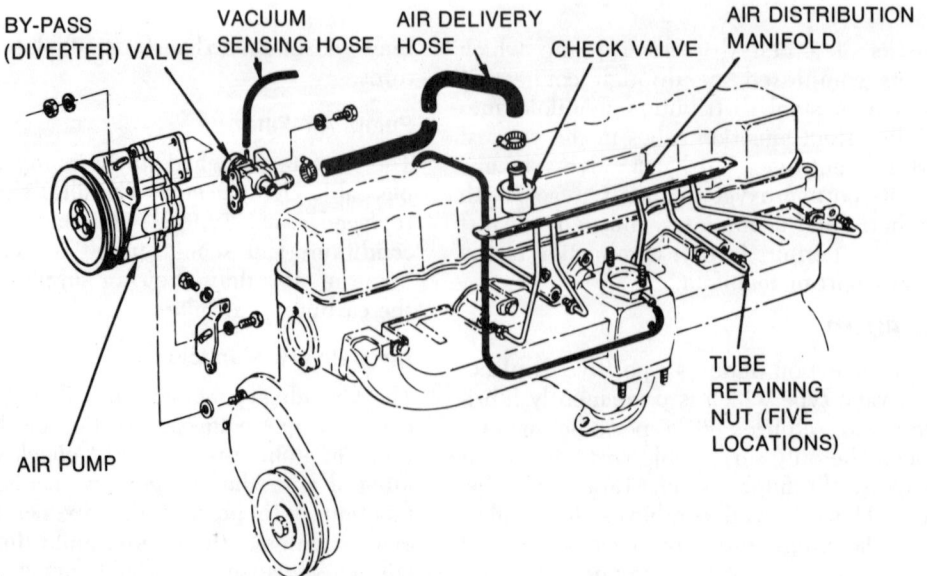

BY-PASS (DIVERTER) VALVE

VACUUM SENSING HOSE

AIR DELIVERY HOSE

CHECK VALVE

AIR DISTRIBUTION MANIFOLD

TUBE RETAINING NUT (FIVE LOCATIONS)

AIR PUMP

Air injection system: 1971–79 6-232, 258

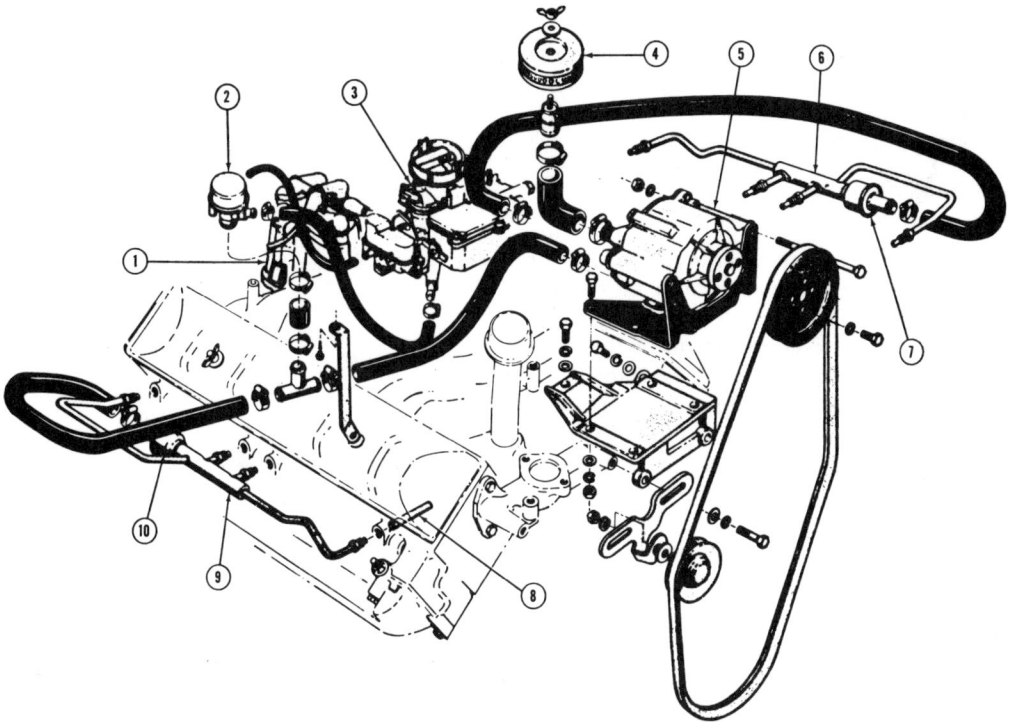

1. Distributor
2. Anti-backfire (gulp) valve
3. Carburetor
4. Filter
5. Air pump
6. Air delivery distribution manifold
7. Check valve
8. Air injection tube(s)
9. Air delivery distribution manifold
10. Check valve

Air injection system: 8-327

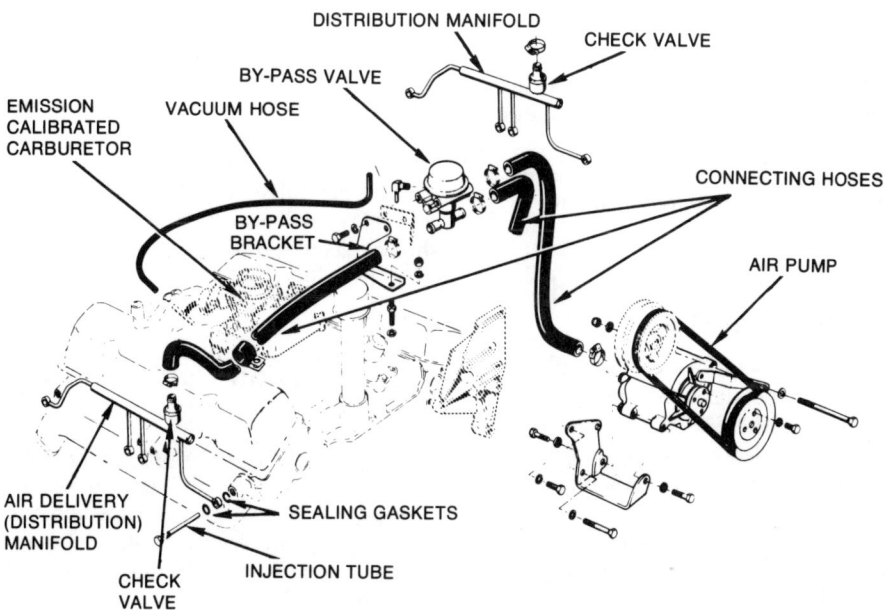

DISTRIBUTION MANIFOLD

CHECK VALVE

BY-PASS VALVE

VACUUM HOSE

EMISSION CALIBRATED CARBURETOR

CONNECTING HOSES

BY-PASS BRACKET

AIR PUMP

AIR DELIVERY (DISTRIBUTION) MANIFOLD

SEALING GASKETS

CHECK VALVE

INJECTION TUBE

Air injection system: 8-304, 360, 401

Air Injection Tubes

The air injection tubes are inserted into the exhaust ports. The tubes project into the exhaust ports, directing air into the vicinity of the exhaust valve.

Anti-Backfire Valve

The anti-backfire diverter valve prevents engine backfire by briefly interrupting the air being injected into the exhaust manifold during periods of deceleration or rapid throttle closure. On the F-head and all of the 1971 and later American Motors engines the valve opens when a sudden increase in manifold vacuum overcomes the diaphragm spring tension. With the valve in the open position the air flow from the air pump is directed to the atmosphere.

On the V6-225, 1966–70 232 Six and 327 V8, the anti-backfire valve is what is commonly called a gulp valve. During rapid deceleration the valve is opened by the sudden high vacuum condition in the intake manifold and gulps air into the intake manifold.

Both of these valves prevent backfiring in the exhaust manifold. Both valves also prevent an over rich fuel mixture from being burned in the exhaust manifold, which would cause backfiring and possible damage to the engine.

CARBURETOR

The carburetors used on engines equipped with emission controls have specific flow characteristics that differ from the carburetors used on vehicles not equipped with emission control devices. The carburetors are identified by number. The correct carburetor should be used when replacement is necessary.

A carburetor dashpot is used on the F-head to control throttle closing speed.

Thermostatically Controlled Air Cleaner-System (TAC)

This system consists of a heat shroud which is integral with the right side exhaust manifold (left side on the 350 V8), a hot air hose and a special air cleaner assembly equipped with a thermal sensor and a vacuum motor and air valve assembly.

The thermal sensor incorporates an air bleed valve which regulates the amount of vacuum applied to the vacuum motor, con-

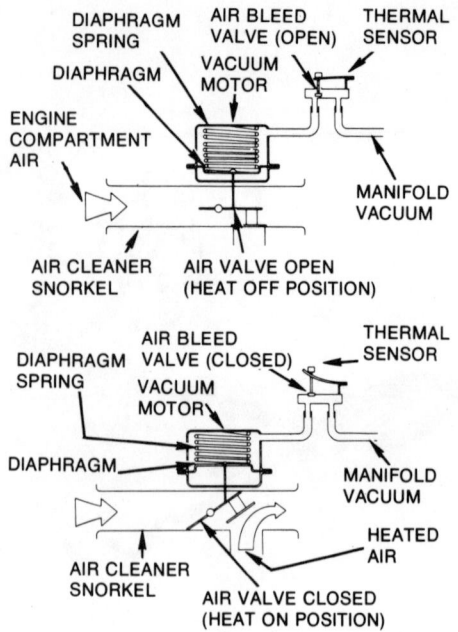

Thermostatically controlled air cleaner (upper open, lower closed)

trolling the air valve position to supply either heated air from the exhaust manifold or air from the engine compartment.

During the warm-up period when underhood temperatures are low, the air bleed valve is closed and sufficient vacuum is applied to the vaccum motor to hold the air valve in the closed (heat on) position.

As the temperature of the air entering the air cleaner approaches approximately 115° F, the air bleed valve opens to decrease the amount of vacuum applied to the vacuum motor. The diaphragm spring in the vacuum motor then moves the air valve into the open (heat off) position, allowing only underhood air to enter the air cleaner.

The air valve in the air cleaner will also open, regardless of air temperature, during heavy acceleration to obtain maximum air flow through the air cleaner.

Transmission Controlled Spark System

The purpose of this system is to reduce the emission of oxides of nitrogen by lowering the peak combustion pressure and temperature during the power stroke.

The system incorporates the following components:

55 WAYS TO IMPROVE FUEL ECONOMY

CHILTON'S
FUEL ECONOMY
& TUNE-UP TIPS

Tune-Up • Spark Plug Diagnosis • Emission Controls

Fuel System • Cooling System • Tires and Wheels

General Maintenance

CHILTON'S FUEL ECONOMY & TUNE-UP TIPS

Fuel economy is important to everyone, no matter what kind of vehicle you drive. The maintenance-minded motorist can save both money and fuel using these tips and the periodic maintenance and tune-up procedures in this Repair and Tune-Up Guide.

There are more than 130,000,000 cars and trucks registered for private use in the United States. Each travels an average of 10-12,000 miles per year, and, in total they consume close to 70 billion gallons of fuel each year. This represents nearly 2/3 of the oil imported by the United States each year. The Federal government's goal is to reduce consumption 10% by 1985. A variety of methods are either already in use or under serious consideration, and they all affect your driving and the cars you will drive. In addition to "down-sizing", the auto industry is using or investigating the use of electronic fuel delivery, electronic engine controls and alternative engines for use in smaller and lighter vehicles, among other alternatives to meet the federally mandated Corporate Average Fuel Economy (CAFE) of 27.5 mpg by 1985. The government, for its part, is considering rationing, mandatory driving curtailments and tax increases on motor vehicle fuel in an effort to reduce consumption. The government's goal of a 10% reduction could be realized — and further government regulation avoided — if every private vehicle could use just 1 less gallon of fuel per week.

How Much Can You Save?

Tests have proven that almost anyone can make at least a 10% reduction in fuel consumption through regular maintenance and tune-ups. When a major manufacturer of spark plugs sur-

TUNE-UP

1. Check the cylinder compression to be sure the engine will really benefit from a tune-up and that it is capable of producing good fuel economy. A tune-up will be wasted on an engine in poor mechanical condition.

2. Replace spark plugs regularly. New spark plugs alone can increase fuel economy 3%.

3. Be sure the spark plugs are the correct type (heat range) for your vehicle. See the Tune-Up Specifications.

Heat range refers to the spark plug's ability to conduct heat away from the firing end. It must conduct the heat away in an even pattern to avoid becoming a source of pre-ignition, yet it must also operate hot enough to burn off conductive deposits that could cause misfiring.

The heat range is usually indicated by a number on the spark plug, part of the manufacturer's designation for each individual spark plug. The numbers in bold-face indicate the heat range in each manufacturer's identification system.

Manufacturer	Typical Designation
AC	R **45** TS
Bosch (old)	WA **145** T30
Bosch (new)	HR **8** Y
Champion	RBL **15** Y
Fram/Autolite	**415**
Mopar	P-**62** PR
Motorcraft	BRF-**42**
NGK	BP **5** ES-15
Nippondenso	W **16** EP
Prestolite	14GR **5** 2A

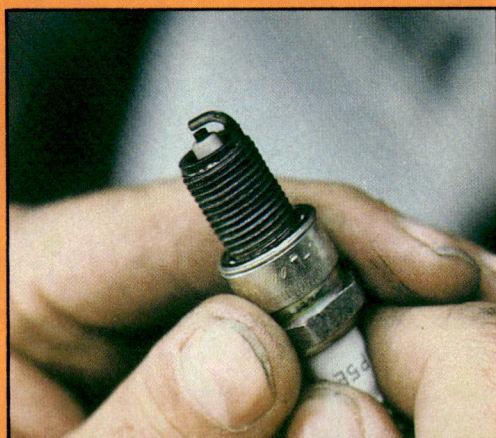

Periodically, check the spark plugs to be sure they are firing efficiently. They are excellent indicators of the internal condition of your engine.

On AC, Bosch (new), Champion, Fram/Autolite, Mopar, Motorcraft and Prestolite, a higher number indicates a hotter plug. On Bosch (old), NGK and Nippondenso, a higher number indicates a colder plug.

4. Make sure the spark plugs are properly gapped. See the Tune-Up Specifications in this book.

5. Be sure the spark plugs are firing efficiently. The illustrations on the next 2 pages show you how to "read" the firing end of the spark plug.

6. Check the ignition timing and set it to specifications. Tests show that almost all cars

veyed over 6,000 cars nationwide, they found that a tune-up, on cars that needed one, increased fuel economy over 11%. Replacing worn plugs alone, accounted for a 3% increase. The same test also revealed that 8 out of every 10 vehicles will have some maintenance deficiency that will directly affect fuel economy, emissions or performance. Most of this mileage-robbing neglect could be prevented with regular maintenance.

Modern engines require that all of the functioning systems operate properly for maximum efficiency. A malfunction anywhere wastes fuel. You can keep your vehicle running as efficiently and economically as possible, by being aware of your vehicles operating and performance characteristics. If your vehicle suddenly develops performance or fuel economy problems it could be due to one or more of the following:

PROBLEM	POSSIBLE CAUSE
Engine Idles Rough	Ignition timing, idle mixture, vacuum leak or something amiss in the emission control system.
Hesitates on Acceleration	Dirty carburetor or fuel filter, improper accelerator pump setting, ignition timing or fouled spark plugs.
Starts Hard or Fails to Start	Worn spark plugs, improperly set automatic choke, ice (or water) in fuel system.
Stalls Frequently	Automatic choke improperly adjusted and possible dirty air filter or fuel filter.
Performs Sluggishly	Worn spark plugs, dirty fuel or air filter, ignition timing or automatic choke out of adjustment.

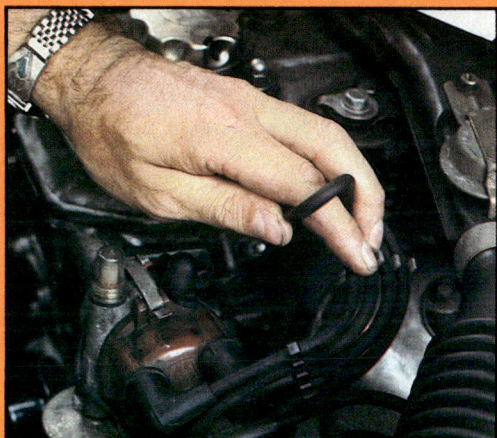

Check spark plug wires on conventional point type ignition for cracks by bending them in a loop around your finger.

Be sure that spark plug wires leading to adjacent cylinders do not run too close together. (Photo courtesy Champion Spark Plug Co.)

have incorrect ignition timing by more than 2°.

7. If your vehicle does not have electronic ignition, check the points, rotor and cap as specified.

8. Check the spark plug wires (used with conventional point-type ignitions) for cracks and burned or broken insulation by bending them in a loop around your finger. Cracked wires decrease fuel efficiency by failing to deliver full voltage to the spark plugs. One misfiring spark plug can cost you as much as 2 mpg.

9. Check the routing of the plug wires. Misfiring can be the result of spark plug leads to adjacent cylinders running parallel to each other and too close together. One wire tends to pick up voltage from the other causing it to fire "out of time".

10. Check all electrical and ignition circuits for voltage drop and resistance.

11. Check the distributor mechanical and/or vacuum advance mechanisms for proper functioning. The vacuum advance can be checked by twisting the distributor plate in the opposite direction of rotation. It should spring back when released.

12. Check and adjust the valve clearance on engines with mechanical lifters. The clearance should be slightly loose rather than too tight.

SPARK PLUG DIAGNOSIS

Normal

APPEARANCE: This plug is typical of one operating normally. The insulator nose varies from a light tan to grayish color with slight electrode wear. The presence of slight deposits is normal on used plugs and will have no adverse effect on engine performance. The spark plug heat range is correct for the engine and the engine is running normally.

CAUSE: Properly running engine.

RECOMMENDATION: Before reinstalling this plug, the electrodes should be cleaned and filed square. Set the gap to specifications. If the plug has been in service for more than 10-12,000 miles, the entire set should probably be replaced with a fresh set of the same heat range.

Oil Deposits

APPEARANCE: The firing end of the plug is covered with a wet, oily coating.

CAUSE: The problem is poor oil control. On high mileage engines, oil is leaking past the rings or valve guides into the combustion chamber. A common cause is also a plugged PCV valve, and a ruptured fuel pump diaphragm can also cause this condition. Oil fouled plugs such as these are often found in new or recently overhauled engines, before normal oil control is achieved, and can be cleaned and reinstalled.

RECOMMENDATION: A hotter spark plug may temporarily relieve the problem, but the engine is probably in need of work.

Incorrect Heat Range

APPEARANCE: The effects of high temperature on a spark plug are indicated by clean white, often blistered insulator. This can also be accompanied by excessive wear of the electrode, and the absence of deposits.

CAUSE: Check for the correct spark plug heat range. A plug which is too hot for the engine can result in overheating. A car operated mostly at high speeds can require a colder plug. Also check ignition timing, cooling system level, fuel mixture and leaking intake manifold.

RECOMMENDATION: If all ignition and engine adjustments are known to be correct, and no other malfunction exists, install spark plugs one heat range colder.

Photos Courtesy Champion Spark Plug Co.

Carbon Deposits

APPEARANCE: Carbon fouling is easily identified by the presence of dry, soft, black, sooty deposits.

CAUSE: Changing the heat range can often lead to carbon fouling, as can prolonged slow, stop-and-start driving. If the heat range is correct, carbon fouling can be attributed to a rich fuel mixture, sticking choke, clogged air cleaner, worn breaker points, retarded timing or low compression. If only one or two plugs are carbon fouled, check for corroded or cracked wires on the affected plugs. Also look for cracks in the distributor cap between the towers of affected cylinders.

RECOMMENDATION: After the problem is corrected, these plugs can be cleaned and reinstalled if not worn severely.

MMT Fouled

APPEARANCE: Spark plugs fouled by MMT (Methycyclopentadienyl Maganese Tricarbonyl) have reddish, rusty appearance on the insulator and side electrode.

CAUSE: MMT is an anti-knock additive in gasoline used to replace lead. During the combustion process, the MMT leaves a reddish deposit on the insulator and side electrode.

RECOMMENDATION: No engine malfunction is indicated and the deposits will not affect plug performance any more than lead deposits (see Ash Deposits). MMT fouled plugs can be cleaned, regapped and reinstalled.

High Speed Glazing

APPEARANCE: Glazing appears as shiny coating on the plug, either yellow or tan in color.

CAUSE: During hard, fast acceleration, plug temperatures rise suddenly. Deposits from normal combustion have no chance to fluff-off; instead, they melt on the insulator forming an electrically conductive coating which causes misfiring.

RECOMMENDATION: Glazed plugs are not easily cleaned. They should be replaced with a fresh set of plugs of the correct heat range. If the condition recurs, using plugs with a heat range one step colder may cure the problem.

Ash (Lead) Deposits

APPEARANCE: Ash deposits are characterized by light brown or white colored deposits crusted on the side or center electrodes. In some cases it may give the plug a rusty appearance.

CAUSE: Ash deposits are normally derived from oil or fuel additives burned during normal combustion. Normally they are harmless, though excessive amounts can cause misfiring. If deposits are excessive in short mileage, the valve guides may be worn.

RECOMMENDATION: Ash-fouled plugs can be cleaned, gapped and reinstalled.

Detonation

APPEARANCE: Detonation is usually characterized by a broken plug insulator.

CAUSE: A portion of the fuel charge will begin to burn spontaneously, from the increased heat following ignition. The explosion that results applies extreme pressure to engine components, frequently damaging spark plugs and pistons.

Detonation can result by over-advanced ignition timing, inferior gasoline (low octane) lean air/fuel mixture, poor carburetion, engine lugging or an increase in compression ratio due to combustion chamber deposits or engine modification.

RECOMMENDATION: Replace the plugs after correcting the problem.

Photos Courtesy Fram Corporation

EMISSION CONTROLS

13. Be aware of the general condition of the emission control system. It contributes to reduced pollution and should be serviced regularly to maintain efficient engine operation.

14. Check all vacuum lines for dried, cracked or brittle conditions. Something as simple as a leaking vacuum hose can cause poor performance and loss of economy.

15. Avoid tampering with the emission control system. Attempting to improve fuel econ-

FUEL SYSTEM

Check the air filter with a light behind it. If you can see light through the filter it can be reused.

Extremely clogged filters should be discarded and replaced with a new one.

18. Replace the air filter regularly. A dirty air filter richens the air/fuel mixture and can increase fuel consumption as much as 10%. Tests show that ⅓ of all vehicles have air filters in need of replacement.

19. Replace the fuel filter at least as often as recommended.

20. Set the idle speed and carburetor mixture to specifications.

21. Check the automatic choke. A sticking or malfunctioning choke wastes gas.

22. During the summer months, adjust the automatic choke for a leaner mixture which will produce faster engine warm-ups.

COOLING SYSTEM

29. Be sure all accessory drive belts are in good condition. Check for cracks or wear.

30. Adjust all accessory drive belts to proper tension.

31. Check all hoses for swollen areas, worn spots, or loose clamps.

32. Check coolant level in the radiator or expansion tank.

33. Be sure the thermostat is operating properly. A stuck thermostat delays engine warm-up and a cold engine uses nearly twice as much fuel as a warm engine.

34. Drain and replace the engine coolant at least as often as recommended. Rust and scale

TIRES & WHEELS

38. Check the tire pressure often with a pencil type gauge. Tests by a major tire manufacturer show that 90% of all vehicles have at least 1 tire improperly inflated. Better mileage can be achieved by over-inflating tires, but never exceed the maximum inflation pressure on the side of the tire.

39. If possible, install radial tires. Radial tires deliver as much as ½ mpg more than bias belted tires.

40. Avoid installing super-wide tires. They only create extra rolling resistance and decrease fuel mileage. Stick to the manufacturer's recommendations.

41. Have the wheels properly balanced.

omy by tampering with emission controls is more likely to worsen fuel economy than improve it. Emission control changes on modern engines are not readily reversible.

16. Clean (or replace) the EGR valve and lines as recommended.

17. Be sure that all vacuum lines and hoses are reconnected properly after working under the hood. An unconnected or misrouted vacuum line can wreak havoc with engine performance.

23. Check for fuel leaks at the carburetor, fuel pump, fuel lines and fuel tank. Be sure all lines and connections are tight.

24. Periodically check the tightness of the carburetor and intake manifold attaching nuts and bolts. These are a common place for vacuum leaks to occur.

25. Clean the carburetor periodically and lubricate the linkage.

26. The condition of the tailpipe can be an excellent indicator of proper engine combustion. After a long drive at highway speeds, the inside of the tailpipe should be a light grey in color. Black or soot on the insides indicates an overly rich mixture.

27. Check the fuel pump pressure. The fuel pump may be supplying more fuel than the engine needs.

28. Use the proper grade of gasoline for your engine. Don't try to compensate for knocking or "pinging" by advancing the ignition timing. This practice will only increase plug temperature and the chances of detonation or pre-ignition with relatively little performance gain.

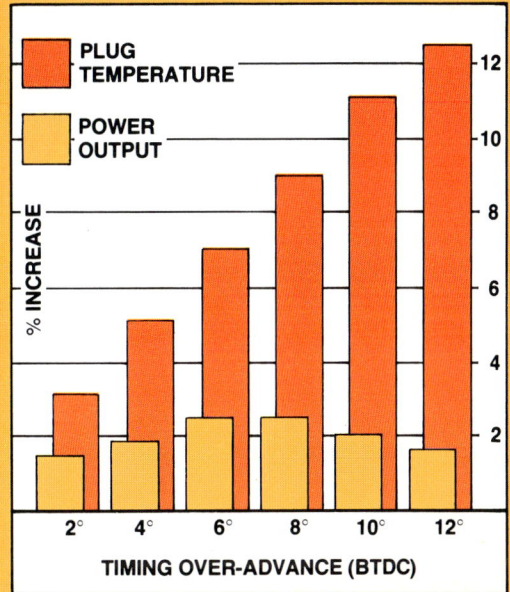

Increasing ignition timing past the specified setting results in a drastic increase in spark plug temperature with increased chance of detonation or preignition. Performance increase is considerably less. (Photo courtesy Champion Spark Plug Co.)

that form in the engine should be flushed out to allow the engine to operate at peak efficiency.

35. Clean the radiator of debris that can decrease cooling efficiency.

36. Install a flex-type or electric cooling fan, if you don't have a clutch type fan. Flex fans use curved plastic blades to push more air at low speeds when more cooling is needed; at high speeds the blades flatten out for less resistance. Electric fans only run when the engine temperature reaches a predetermined level.

37. Check the radiator cap for a worn or cracked gasket. If the cap does not seal properly, the cooling system will not function properly.

42. Be sure the front end is correctly aligned. A misaligned front end actually has wheels going in different directions. The increased drag can reduce fuel economy by .3 mpg.

43. Correctly adjust the wheel bearings. Wheel bearings that are adjusted too tight increase rolling resistance.

Check tire pressures regularly with a reliable pocket type gauge. Be sure to check the pressure on a cold tire.

GENERAL MAINTENANCE

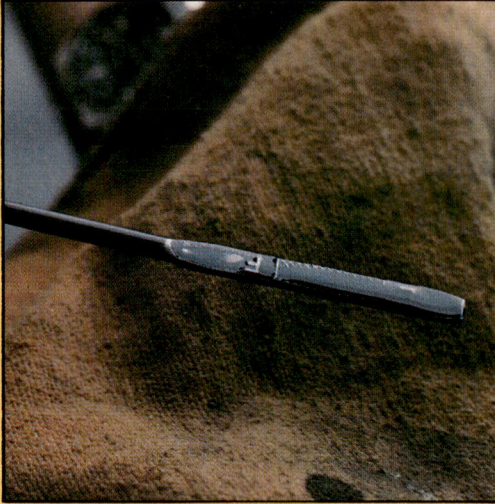

Check the fluid levels (particularly engine oil) on a regular basis. Be sure to check the oil for grit, water or other contamination.

A vacuum gauge is another excellent indicator of internal engine condition and can also be installed in the dash as a mileage indicator.

44. Periodically check the fluid levels in the engine, power steering pump, master cylinder, automatic transmission and drive axle.

45. Change the oil at the recommended interval and change the filter at every oil change. Dirty oil is thick and causes extra friction between moving parts, cutting efficiency and increasing wear. A worn engine requires more frequent tune-ups and gets progressively worse fuel economy. In general, use the lightest viscosity oil for the driving conditions you will encounter.

46. Use the recommended viscosity fluids in the transmission and axle.

47. Be sure the battery is fully charged for fast starts. A slow starting engine wastes fuel.

48. Be sure battery terminals are clean and tight.

49. Check the battery electrolyte level and add distilled water if necessary.

50. Check the exhaust system for crushed pipes, blockages and leaks.

51. Adjust the brakes. Dragging brakes or brakes that are not releasing create increased drag on the engine.

52. Install a vacuum gauge or miles-per-gallon gauge. These gauges visually indicate engine vacuum in the intake manifold. High vacuum = good mileage and low vacuum = poorer mileage. The gauge can also be an excellent indicator of internal engine conditions.

53. Be sure the clutch is properly adjusted. A slipping clutch wastes fuel.

54. Check and periodically lubricate the heat control valve in the exhaust manifold. A sticking or inoperative valve prevents engine warm-up and wastes gas.

55. Keep accurate records to check fuel economy over a period of time. A sudden drop in fuel economy may signal a need for tune-up or other maintenance.

AMBIENT TEMPERATURE
OVERRIDE SWITCH
(CLOSES AT 63°F)

SOLENOID
VACUUM
VALVE
(NO VACUUM
ADVANCE WHEN
ENERGIZED)

BATTERY
FEED
WIRE

SOLENOID
CONTROL
SWITCH
(OPENS AT
APPROX.
25–30
MPH OR
TOP GEAR

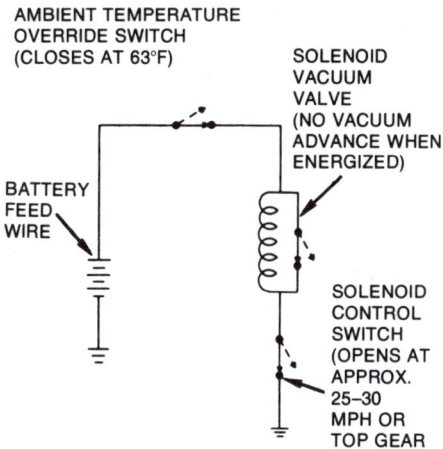

TCS electrical circuit

PCV HOSE

PORTED
VACUUM
"T"
CONNECTION

SOLENOID
VACUUM
VALVE

PORTED
VACUUM
HOSE

MANIFOLD
VACUUM
HOSE

DISTRIBUTOR
VACUUM HOSE

BELOW
160°F

AT 160°F

2 D 1

2 D 1

CLOSED   OPEN   OPEN   CLOSED

COOLANT TEMPERATURE OVERRIDE SWITCH

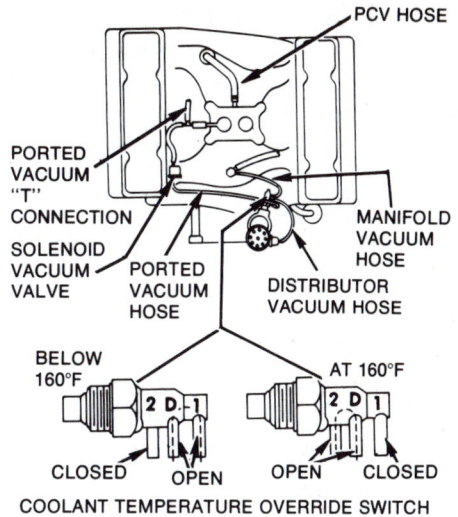

TCS hose routing

Ambient Temperature Override Switch

This switch, located at the firewall, senses ambient temperatures and completes the electrical circuit from the battery to the solenoid vacuum valve when ambient temperatures are above 63° F.

Solenoid Vacuum Valve

This valve is attached to the ignition coil bracket at the right side of the engine (V8 engines) or to a bracket at the rear of the intake manifold (Sixes). When the valve is energized, carburetor vacuum is blocked off and the distributor vacuum line is vented to the atmosphere through a port in the valve, resulting in no vacuum advance. When the valve is deenergized, vacuum is applied to the distributor resulting in normal vacuum advance.

Solenoid Control Switch

This switch is located at the transmission valve body. It opens or closes in relation to car speed and gear range. When the transmission is in high gear, the switch opens and breaks the ground circuit to the solenoid vacuum valve. In lower gear ranges the switch closes and completes the ground circuit to the solenoid vacuum valve. With a manual transmission, the switch is operated by the transmission shifter shaft. With automatic transmissions, the switch is controlled by the speedometer gear speed. Under speeds of 25 mph, the switch is activated.

Coolant Temperature Override Switch

This switch is used only on the 304 V8. It is threaded into the thermostat housing. The switch reacts to coolant temperatures to route either intake manifold or carburetor vacuum to the distributor vacuum advance diaphragm.

When the coolant temperature is below 160° F, intake manifold vacuum is applied through a hose connection to the distributor advance diaphragm, resulting in full vacuum advance.

When the coolant temperature is above 160° F, intake manifold vacuum is blocked off and carburetor vacuum is then applied through the solenoid vacuum valve to the distributor advance diaphragm, resulting in decreased vacuum advance.

The relationship between distributor vacuum advance and the operation of the TCS system and coolant temperature override switch can be determined by referring to the Emission Control Distributor Vacuum Application Chart.

Maintenance and Service

Efficient performance of the exhaust emission control system is dependent upon precise maintenance.

Carburetor

Check the carburetor for the proper application. Check the dashpot for proper operation and adjust as required. When the throttle is released quickly, the arm of the dashpot should fully extend itself and should catch the throttle lever, letting it back to idle position gradually.

Proper idle mixture adjustment is impera-

Emission Controlled Distributor Vacuum Application Chart for Vehicles Equipped with TCS

Manual Transmission (gear)		Automatic Transmission (Vehicle Speed)	Ambient (air) Temperature	Coolant Temperature	Vacuum Applied to Distributor
3-Speed	4-Speed				
1–2	1–2–3	Under 25 mph	Below 63° F	Below 160° F	Manifold
1–2	1–2–3	Under 25 mph	Below 63° F	Above 160° F	Ported
1–2	1–2–3	Under 25 mph	Above 63° F	Above 160° F	None
1–2	1–2–3	Under 25 mph	Above 63° F	Below 160° F	Manifold
3	4	25–30 mph	Below 63° F	Below 160° F	Manifold
3	4	25–30 mph	Below 63° F	Above 160° F	Ported
3	4	25–30 mph	Above 63° F	Above 160° F	Ported
3	4	25–30 mph	Above 63° F	Below 160° F	Manifold

NOTE: *If equipped with thermal vacuum switch (TVS), intake manifold vacuum is applied to the distributor when engine coolant temperature reaches 225° F.*

tive for best exhaust emission control. The idle adjustment should be made with the engine at normal operating temperature and the air cleaner in place. All lights and accessories must be turned off and the transmission must be in neutral. See the tune-up chapter for adjustment procedures.

Distributor

Check the distributor number for proper application. Check the distributor cam dwell angle and point condition and adjust to specifications or replace as required. See the tune-up chapter for procedures.

Anti-Backfire Diverter Valve

On the F-head, the anti-backfire valve remains open except when the throttle is closed rapidly from an open position.

To check the valve for proper operation, accelerate the engine in neutral, allowing the throttle to close rapidly. The valve is operating satisfactorily when no exhaust system backfire occurs. A further check can be made by removing the large hose that runs from the anti-backfire valve to the check valve and accelerating the engine and allowing the throttle to close rapidly. If there is an audible momentary interruption of the flow of air then it can be assumed that the valve is working correctly.

To check the valve on a V6, pre-1971 232 Six and 327 V8, listen for backfire when the throttle is released quickly. If none exists, the valve is doing its job. To check further, remove the large hose that connects the valve with the air pump. Place a finger over the open end of the hose, not the valve, and

accelerate the engine, allowing the throttle to close rapidly. The valve is operating satisfactorily if there is a momentary audible rush of air.

Check Valve

The check valve in the air distribution manifold prevents the reverse flow of exhaust gases to the pump in the event the pump should become inoperative or should exhaust pressure ever exceed the pump pressure.

To check this valve for proper operation, remove the air supply hose from the pump at the distribution manifold. With the engine running, listen for exhaust leakage where the check valve is connected to the distribution manifold. If leakage is audible, the valve is not operating correctly.

Air Pump

Check for the proper drive belt tension and adjust as necessary. Do not pry on the die cast pump housing. Check to see if the pump is discharging air. Remove the air outlet hose at the pump. With the engine running, air should be felt at the pump outlet opening.

REMOVAL AND INSTALLATION
Air Pump

1. Loosen the air pump adjusting bracket bolts.
2. Remove the drive belt.
3. Remove the air pump intake and discharge hoses.
4. Remove the air pump from the engine.
5. To install, reverse the above procedure.

Anti-backfire Valve

To remove the anti-backfire valve disconnect the hoses and bracket-to-engine attaching screws. Install in the reverse order of removal.

Air Distribution Manifold and Air Injection Tubes

It is necessary to remove the exhaust manifold only on the F-head prior to removing the air distribution manifold and the air injection tubes. On all the other engines, these components can be removed with the manifolds on the engine.

1. Disconnect the air delivery hose from the air injection manifold. Remove the exhaust manifold on the F-head.

2. Remove the air distribution manifold from the air injection tubes on the F-head only.

3. Unscrew the air injection tube from the exhaust manifold or the head. Some resistance may be encountered because of the normal build up of carbon. The application of heat may be helpful in removing the air injection tubes.

4. Install in the reverse order of removal.
NOTE: *There are two lengths of tubes used with the F-head. The shorter tubes are installed in numbers 1 and 4 cylinders. The air injection tubes must be installed on the exhaust manifold prior to installing the exhaust manifold on the engine.*

Exhaust Gas Recirculation (EGR) System

The EGR system consists of a diaphragm actuated flow control valve (EGR valve), coolant temperature override switch, low temperature vacuum signal modulator, high temperature vacuum signal modulator, and connecting hoses. This system is only installed on the 1973 and 1974 engines.

All 1977 and later California units have a back pressure sensor which modulates EGR signal vacuum according to the rise or fall of exhaust pressure in the manifold. A restrictor plate is not used in these applications.

The purpose of the EGR system is to limit the formation of nitrogen oxides by diluting the fresh air intake charge with a metered amount of exhaust gas, thereby reducing the peak temperatures of the burning gases in the combustion chambers.

EGR Valve

The EGR valve is mounted on a machined surface at the rear of the intake manifold on the V8s and on the side of the intake manifold on the sixes.

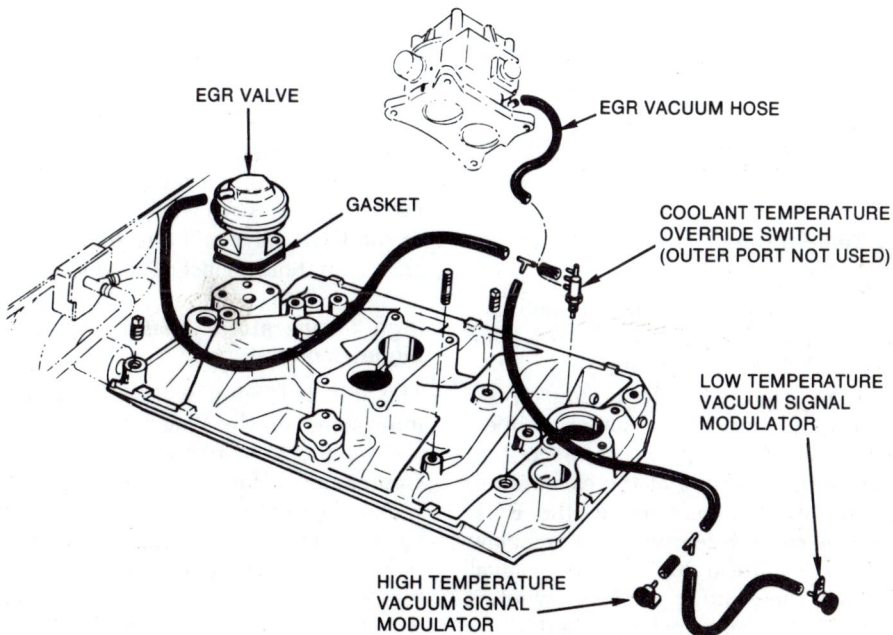

EGR system: 1973–74 8-304, 360, 401

HOSE CONNECTION
TO EGR PORT
AT CARBURETOR

COOLANT TEMPERATURE
OVERRIDE SWITCH
(OUTER PORT NOT USED)

DISCHARGE
PASSAGE

GASKET

EGR VALVE

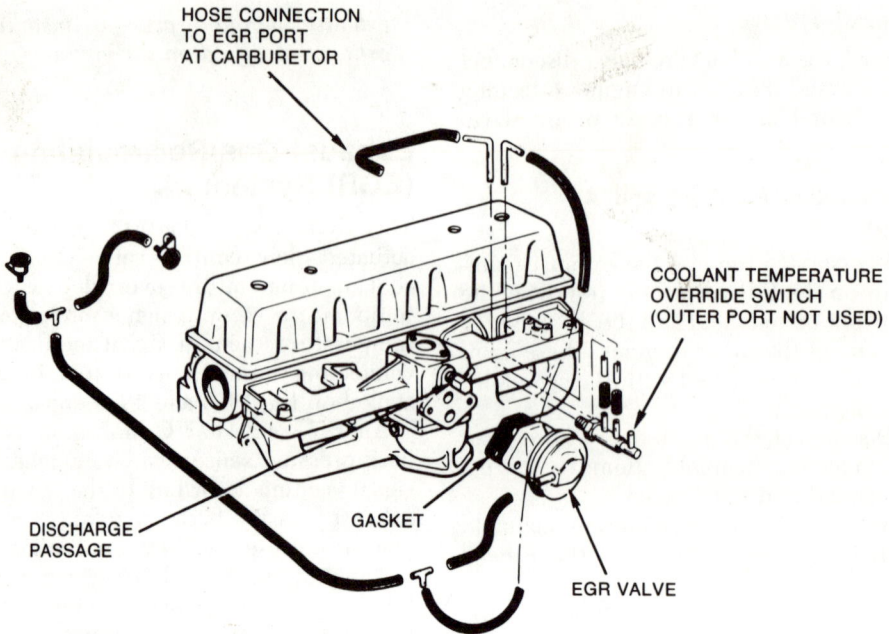

EGR system 1973–74 6-232, 258

The valve is held in a normally closed position by a coil spring located above the diaphragm. A special fitting is provided at the carburetor to route ported (above the throttle plates) vacuum through hose connections to a fitting located above the diaphragm on the valve. A passage in the intake manifold directs exhaust gas from the exhaust crossover passage (V8) or from below the riser area (Sixes) to the EGR valve. When the diaphragm is actuated by vacuum, the valve opens and meters exhaust gas through another passage in the intake manifold to the floor of the intake manifold below the carburetor.

Coolant Temperature Override Switch

This switch is located in the intake manifold at the coolant passage adjacent to the oil filler tube on the V8s or at the left side of the engine block (formerly the drain plug) on the Sixes. The outer port of the switch is open and not used. The inner port is connected by a hose to the EGR fitting at the carburetor. The center port is connected to the EGR valve. When coolant temperature is below 115° F, (160° F on the 304 V8 with manual transmission), the center port of the switch is closed and no vacuum signal is applied to the EGR valve, therefore, no exhaust gas will flow through the valve. When the coolant

temperature reaches 115° F both the center port and the inner port of the switch are open and a vacuum signal is applied to the EGR valve. This vacuum signal is, however, subject to regulation by the low and high temperature signal modulators.

Low Temperature Vaccum Signal Modulator

This unit is located just to the right of the radiator behind the grill opening. The low temperature vacuum signal modulator vacuum hose is connected by a plastic T-fitting to the EGR vacuum signal hose. The modulator is open when ambient temperatures are below 60° F. This causes a weakened vacuum signal to the EGR valve and a resultant decrease in the amount of exhaust gas being recirculated.

High Temperature Vacuum Signal Modulator

This unit is located at the right front fender inner panel on the Wagoneer and Cherokee, and on the front of the battery tray on the Commando. The high temperature vacuum signal modulator is connected to the EGR vacuum signal hose by a plastic T-fitting. The modulator opens when the underhood air temperatures reach 115° F and it causes a weakened vacuum signal to the EGR valve, thus reducing the amount of exhaust gases being recirculated.

Electric Assist Choke

An electric assist choke is used on 4 bbl carburetors to more accurately match the choke operation to engine requirements. It provides extra heat to the choke bimetal spring to speed up the choke valve opening after the underhood air temperature reaches 95° F ± 15° F. Its purpose is to reduce the emission of carbon monoxide (CO) during the engine's warm-up period.

A special AC terminal is provided at the alternator to supply a 7 volt power source for the electric choke. A thermostatic switch within the choke cover closes when the underhood air temperature reaches 95° F ± 15° F and allows current to flow to a ceramic heating element. The circuit is completed through the choke cover ground strap and choke housing to the engine. As the heating element warms up, heat is absorbed by an attached metal plate which in turn heats the choke bimetal spring.

After the engine is turned off, the thermostatic switch remains closed until the underhood temperature drops below approximately 65° F. Therefore, the heating element will immediately begin warming up when the engine is restarted, if the underhood temperature is above 65° F.

Controlled Combustion System—350 V8

The controlled combustion system used on the 350 V8 engines limits the hydrocarbons and carbon monoxide emissions from the exhaust. The system includes an engine designed for low emissions, lean carburetor calibration at idle and part-throttle, plus lean choke calibration.

The engine has "ported" spark advance, with the vacuum take-off just above the throttle valve so that there is no vacuum advance at closed throttle, but vacuum advance begins as soon as the throttle is opened slightly. To reduce emissions at idle and at lower engine speeds, the engine timing is such that the distributor will not have centrifugal advance until about 850 rpm.

The lean carburetion is possible because of the heated air system (thermostatically controlled air cleaner). With the heated air system operating, inlet air temperature is around 115° F after the first few minutes of operation. This makes the use of lean carburetor calibration possible and the engine still responds well in cold weather.

Fuel Tank Vapor Emission Control System

A closed fuel tank system is used on all models through 1974, some models 1975–78, and all 1979 models, to route raw fuel vapor from the fuel tank into the PCV system (sixes) or air cleaner snorkle (V8s) where it is burned along with the fuel-air mixture. The system prevents raw fuel vapors from entering the atmosphere.

The fuel vapor system consists of internal fuel tank venting, a vacuum-pressure fuel tank filler cap, an expansion tank or charcoal filled canister, liquid limit fill valve, and internal carburetor venting.

Fuel vapor pressure in the fuel tank forces the vapor through vent lines to the expansion tank or charcoal filled storage canister. The vapor then travels through a single vent line to the limit fill valve which regulates the vapor flow to the valve cover or air cleaner.

The fuel tank vent line is routed through the limit fill valve to the valve cover on the left side on the 1972 V8s. On the 1973 Sixes, it travels to the intake manifold and on the V8s it is routed to the carburetor air cleaner.

LIMIT FILL VALVE

This valve is essentially a combination vapor flow regulator and pressure relief valve. It regulates vapor flow from the fuel tank vent line into the valve cover. The valve consists of a housing, a spring loaded diaphragm and a diaphragm cover. As tank vent pressure increases, the diaphragm lifts permitting vapor to flow through. The pressure at which this occurs is 4–6 in. of water column. This action regulates the flow of vapors under severe conditions but generally prohibits the flow of vapor during normal temperature operation, thus minimizing driveability problems.

Catalytic Converter

All 1977–78 California models and all 1979 models are equipped with a catalytic converter. Most models use a single pellet-filled unit, while some California models use a single monolithic type unit.

The pellet type contains beads of alumina coated with platinum and palladium, contained in a stainless steel canister. A plug is provided in the unit for replacement of the beads if they become fouled. The monolithic unit uses an extruded core resembling a honeycomb. The core layers are coated with platinum and palladium. This unit is not serviceable.

Vacuum Throttle Modulating System (VTM)

This system is designed to reduce the level of hydrocarbon emission during rapid throttle closure at high speed. It is used on some 49 state and all California Wagoneer and Cherokee models, with a V-8 engine.

The system consists of a deceleration valve located at the right front of the intake manifold, and a throttle modulating diaphragm located at the carburetor base. The valve and the diaphagm are connected by a vacuum hose and the valve is connected to direct manifold vacuum. During deceleration, manifold vacuum acts to delay, slightly, the closing of the throttle plate.

To Adjust:

1. Run the engine to normal operating temperature and set the idle speed to specification. Shut off the engine.

2. Position the throttle lever against the curb idle adjusting screw.

3. Measure the clearance between the throttle modulating diaphragm plunger and the throttle lever. A clearance of $1/16$ inch should exist.

4. Adjust the clearance, if necessary, by loosening the jam nut and turning the diaphragm assembly.

FUEL SYSTEM

Fuel Pump

REMOVAL AND INSTALLATION ALL ENGINES

1. Disconnect the inlet and outlet fuel lines.

2. Remove the two fuel pump body attaching nuts and lockwashers.

3. Pull the pump and gasket free of the enine. Make sure that the mating surfaces of the fuel pump and the engine are clean.

4. Cement a new gasket to the mounting flange of the fuel pump.

5. Position the fuel pump on the engine block so that the lever of the fuel pump rests on the fuel pump cam of the camshaft.

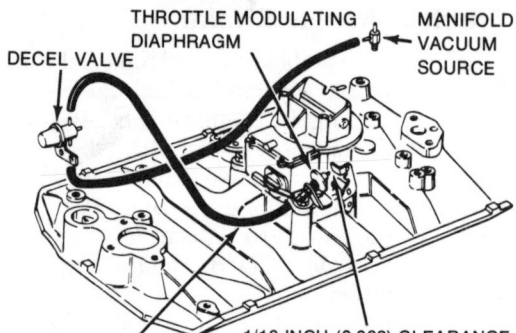

Typical vacuum throttle modulating system

6. Secure the fuel pump to the block with the two cap screws and lock washers.

7. Connect the intake and outlet fuel lines to the fuel pump.

FUEL PUMP TESTING

Volume Check

Disconnect the fuel line from the carburetor. Place the open end in a suitable container. Start the engine and operate it at normal idle speed. The pump should deliver at least one quart in one minute.

Pressure Check

Disconnect the fuel line at the carburetor. Install a T-fitting on the open end of the fuel line and refit the line to the carburetor. Plug a pressure gauge into the remaining opening

1. Fuel outlet   2. Vapor return   3. Fuel inlet

6-225 and 8-350 fuel pump; the unit is not serviceable

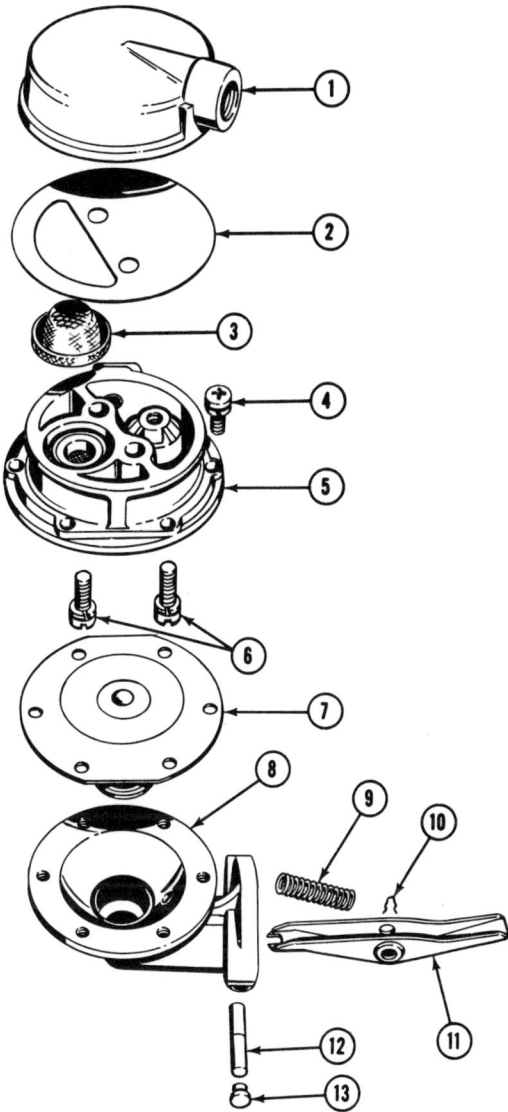

1. Housing cover
2. Air dome diaphragm
3. Strainer
4. Screw and washer
5. Housing
6. Cover screw and lockwashers
7. Main diaphragm
8. Pump body
9. Cam lever return spring
10. Pin retainer
11. Cam lever
12. Cam lever pin
13. Lever seal shaft plug

4-134 and 8-327 fuel pump

1. Ball	10. Valve housing
2. Bowl	11. Valve assembly
3. Spring	12. Screws
4. Filter	13. Diaphragm and oil seal
5. Gasket	14. Pump body
6. Pump body	15. Cam lever spring
7. Gasket	16. Cam lever
8. Valve assembly	17. Gasket
9. Screws	18. Cam lever pin and plug

1966–70 6-232 fuel pump, used on some 4-134 engines

of the T-fitting. The hose leading to the pressure gauge should not be any longer than 6 inches. Start the engine and let it run at idle speed. Pressure readings are given in the Tune-Up Specifications Chart.

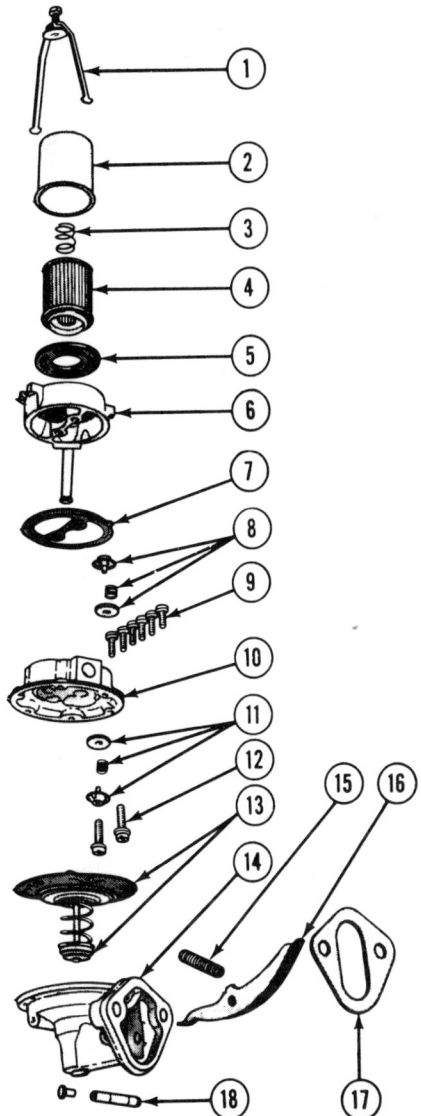

Carburetors

REMOVAL AND INSTALLATION ALL ENGINES

To remove the carburetor from any engine, first remove the air cleaner from the top of the carburetor. Remove all lines and hoses, noting their positions to facilitate installation.

Remove all throttle and choke linkage at the carburetor. Remove the carburetor attaching nuts which hold it to the intake manifold. Lift the carburetor from the engine along with the carburetor base gasket. Discard the gasket. Install the carburetor in the reverse order of removal, using a new base gasket.

OVERHAUL ALL TYPES

Efficient carburetion depends greatly on careful cleaning and inspection during overhaul since dirt, gum, water, or varnish in or on the carburetor parts are often responsible for poor performance.

Overhaul your carburetor in a clean, dust-free area. Carefully disassemble the carburetor, referring often to the exploded views. Keep all similar and look-alike parts segregated during disassembly and cleaning to avoid accidental interchange during assembly. Make a note of all jet sizes.

When the carburetor is disassembled, wash all parts (except diaphragms, electric choke units, pump plunger, and any other plastic, leather, fiber, or rubber parts) in clean carburetor solvent. Do not leave parts in the solvent any longer than is necessary to sufficiently loosen the deposits. Excessive

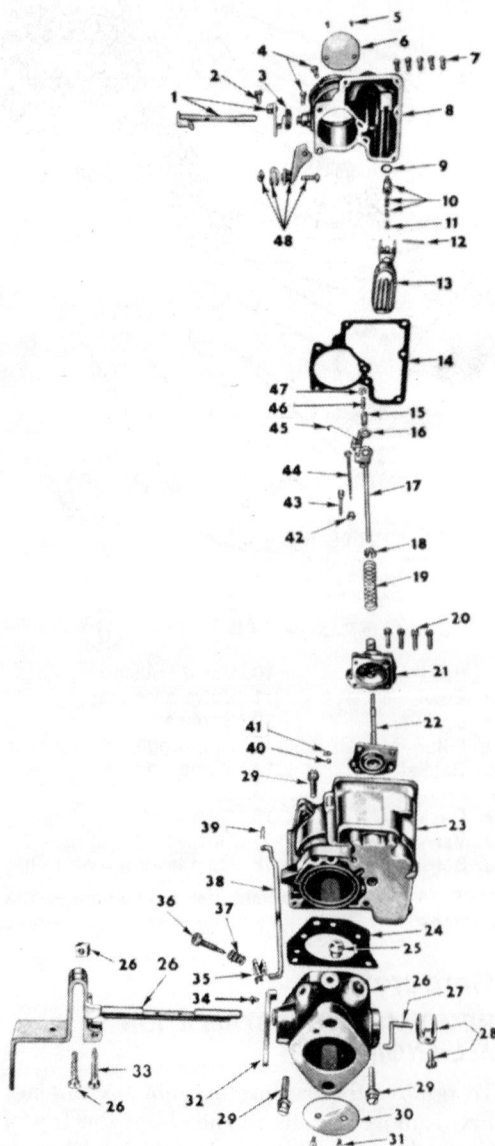

1. Choke shaft and lever
2. Screw
3. Choke lever spring
4. Screw and washer
5. Choke valve screw
6. Choke valve
7. Screw and washer
8. Air horn
9. Needle seat gasket
10. Needle spring and seat
11. Needle pin
12. Float pin
13. Float
14. Gasket
15. Pump spring
16. Metering rod arm
17. Pump link
18. Pump spring retainer
19. Vacuum diaphragm spring
20. Screw and washer
21. Diaphragm housing
22. Diaphragm
23. Body
24. Gasket
25. Idle port plug
26. Throttle body lever and shaft assembly
27. Pump link connector
28. Throttle shaft arm
29. Screw and washer
30. Throttle valve
31. Throttle valve screw
32. Fast idle arm
33. Adjusting screw
34. Body flange plug
35. Clevis clip
36. Idle adjusting screw
37. Idle screw spring
38. Fast idle connector rod
39. Pin spring
40. Ball check valve
41. Ball check valve retainer ring
42. Metering rod jet
43. Low speed jet
44. Metering rod
45. Metering rod spring
46. Inner pump spring
47. Pump spring retainer
48. Bracket and clamp assembly (choke and throttle)

Carter YF exploded view

cleaning may remove the special finish from the float bowl and choke valve bodies, leaving these parts unfit for service. Rinse all parts in clean solvent and blow them dry with compressed air or allow them to air dry. Wipe clean all cork, plastic, leather, and fiber parts with a clean, lint-free cloth.

Blow out all passages and jets with compressed air and be sure that there are no restrictions or blockages. Never use wire or similar tools to clean jets, fuel passages, or air bleeds. Clean all jets and valves separately to avoid accidental interchange.

Check all parts for wear or damage. If wear or damage is found, replace the defective parts. Especially check the following:

1. Check the float needle and seat for wear. If wear is found, replace the complete assembly.

2. Check the float hinge pin for wear and the float(s) for dents or distortion. Replace the float if fuel has leaked into it.

3. Check the throttle and choke shaft bores for wear or an out-of-round condition. Damage or wear to the throttle arm, shaft, or shaft bore will often require replacement of the throttle body. These parts require a close tolerance of fit; wear may allow air leakage, which could affect starting and idling.

NOTE: *Throttle shafts and bushings are not included in overhaul kits. They can be purchased separately.*

4. Inspect the idle mixture adjusting needles for burrs or grooves. Any such condition requires replacement of the needle, since you will not be able to obtain a satisfactory idle.

5. Test the accelerator pump check valves. They should pass air one way but not the other. Test for proper seating by blowing and sucking on the valve. Replace the valve if necessary. If the valve is satisfactory, wash the valve again to remove breath moisture.

6. Check the bowl cover for warped surfaces with a straightedge.

7. Closely inspect the valves and seats for wear and damage, replacing as necessary.

8. After the carburetor is assembled, check the choke valve for freedom of operation.

Carburetor overhaul kits are recommended for each overhaul. These kits contain

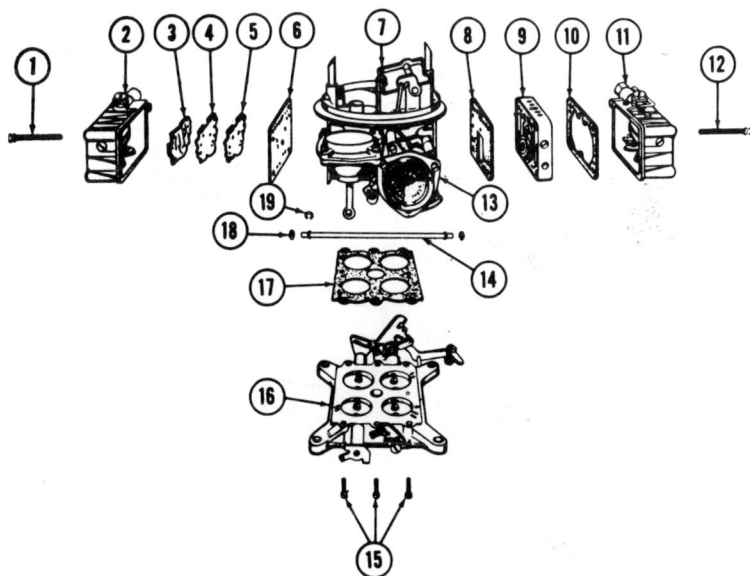

1. Secondary fuel bowl screw and gasket
2. Secondary fuel bowl
3. Secondary fuel bowl gasket
4. Secondary metering body
5. Secondary metering body gasket
6. Secondary metering plate
7. Secondary metering body plate gasket
8. Primary metering body gasket
9. Primar metering body
10. Primary fuel bowl gasket
11. Primary fuel bowl
12. Primary fuel bowl screw and gasket
13. Main body
14. Fuel line tubing
15. Throttle body-to-main body screw and lockwasher
16. Throttle body
17. Throttle body-to-main body gasket
18. Fuel line tubing O-ring
19. Secondary diaphragm rod retainer

Holley 4160 exploded view

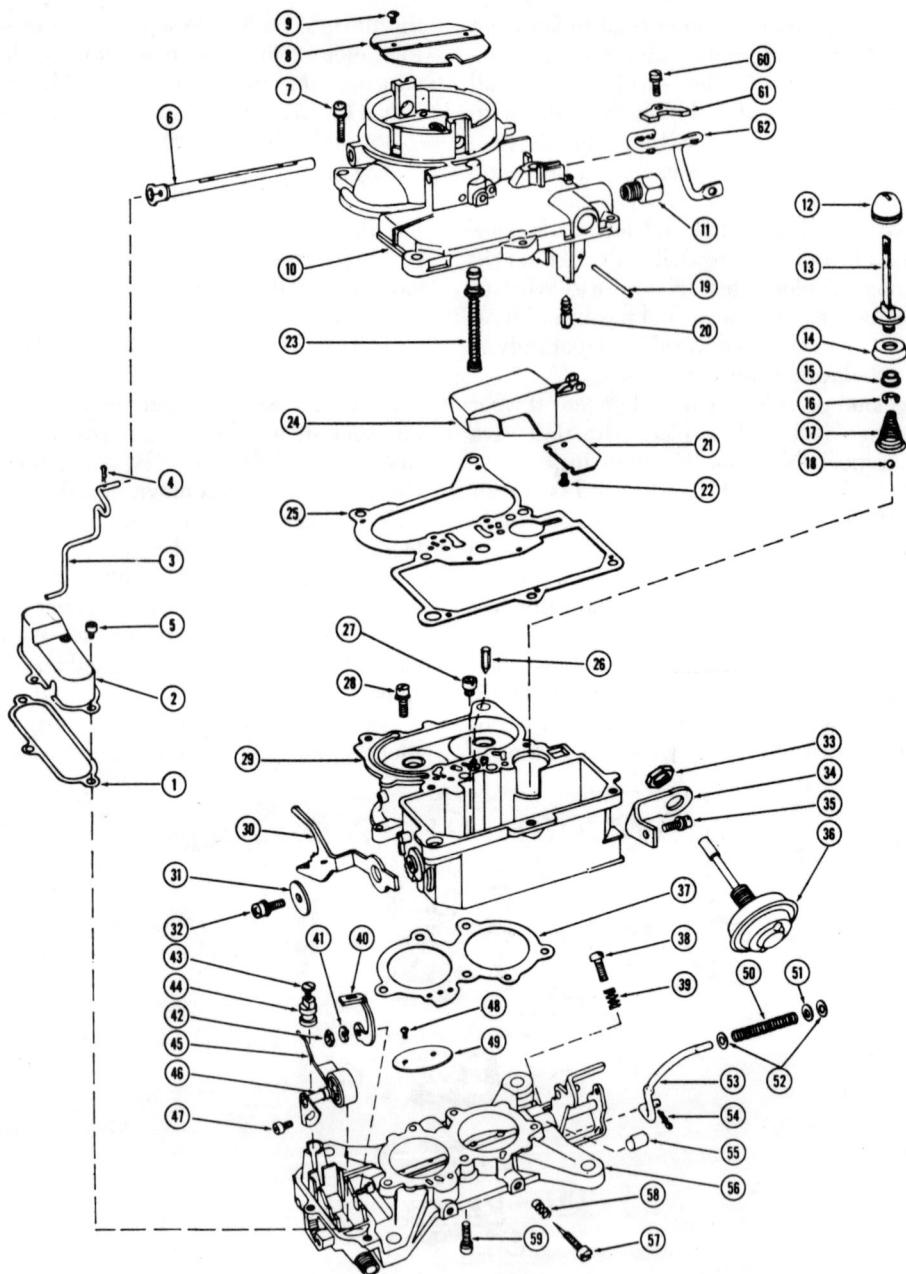

1. Thermostat housing cover gasket
2. Thermostat cover and guide assembly
3. Choke rod
4. Choke rod retainer
5. Thermostat housing cover screw
6. Choke shaft and lever assembly
7. Air horn-to-main body screw
8. Choke plate
9. Choke plate screw
10. Air horn and plugs assembly
11. Fuel inlet fitting
12. Acceleration pump stem seal
13. Acceleration pump assembly
14. Acceleration pump cup
15. Acceleration pump cup liner
16. Acceleration pump cup retainer
17. Acceleration pump return spring
18. Acceleration pump inlet valve
19. Float hinge pin
20. Fuel inlet needle assembly
21. Fuel bowl baffle
22. Fuel bowl baffle screw
23. Power valve piston assembly
24. Float assembly
25. Main body gasket
26. Pump discharge valve
27. Main jet
28. Throttle body-to-main body screw and lockwasher

Holley 2209 exploded view

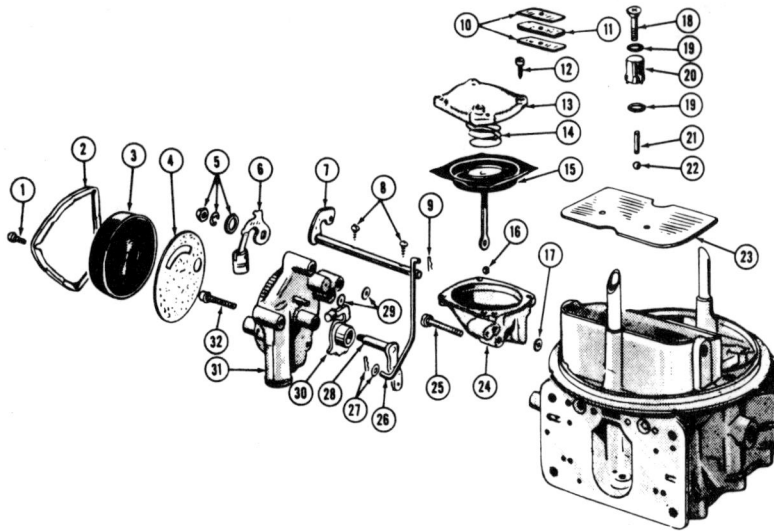

1. Thermostat housing cover screw
2. Thermostat housing cover clamp
3. Thermostat housing cover
4. Thermostat housing cover gasket
5. Choke shaft nut, lockwasher and spacer
6. Choke lever link and piston assembly
7. Choke shaft
8. Choke plate screw
9. Choke rod retainer (upper)
10. Choke rod seal retainer
11. Choke rod felt seal
12. Diaphragm cover screw and lockwasher
13. Diaphragm cover
14. Diaphragm spring
15. Diaphragm assembly
16. Diaphragm check ball
17. Diaphragm housing gasket
18. Pump discharge nozzle screw
19. Pump discharge nozzle gasket
20. Pump discharge nozzle
21. Pump check weight
22. Pump check ball
23. Choke plate
24. Diaphragm housing
25. Secondary diaphragm housing screw and lockwasher
26. Choke rod
27. Lower choke rod retainer and washer
28. Choke housing shaft
29. Choke housing gasket
30. Fast idle cam
31. Choke housing
32. Choke housing screw and lockwasher

Holley 4160 main body exploded view

29. Main body and plugs assembly
30. Fast idle cam
31. Fast idle cam washer
32. Fast idle cam screw
33. Dashpot nut
34. Dashpot bracket
35. Dashpot bracket screw and lockwasher
36. Dashpot assembly
37. Throttle body gasket
38. Throttle stop screw
39. Throttle stop screw spring
40. Fast idle and dechoke lever
41. Lockwasher
42. Nut
43. Choke piston screw
44. Choke piston
45. Choke piston link
46. Choke thermostat assembly
47. Choke thermostat assembly lockscrew
48. Throttle plate screw
49. Throttle plate
50. Pump drive spring
51. Pump link washer
52. Pump drive spring retainer
53. Pump operating link
54. Pump operating link retainer
55. Throttle shaft bearing ribbon
56. Throttle body and shaft assembly
57. Idle adjusting needle
58. Idle adjusting needle spring
59. Throttle body-to-main body screw and lockwasher
60. Clamp retainer screw
61. Pump rod clamp
62. Pump rod

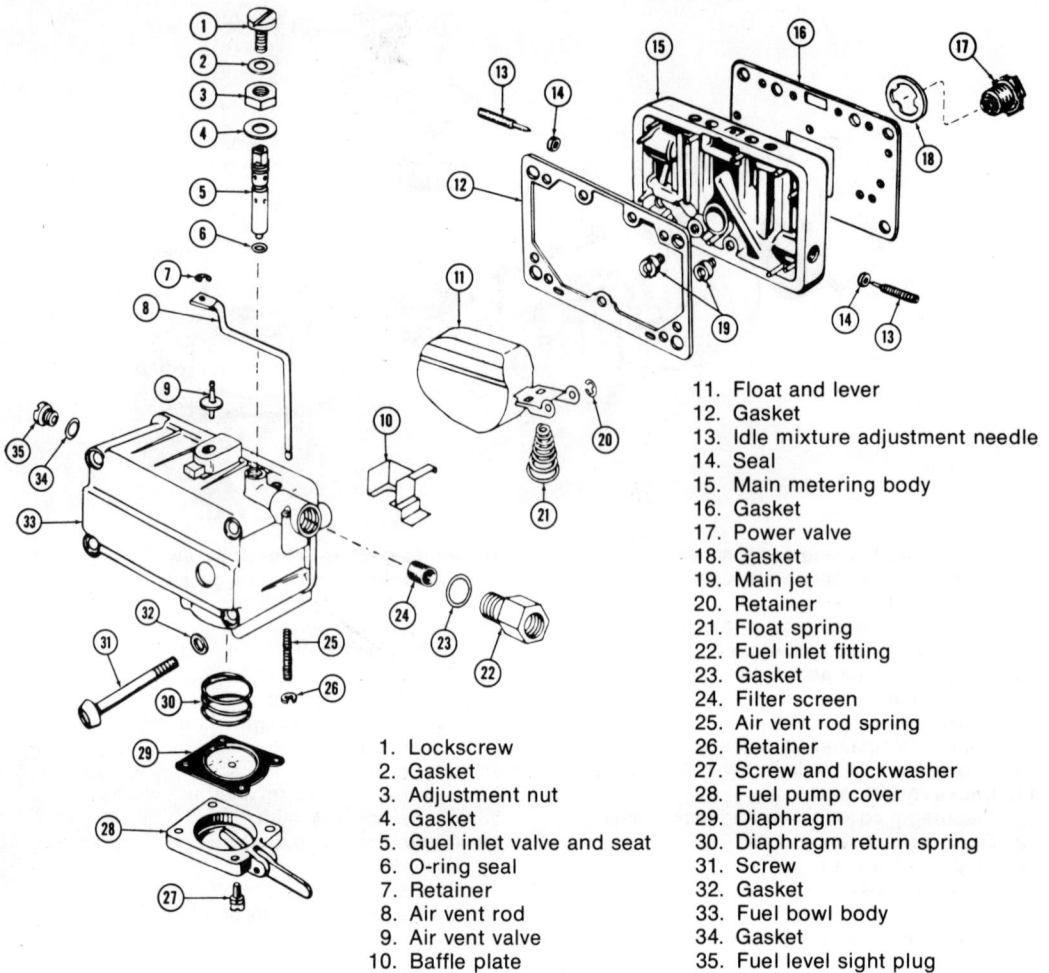

11. Float and lever
12. Gasket
13. Idle mixture adjustment needle
14. Seal
15. Main metering body
16. Gasket
17. Power valve
18. Gasket
19. Main jet
20. Retainer
21. Float spring
22. Fuel inlet fitting
23. Gasket
24. Filter screen
25. Air vent rod spring
26. Retainer
27. Screw and lockwasher
28. Fuel pump cover
29. Diaphragm
30. Diaphragm return spring
31. Screw
32. Gasket
33. Fuel bowl body
34. Gasket
35. Fuel level sight plug

1. Lockscrew
2. Gasket
3. Adjustment nut
4. Gasket
5. Guel inlet valve and seat
6. O-ring seal
7. Retainer
8. Air vent rod
9. Air vent valve
10. Baffle plate

Holley 4160 main fuel bowl and metering body

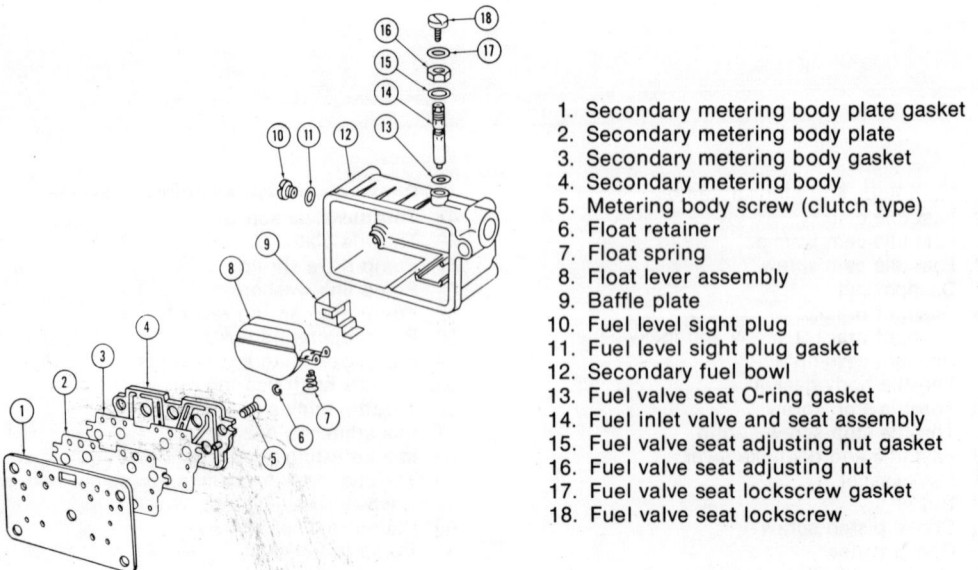

1. Secondary metering body plate gasket
2. Secondary metering body plate
3. Secondary metering body gasket
4. Secondary metering body
5. Metering body screw (clutch type)
6. Float retainer
7. Float spring
8. Float lever assembly
9. Baffle plate
10. Fuel level sight plug
11. Fuel level sight plug gasket
12. Secondary fuel bowl
13. Fuel valve seat O-ring gasket
14. Fuel inlet valve and seat assembly
15. Fuel valve seat adjusting nut gasket
16. Fuel valve seat adjusting nut
17. Fuel valve seat lockscrew gasket
18. Fuel valve seat lockscrew

Holley 4160 secondary fuel bowl and metering body

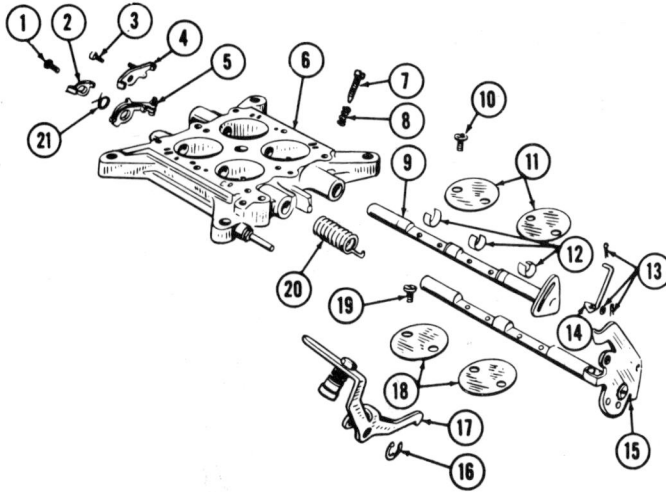

1. Screw and lockwasher
2. Fast idle pick-up lever
3. Screw and lockwasher
4. Secondary diaphragm lever
5. Fast idle cam lever
6. Throttle body
7. Idle speed adjustment screw
8. Spring
9. Secondary throttle shaft assembly
10. Screw
11. Secondary throttle plate
12. Throttle shaft sleeve
13. Retainer pins and washers
14. Throttle connecting rod
15. Primary throttle shaft and lever assembly
16. Retainer
17. Pump lever assembly
18. Primary throttle plate
19. Screw
20. Return spring
21. Lever spring

Holley 4160 throttle body

all gaskets and new parts to replace those that deteriorate most rapidly. Failure to replace all parts supplied with the kit (especially gaskets) can result in poor performance later.

Some carburetor manufacturers supply overhaul kits of three basic types: minor repair; major repair; and gasket kits. Basically, they contain the following:

Minor Repair Kits
    All gaskets
    Float needle valve
    Volume control screw
    All diaphragms
    Spring for the pump diaphragm
Major Repair Kits
    All jets and gaskets
    All diaphragms
    Float needle valve
    Volume control screw
    Pump ball valve
    Main jet carrier
    Float
    Complete intermediate rod
    Intermediate pump lever
    Complete injector tube
    Some cover hold-down screws and washers
Gasket Kits
    All gaskets
After cleaning and checking all compo-

nents, reassemble the carburetor, using new parts and referring to the exploded view. When reassembling, make sure that all screws and jets are tight in their seats, but do not overtighten, as the tips will be distorted. Tighten all screws gradually, in rotation. Do not tighten needle valves into their seats; uneven jetting will result. Always use new gaskets. Be sure to adjust the float level when reassembling.

FLOAT AND FUEL LEVEL ADJUSTMENT
F-Head—Carter Model YF

1. Remove and invert the bowl cover.
2. Remove the bowl cover gasket.

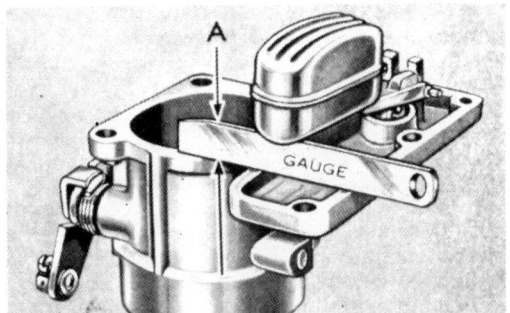

Float level adjustment: Carter YF, 4-134 engine

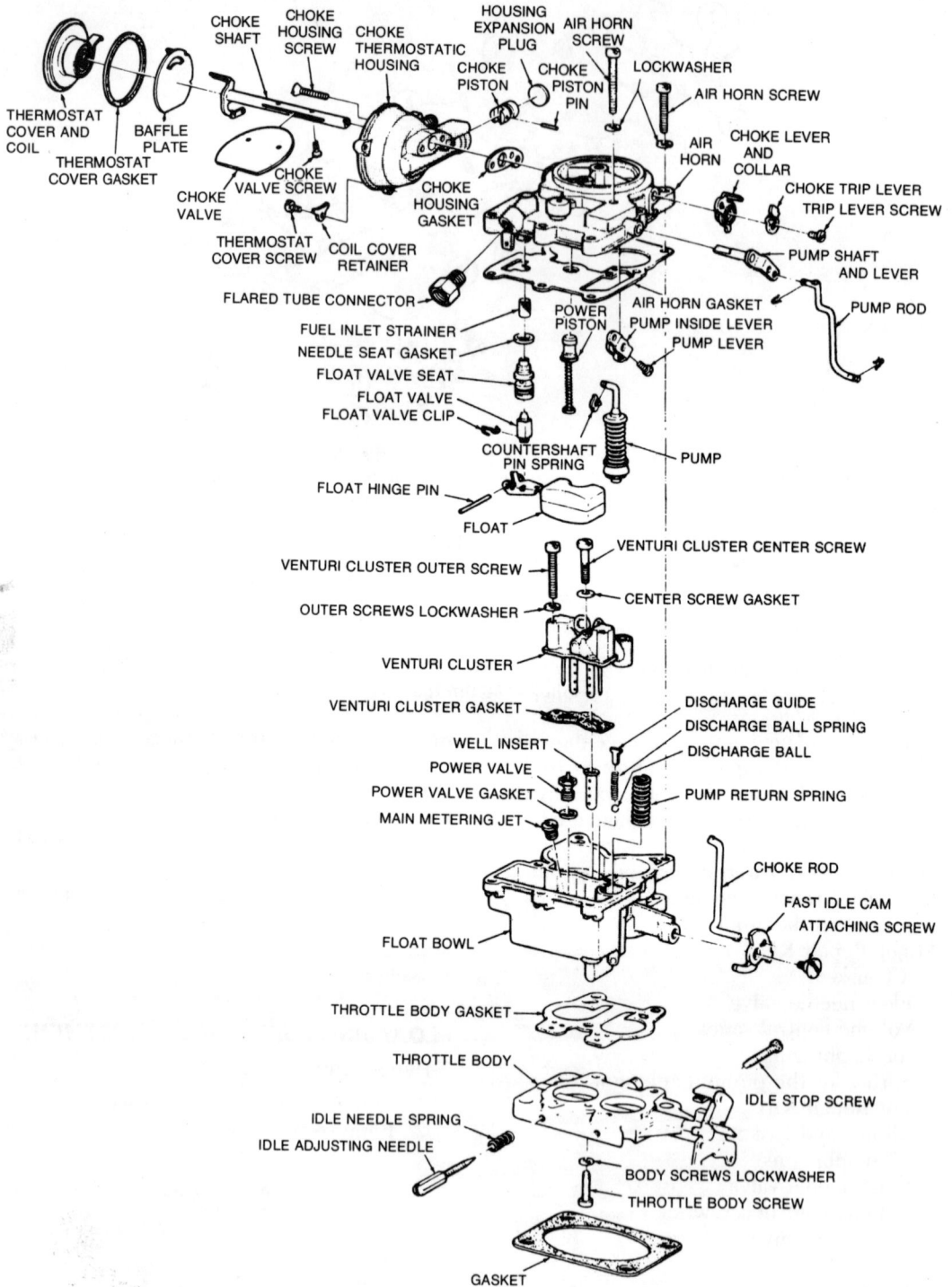

CHOKE SHAFT
CHOKE HOUSING SCREW
CHOKE THERMOSTATIC HOUSING
HOUSING EXPANSION PLUG
AIR HORN SCREW
CHOKE PISTON
CHOKE PISTON PIN
LOCKWASHER
AIR HORN SCREW
THERMOSTAT COVER AND COIL
BAFFLE PLATE
THERMOSTAT COVER GASKET
CHOKE VALVE SCREW
CHOKE VALVE
AIR HORN
CHOKE LEVER AND COLLAR
CHOKE HOUSING GASKET
CHOKE TRIP LEVER
TRIP LEVER SCREW
THERMOSTAT COVER SCREW
COIL COVER RETAINER
PUMP SHAFT AND LEVER
FLARED TUBE CONNECTOR
AIR HORN GASKET
PUMP ROD
FUEL INLET STRAINER
POWER PISTON
PUMP INSIDE LEVER
NEEDLE SEAT GASKET
PUMP LEVER
FLOAT VALVE SEAT
FLOAT VALVE
FLOAT VALVE CLIP
COUNTERSHAFT PIN SPRING
PUMP
FLOAT HINGE PIN
FLOAT
VENTURI CLUSTER OUTER SCREW
VENTURI CLUSTER CENTER SCREW
OUTER SCREWS LOCKWASHER
CENTER SCREW GASKET
VENTURI CLUSTER
VENTURI CLUSTER GASKET
DISCHARGE GUIDE
WELL INSERT
DISCHARGE BALL SPRING
POWER VALVE
DISCHARGE BALL
POWER VALVE GASKET
PUMP RETURN SPRING
MAIN METERING JET
CHOKE ROD
FAST IDLE CAM
ATTACHING SCREW
FLOAT BOWL
THROTTLE BODY GASKET
THROTTLE BODY
IDLE STOP SCREW
IDLE NEEDLE SPRING
IDLE ADJUSTING NEEDLE
BODY SCREWS LOCKWASHER
THROTTLE BODY SCREW
GASKET

Rochester 2GV exploded view

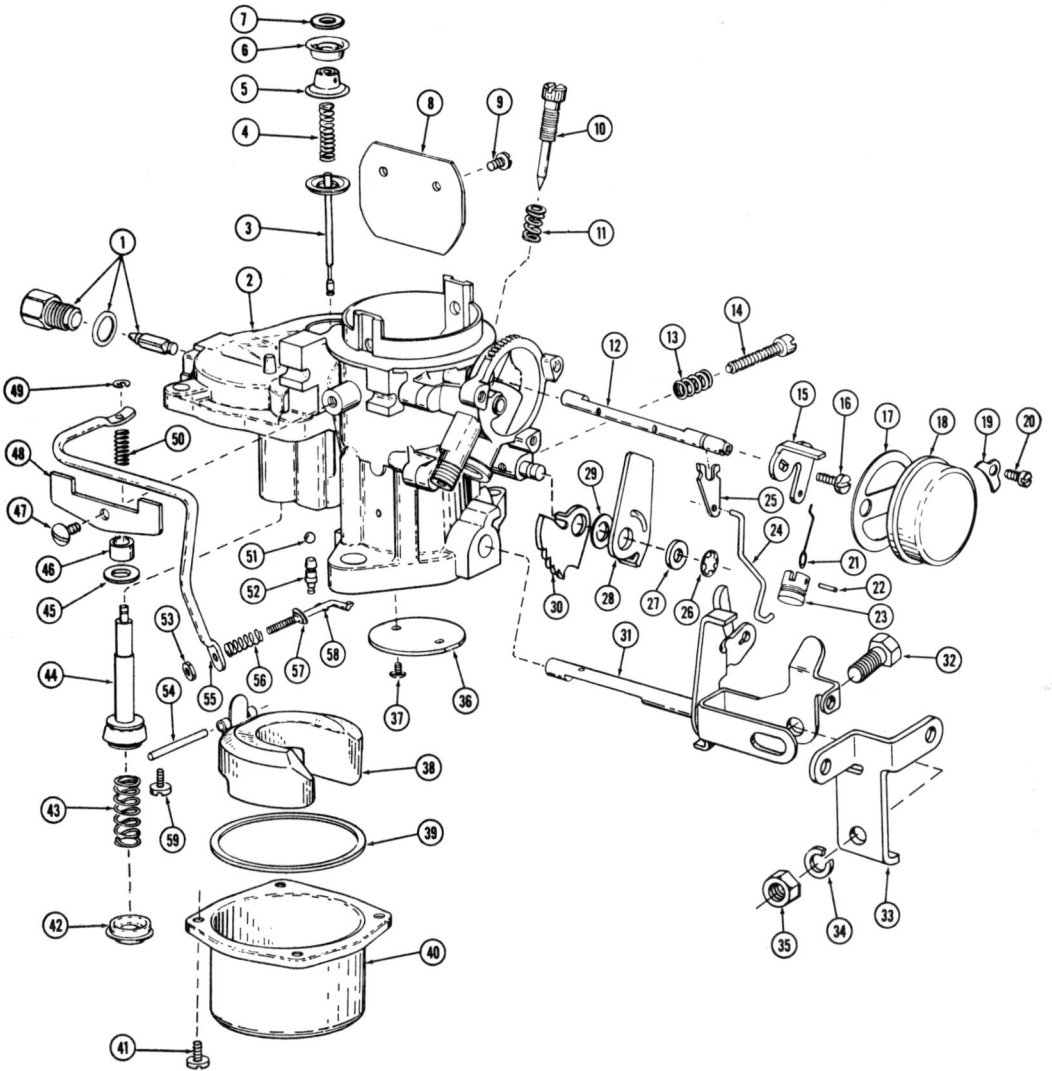

1. Needle and seat assembly
2. Carburetor body
3. Fuel enrichment rod and diaphragm assembly
4. Diaphragm spring
5. Spring retainer
6. Diaphragm cover
7. Cover washer
8. Choke valve plate
9. Screw
10. Idle mixture adjustment screw
11. Spring
12. Choke shaft
13. Spring
14. Curb idle speed adjustment screw
15. Choke piston lever
16. Screw
17. Gasket
18. Thermostatic spring and housing assembly
19. Retainer
20. Screw
21. Choke piston link
22. Pin
23. Choke piston
24. Fast idle connector rod
25. Choke fast idle lever
26. Retaining ring
27. Flat washer
28. Fast idle weight
29. Nonmetallic washer
30. Fast idle cam
31. Throttle shaft and lever assembly
32. Capscrew
33. Throttle linkage adapter
34. Lockwasher
35. Nut
36. Throttle valve plate
37. Screw
38. Float
39. Gasket
40. Fuel bowl
41. Screw
42. Pump retainer
43. Lower spring
44. Accelerator pump piston assembly
45. Flat washer
46. Bushing
47. Screw
48. Arm retainer
49. Retaining ring
50. Upper spring
51. Accelerator pump intake check ball
52. Check ball seat
53. Adjustment nut
54. Pin
55. Accelerator pump arm
56. Spring
57. Spring retainer
58. Accelerator pump connector link
59. Screw

Carter RBS: 1966–70 6-232

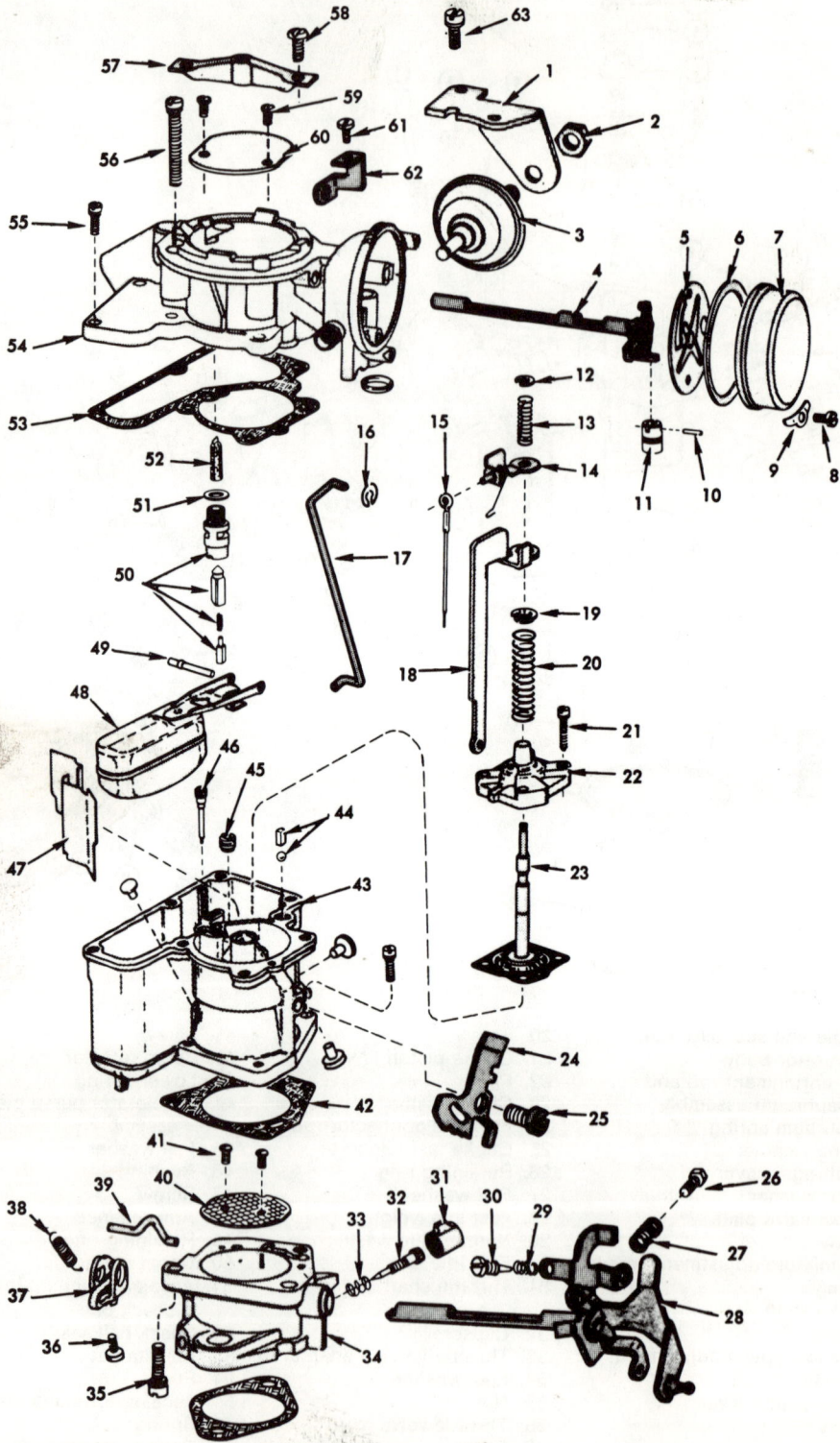

1971 and later Carter YF

3. Allow the weight of the float to rest on the needle and spring. Be sure that there is no compression of the spring other than by the weight of the float.

4. Adjust the level by bending the float arm lip that contacts the needle (not the arm) to provide $17/64$ in. of clearance on models made during and after 1968. On models prior to 1968 the clearance is to be set at $5/16$ in.

V6-225—Rochester Model 2G

The procedure for adjusting the float level of the two barrel carburetor installed on the V6 is the same as the procedure for the F-head up to Step 4.

The actual measurement is taken from the air horn gasket to the lip at the toe of the float. This distance should be $15/32$ in. To adjust the float level, bend the float arm as required.

The float drop adjustment is accomplished in the following manner: With the bowl cover turned in the upright position, measure the distance from the gasket to the notch at the toe of the float. Bend the tang as required to obtain a measurement of $17/32$ in.

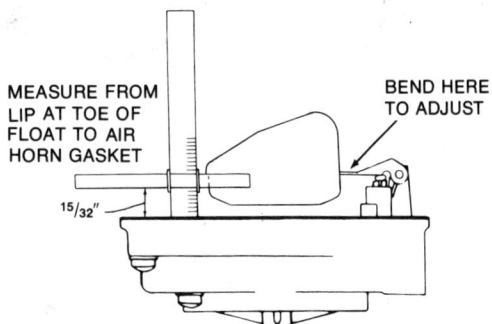

MEASURE FROM LIP AT TOE OF FLOAT TO AIR HORN GASKET

BEND HERE TO ADJUST

$15/32''$

Float level adjustment; Rochester 2GV, 6-225 and 8-350 engines

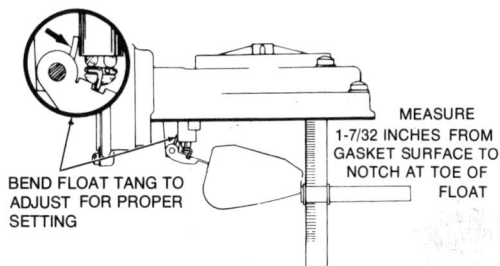

BEND FLOAT TANG TO ADJUST FOR PROPER SETTING

MEASURE 1-7/32 INCHES FROM GASKET SURFACE TO NOTCH AT TOE OF FLOAT

Float drop adjustment: Rochester 2GV, 6-225 and 8-350 engines

1. Dashpot bracket
2. Dashpot lock nut
3. Dashpot
4. Choke shaft and lever assembly
5. Baffle plate
6. Choke cover gasket
7. Choke cover
8. Choke cover retaining screw (3)
9. Choke cover retainer (3)
10. Choke piston pin
11. Choke piston
12. Upper pump spring retainer
13. Upper pump spring
14. Metering rod arm and spring
15. Metering rod
16. Choke rod retaining clip
17. Choke rod
18. Pump lifter link
19. Lower pump spring retainer
20. Lower pump spring
21. Pump housing retaining screw (4)
22. Pump housing
23. Pump diaphragm assembly
24. Fast idle cam
25. Fast idle cam retaining screw
26. Curb idle speed adjusting screw
27. Curb idle screw spring
28. Throttle shaft and lever assembly
29. Fast idle screw spring
30. Fast idle speed adjusting screw
31. Idle limiter cap
32. Idle mixture screw

33. Idle mixture screw spring
34. Throttle body
35. Throttle body retaining screw (3)
36. Throttle shaft arm set screw
37. Throttle shaft arm
38. Throttle shaft return spring
39. Pump connector link
40. Throttle valve
41. Throttle valve retaining screw (2)
42. Throttle body gasket
43. Main body
44. Pump discharge check ball and weight
45. Metering rod jet
46. Low speed jet
47. Fuel bowl baffle
48. Float and lever assembly
49. Float pin
50. Needle and seat assembly
51. Needle seat gasket
52. Screen
53. Air horn gasket
54. Air horn
55. Short air horn retaining screw (3)
56. Long air horn retaining screw (3)
57. Air cleaner bracket
58. Air cleaner bracket retaining screw (2)
59. Choke valve retaining screw (2)
60. Choke valve
61. Choke lever retaining screw
62. Choke lever
63. Dashpot bracket retaining screw

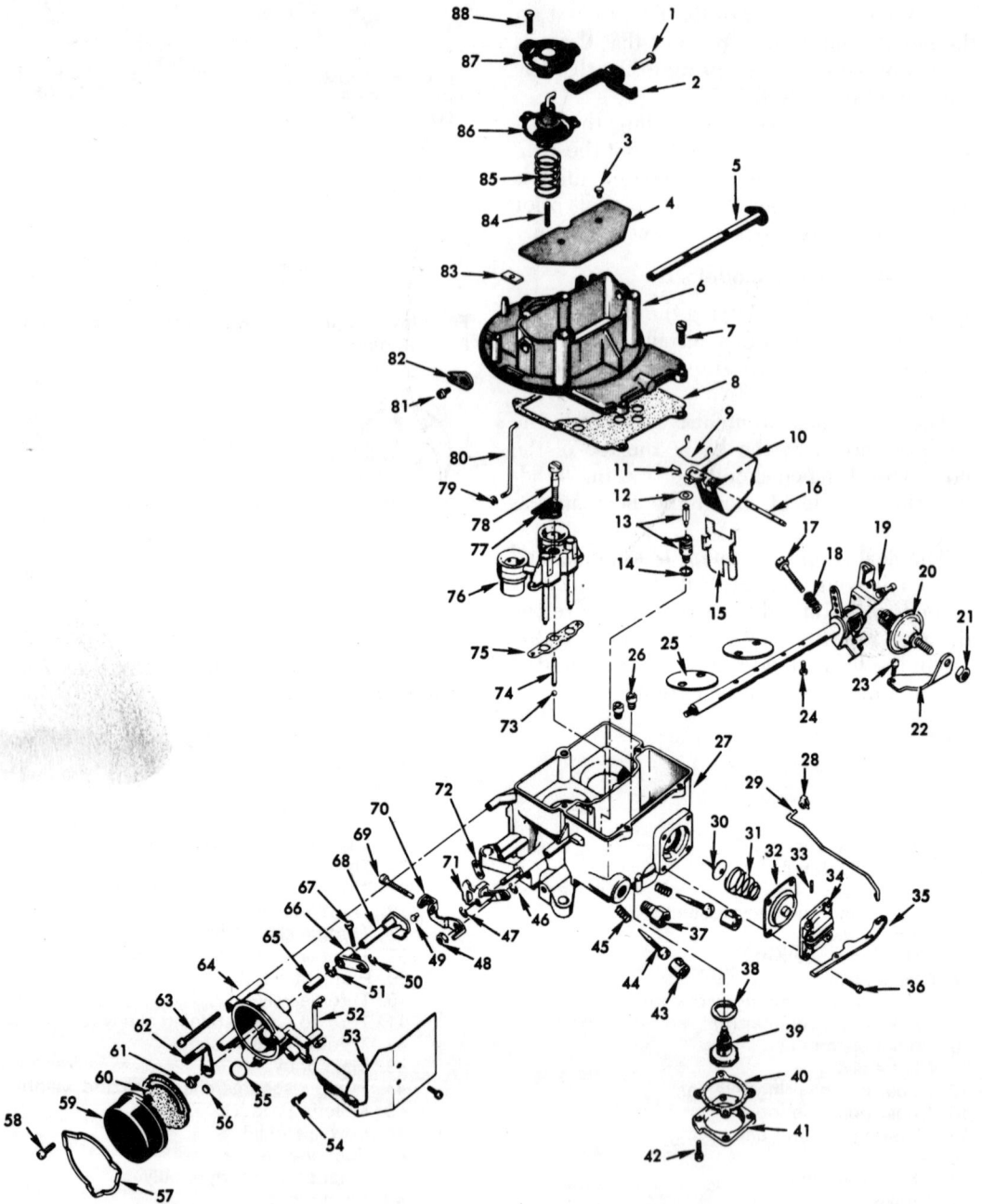

1. Pivot pin
2. Modulator arm
3. Choke valve retaining screw (2)
4. Choke valve
5. Choke shaft
6. Air horn
7. Air horn retaining screw (4)
8. Air horn gasket
9. Float shaft retainer
10. Float and lever assembly
11. Needle retaining clip
12. Deflector
13. Needle and seat assembly
14. Needle seat gasket
15. Fuel bowl baffle
16. Float shaft
17. Curb idle adjusting screw
18. Curb idle adjusting screw spring
19. Throttle shaft and lever assembly
20. Dashpot
21. Dashpot locknut
22. Dashpot bracket
23. Dashpot bracket retaining screw
24. Throttle valve retaining screw (4)
25. Throttle valve (2)
26. Main jet (2)

Autolite 2100 exploded view: 8-304, 360

27. Main body
28. Pump rod retainer
29. Pump rod
30. Elastomer valve
31. Pump return spring
32. Pump diaphragm
33. Pump lever pin
34. Pump cover
35. Pump lever
36. Pump cover retaining screw (4)
37. Fuel inlet fitting
38. Power valve gasket
39. Power valve
40. Power valve cover gasket
41. Power valve cover
42. Power valve cover retaining screw (4)
43. Idle limiter cap (2)
44. Idle mixture screw (2)
45. Idle mixture screw spring (2)
46. Retainer
47. Retainer
48. Fast idle lever retaining nut
49. Fast idle lever pin
50. Retainer
51. Retainer
52. Fast idle cam rod
53. Choke shield
54. Choke shield retaining screw (2)
55. Piston passage plug
56. Heat passage plug
57. Choke cover retaining clamp
58. Choke retaining screw (3)
59. Choke cover
60. Choke cover gasket
61. Thermostat lever retaining screw
62. Thermostat lever
63. Choke housing retaining screw (3)
64. Choke housing
65. Choke shaft bushing
66. Fast idle cam lever
67. Fast idle cam lever adjusting screw
68. Thermostatic choke shaft
69. Fast idle speed adjusting screw
70. Fast idle lever
71. Fast idle cam
72. Choke housing gasket
73. Pump discharge check ball
74. Pump discharge weight
75. Booster venturi gasket
76. Booster venturi assembly
77. Air distribution plate
78. Pump discharge screw
79. Retainer
80. Choke rod
81. Choke lever retaining screw
82. Choke lever
83. Choke rod seal
84. Stop screw
85. Modulator return spring
86. Modulator diaphragm assembly
87. Modulator cover
88. Modulator retaining screw (3)

232, 258 Sixes—Carter Model YF

NOTE: *This procedure applies to 1971 and later engines. See below for 1966–69 6-232.*

Remove and invert the air horn assembly and remove the gasket. Measure the distance between the top of the float at the free end, and the air horn casting. The measurement should be $^{29}/_{64}$ in. Adjust by bending the float lever.

NOTE: *The fuel inlet needle must be held off its seat while bending the float lever in order to prevent damage to the needle and seat.*

To adjust the float drop, hold the air horn in the upright position and measure the distance between the top of the float, at the extreme outer end, and the air horn casting. The measurement should be 1¼ in. Adjust by bending the tab at the rear of the float lever.

6-258 2BBL-Carter BBD

1. Remove the air horn from the carburetor.

2. Apply light finger pressure to the vertical float tab to exert GENTLE pressure against the inlet needle.

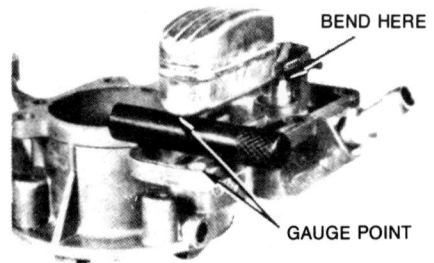

Float level adjustment: Carter YF, 1971 and later 6-232, 258

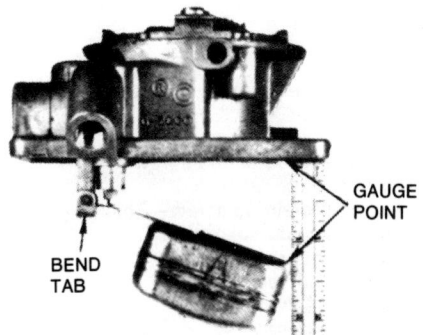

Float drop adjustment: Carter YF, 1971 and later 6-232, 258

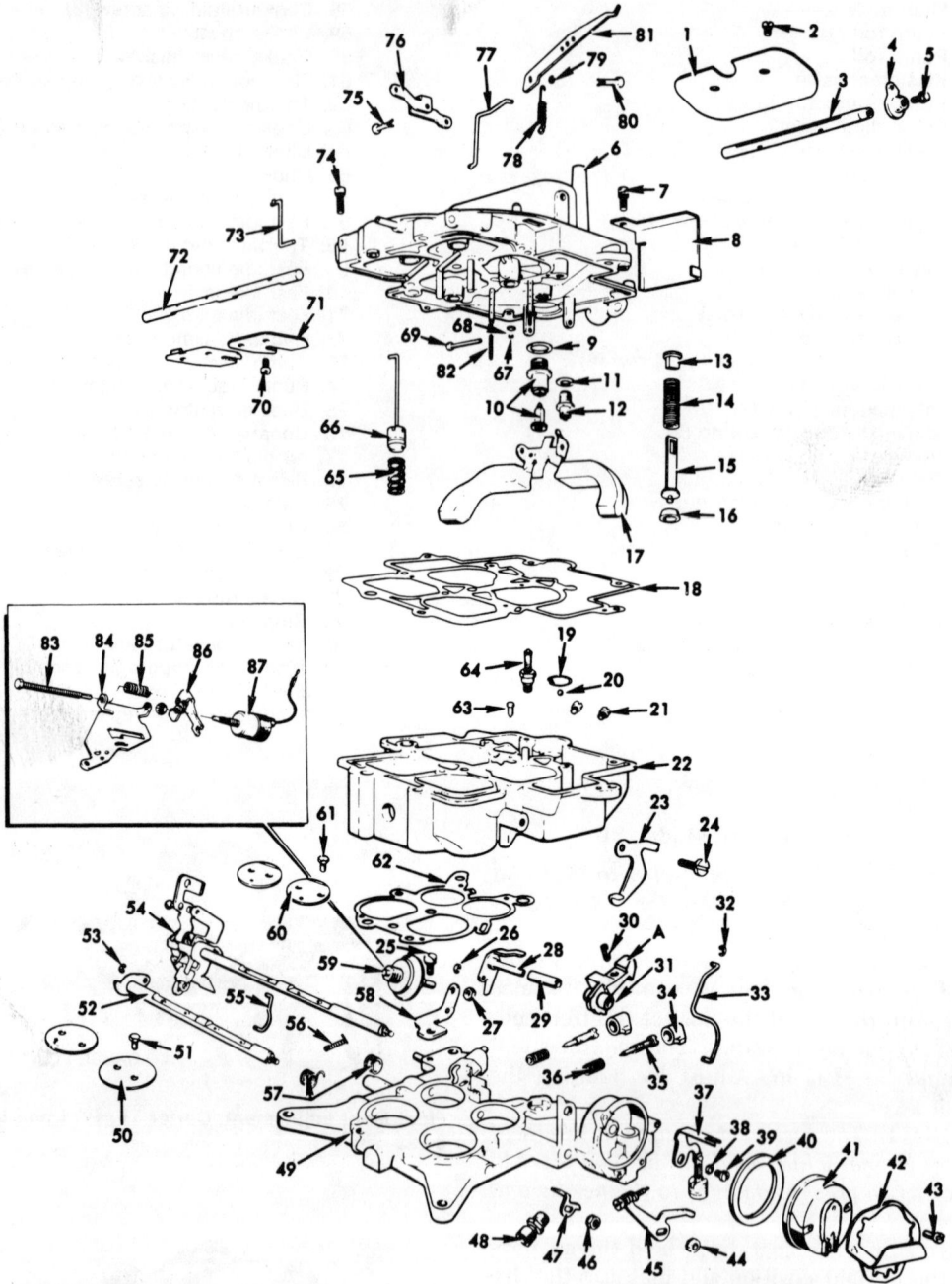

1. Choke valve
2. Choke valve retaining screw (2)
3. Choke shaft
4. Choke lever
5. Choke lever retaining screw
6. Air horn
7. Air horn retaining screw—Phillips (10)
8. Choke shield
9. Needle seat gasket
10. Needle and seat assembly
11. Auxiliary valve gasket
12. Auxiliary valve
13. Pump spring retainer

14. Pump spring
15. Pump piston
16. Pump rubber cap
17. Float and lever assembly
18. Air horn gasket
19. Inlet check valve retainer
20. Inlet check valve
21. Main jet (2)
22. Main body
23. Secondary throttle lockout lever
24. Lockout lever retaining screw
25. Dashpot bracket retaining screw
26. Choke rod lower retainer

Autolite 4300 exploded view: 8-360, 401

27. Dashpot lockout
28. Thermostatic choke shaft
29. Thermostatic choke shaft bushing
30. Fast idle cam adjusting screw
31. Fast idle cam
32. Choke rod upper retainer
33. Choke rod
34. Idle limiter cap (2)
35. Idle mixture screw (2)
36. Idle mixture screw spring (2)
37. Choke piston and lever assembly
38. Choke piston lever retaining washer
39. Choke piston lever retaining screw
40. Choke cover gasket
41. Choke cover (electric)
42. Choke cover retaining clamp
43. Choke cover retaining screw (3)
44. Fast idle lever retaining nut
45. Fast idle lever
46. Secondary stop lever retaining nut
47. Secondary throttle stop lever
48. Choke heat inlet fitting
49. Throttle body
50. Secondary throttle valve (2)
51. Throttle valve retaining screw (4)
52. Secondary throttle shaft
53. Secondary link retainer
54. Primary throttle shaft assembly
55. Secondary throttle link
56. Curb idle adjusting screw
57. Throttle return spring

58. Dashpot bracket
59. Dashpot
60. Primary throttle valve (2)
61. Throttle valve retaining screw
62. Throttle body gasket
63. Pump discharge needle
64. Power valve
65. Dampener spring
66. Dampener piston
67. Pump air bleed retainer
68. Pump air bleed (Viton disc)
69. Float pin
70. Air valve retaining screw (4)
71. Air valve (2)
72. Air valve shaft
73. Air valve link
74. Air horn retaining screw
75. Pivot pin
76. Air valve lever
77. Pump rod
78. Pump return spring
79. Pump rod washer
80. Pivot pin
81. Pump lever
82. Atomizing screen
83. Adjusting screw
84. Mounting bracket
85. Adjusting screw spring
86. Carriage
87. Electric solenoid

3. Lay a straight edge across the float bowl and measure the gap between the straight edge and the top of the float at its highest point. The gap should be ¼ inch.

4. If adjustment is necessary, remove the float and bend the lower tab. Replace the float and check the gap.

304, 360 V8—Autolite Model 2100

With the air horn assembly and the gasket removed, raise the float by pressing down on the float tab until the fuel inlet needle is lightly seated. Using a T-scale, measure the distance from the fuel bowl machined surface to either corner of the float at the free end. The measurement should be ⅜ inch for 1966–75 models and $^{15}/_{16}$ inch for 1976–79 models. To adjust bend the float tab and hold the fuel inlet needle off its seat in order to prevent damage to the seat and the tip of the needle.

1966–69 6-232—Carter RBS Series

With the carburetor inverted, remove the 4 screws which secure the fuel bowl to the fuel bowl flange on the carburetor body. Remove the fuel bowl gasket. With only the weight of

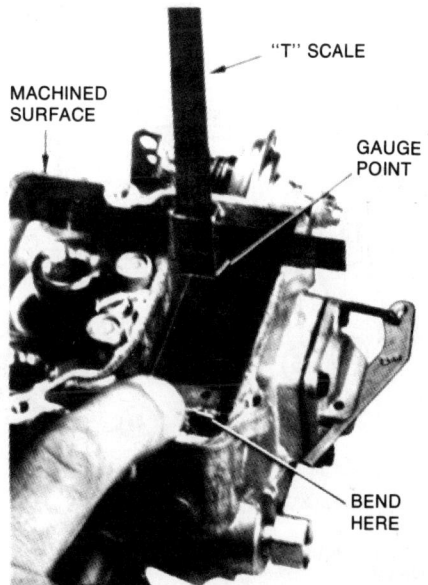

Float level adjustment: Autolite 2100, 8-304, 360

the float pressing the intake valve needle into the seat, measure the clearance between the fuel bowl flange of the carburetor body and each end of the float. This clearance should

THIS
CARBURETOR.

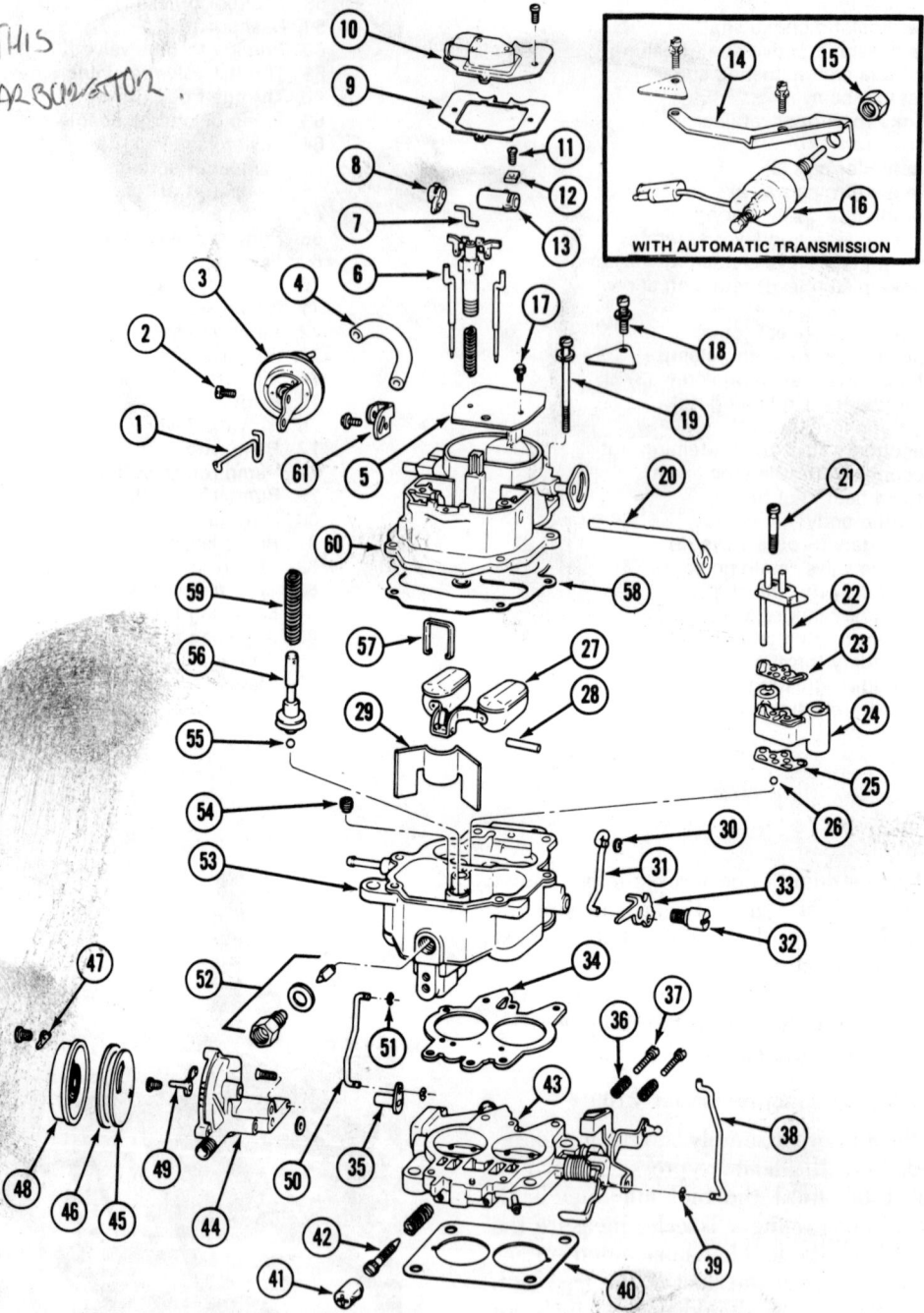

WITH AUTOMATIC TRANSMISSION

1. Diaphragm connector link
2. Screw
3. Choke vacuum diaphragm
4. Hose
5. Valve
6. Metering rod
7. S-Link
8. Pump arm
9. Gasket
10. Rollover check valve
11. Screw

12. Lock
13. Rod lifter
14. Bracket
15. Nut
16. Solenoid
17. Screw
18. Air horn retaining screw (short)
19. Air horn retaining screw (long)
20. Pump lever
21. Venturi cluster screw
22. Idle fuel pick-up tube

Carter BBD exploded view: 6-258

23. Gasket
24. Venturi cluster
25. Gasket
26. Check ball (small)
27. Float
28. Fulcrum pin
29. Baffle
30. Clip
31. Choke link
32. Screw
33. Fast idle cam
34. Gasket
35. Thermostatic choke shaft
36. Spring
37. Screw
38. Pump link
39. Clip
40. Gasket
41. Limiter cap
42. Screw

43. Throttle body
44. Choke housing
45. Baffle
46. Gasket
47. Retainer
48. Choke coil
49. Lever
50. Choke rod
51. Clip
52. Needle and seat assembly
53. Main body
54. Main metering jet
55. Check ball (large)
56. Accelerator pump plunger
57. Fulcrum pin retainer
58. Gasket
59. Spring
60. Air horn
61. Lever

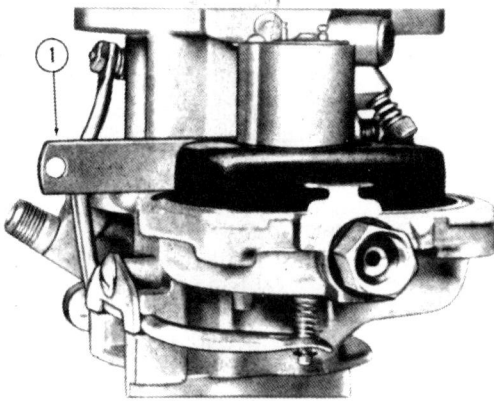

1. 21/32 in. gauge

Float level adjustment: Carter RBS, 1966–69 6-232

be $^{21}/_{32}$ in. If necessary, bend the lip of the float which touches the intake valve needle to obtain the correct clearance. Take care not to press the needle into the seat. Install the fuel bowl gasket on the fuel bowl flange, then install the fuel bowl.

360 and 401 V8s—Autolite Model 4300 (4 bbl)

Invert the air horn assembly and remove the gasket. Use a "T" scale to measure the distance from the float pontoons to the air horn casting. Position the horizontal scale over the flat surface of both float pontoons at the free ends and parallel to the air horn casting. Hold the lower end of the vertical scale in full contact with the smooth area of the air horn casting, located midway between the main discharge nozzles.

NOTE: *Do not allow the end of the vertical*

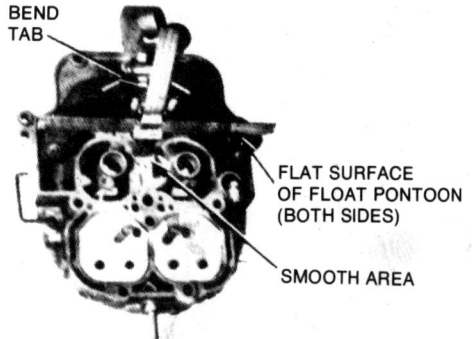

Float level adjustment: Autolite 4300, 8-360, 401

scale to come in contact with any gasket sealing ridge while measuring the float setting.

The free end of each float pontoon should just touch the horizontal scale. If one pontoon is lower than the other, twist the float and lever assembly slightly to align them. The measurement should be $^{13}/_{16}$ in.

Adjust the float level by bending the tab which contacts the fuel inlet needle.

8-360, 401–Motorcraft 4350

1. Invert the air horn assembly and remove the gasket.

2. Measure the distance from the floats to the air horn rim using a T-scale. Position the horizontal scale over the flat surface of both floats at the free ends, parallel to the air horn casting. Hold the lower end of the vertical scale in full contact with the smooth area of the casting, midway between the main discharge nozzles.

DAMPER
LINK

PLUGS

SCREW

AIR VALVE
PLATE

CHOKE PLATE

SCREW

LEVER

AIR VALVE
SHAFT

CHOKE PLATE
SHAFT

SCREW
AIR HORN

SPACER

VACUUM PISTON
LIMITER LEVER

SHAFT

METERING
ROD

ACCELERATING PUMP ARM

ACCELERATING PUMP
SPRING RETAINER

GASKET

ACCELERATING PUMP
LEVER AND ROD

AUXILIARY
VALVE
ASSEMBLY

PUMP SPRING

ACCELERATING PUMP
THROTTLE
LINK

PUMP PISTON

PUMP PISTON CUP

AIR VALVE DAMPER
PISTON AND ROD

RETAINER

METERING RODS
AND YOKE

FUEL INLET VALVE
AND SEAT

FLOAT AND
LEVER
ASSEMBLY

MAIN JETS

GASKET

BALL CHECK
RETAINER

VACUUM PISTON

DISHARGE CHECK BALL

SPRING

MAIN BODY

INLET BALL
CHECK

VALVE

PRIMARY
THROTTLE PLATE

VACUUM PISTON
CYLINDER

PRIMARY THROTTLE SHAFT
AND LEVER ASSEMBLY

SECONDARY
THROTTLE
LOCKOUT LEVER

GASKET

LINK

SCREW

GASKET

SCREW

AUTOMATIC CHOKE
SHAFT AND LEVER

RETAINER

RETAINER

CAM ADJUSTING SCREW

SECONDARY THROTTLE
SHAFT AND LEVER (L.H.)

BUSHING

FAST IDLE CAM

SECONDARY
THROTTLE
PLATES

CAP

CHOKE CONTROL ROD

IDLE
SPEED
SCREW

SCREW

RETURN SPRING

SPRING

THROTTLE BODY

LEVER

CHOKE HEAT
CONNECTION

BAFFLE

CHOKE
COVER

CHOKE
DIAPHRAGM

RETAINER

SCREW

SCREW

FAST IDLE
LEVER

RETURN
SPRING

SECONDARY THROTTLE
SHAFT AND LEVER (R.H.)

FAST IDLE SPEED
ADJUSTING SCREW

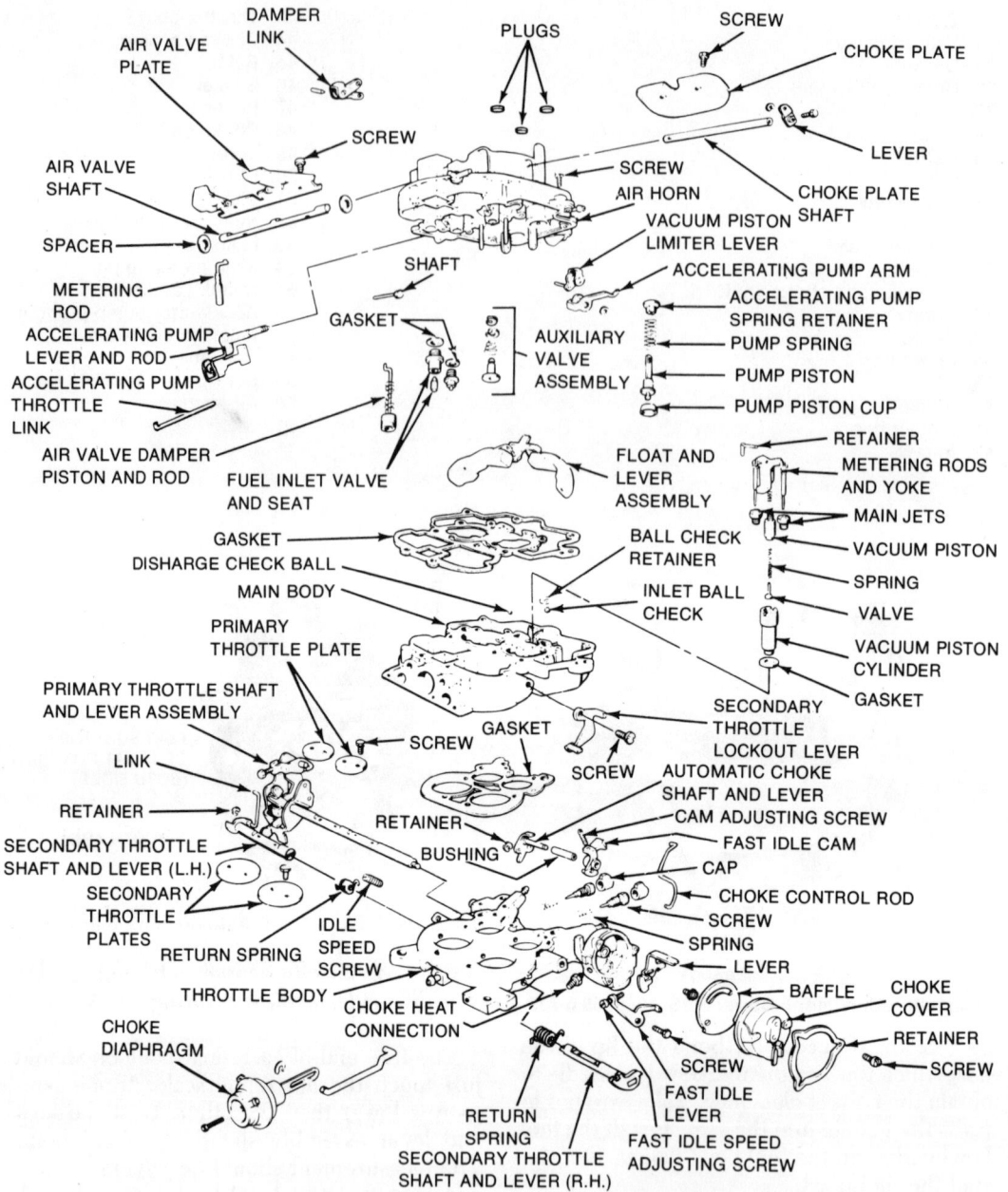

Motorcraft 4350 exploded view: 8-360, 401

CAUTION: *Do not allow the end of the vertical scale to contact any gasket sealing ridge while measuring the float setting.*

3. The free end of the floats should just touch the horizontal scale. Float-to-air horn casting distance should be $^{29}/_{64}$ inch. Bend the vertical tab on the float arm to adjust the distance.

327 V8—Holley Model 2209 (2 BBL)

With the air horn of the carburetor removed from the main body, invert the air horn with

Float level adjustment: Holley 2209, 8-327

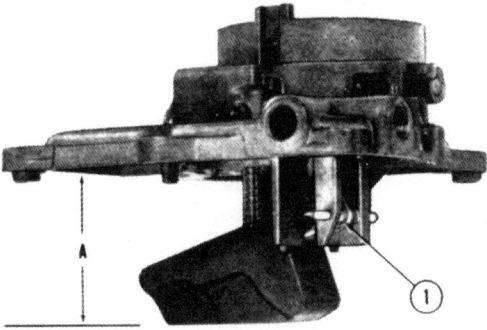

A. Bottom of float is parallel with bottom of gasket surface
1. Stop tab

Float drop adjustment: Holley 2209, 8-327

the float assembly attached and measure the distance between the float and the air horn. The distance should be $9/16$ in. Adjust the distance by bending the tab at the float hinge.

To adjust the float drop, turn the air horn right side up and let the float hang down freely. The bottom surface of the float should be parallel with the gasket mating surface of the air horn. Adjust the position of the float by bending the float drop tang.

327 V8—Holley Model 4160 (4 bbl)

To perform a preliminary dry float adjustment on both the primary and secondary fuel bowl float assemblies, invert the fuel bowl and let the float rest on the fuel inlet valve and seat assembly. The fuel inlet valve and seat assembly can be rotated until the float is parallel with the fuel bowl floor (actually the top of the fuel bowl chamber inverted). Note that this is an initial dry float setting which must be rechecked with the carburetor assembled and on the engine to obtain the proper wet fuel level.

This carburetor has an externally adjustable needle and seat assembly that allows the fuel level to be checked and adjusted without removing the carburetor from the engine.

With the engine running, remove the sight plug from the carburetor bowl on the side opposite from the fuel inlet. If the fuel level is too high, excess fuel will run out through the sight hole. The correct fuel level is just up to the threads at the bottom of the sight hole.

To adjust the level, loosen the top lockscrew on the needle and seat assembly and adjust with the lower nut until the float maintains the fuel at the correct level. Tighten the top lockscrew without disturbing the adjusting nut.

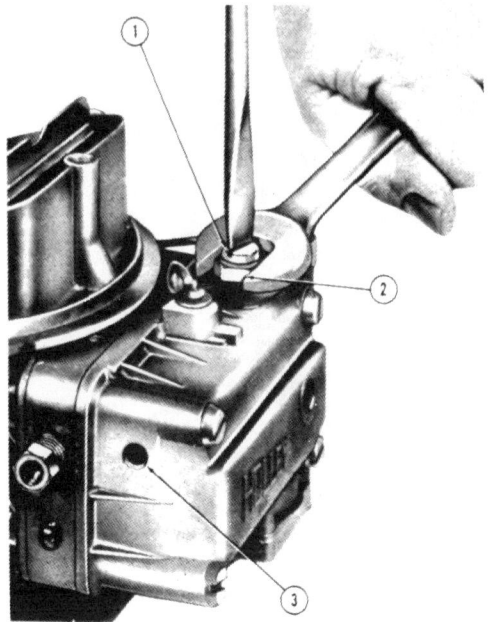

1. Adjustment screw
2. Locknut
3. Fuel level observation port

Float level adjustment: Holley 4160, 8-327

The above procedure applies to both the primary and secondary fuel bowls.

350 V8—Rochester Model 2GV (2 bbl)

The float level and float drop on the Rochester Model 2GV is adjusted in the same manner as for the Model 2G installed on the V6 engine. Refer to the procedure given for that engine. The float level measurement is the same for both carburetors. The float drop measurement for the Model 2GV on the 350 V8 is $1\frac{3}{4}$ in.

FAST IDLE ADJUSTMENT (AIR CLEANER OFF)

F-Head

With the choke held in the wide open position, the lip on the fast-idle rod should contact the boss on the body casting. Adjust it by bending the fast idle link at the offset in the link.

V6 and 350 V8

No fast idle speed adjustment is required. Fast idle is controlled by the curb idle speed adjustment screw. If the curb idle speed is set correctly and the choke rod is properly adjusted, fast idle speed will be correct.

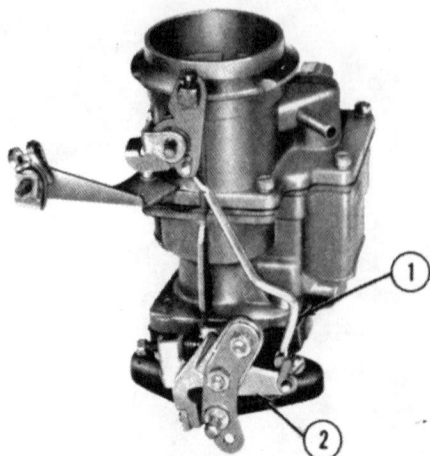

1. Fast idle connector rod
2. Fast idle link

Fast idle adjustment: Carter YF, 4-134

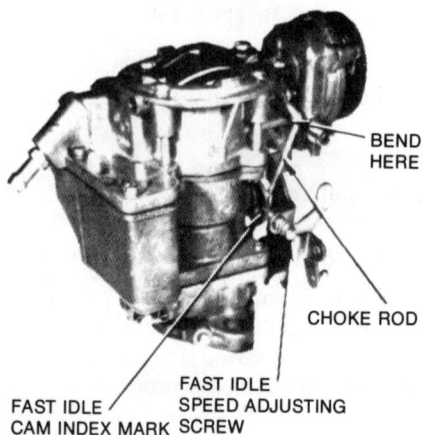

BEND HERE

CHOKE ROD

FAST IDLE
SPEED ADJUSTING
SCREW

FAST IDLE
CAM INDEX MARK

Fast idle cam linkage adjustment: Carter YF, 1971 and later 6-232, 258

FAST IDLE CAM
LEVER SCREW

SECOND STEP
OF CAM

Fast idle adjustment: Autolite 2100, 8-304, 360

1966–69 6-232

Fast idle adjustment should only be made when the carburetor is removed from the engine. With the carburetor body inverted, rotate the fast idle cam and fast idle weight to the extreme counterclockwise position. The cam should press against the fast idle prong of the throttle shaft and lever assembly to hold the throttle valve slightly open. Bend this prong, if necessary, to obtain 0.023 in. of clearance between the throttle valve plate and the idle port side of the carburetor bore, with the cam in this position. With the choke valve closed tightly and the fast idle connector rod at the end of the slot in the fast idle weight, bend the offset portion of the connector rod to align the index mark of the cam with the fast idle prong of the throttle shaft and lever assembly.

1971 and Later 6-232, 258 1-bbl

Partially open the throttle and close the choke valve to rotate the fast idle cam into the cold start position. While holding the choke valve closed, release the throttle. With the fast idle cam in this position, the fast idle adjusting screw must be aligned with the index mark at the back side of the cam. Adjust by bending the choke rod at its upper angle.

6-258 2-bbl—Carter BBD

1. Loosen the choke housing cover and turn it ¼ turn right. Tighten one screw.
2. Slightly open the throttle and place the fast idle screw on the second cam step.
3. Measure the distance between the choke plate and the air horn wall. The gap should be $3/32$ inch.
4. If adjustment is necessary, bend the fast idle cam link down to increase measurement or up to decrease it.
5. Loosen the choke cover screw and return the cap to its original setting. Tighten the cover screws.

304, 360 and 401 V8s (2 bbl and 4 bbl)

Set the fast idle speed with the engine at operating temperature and the fast idle speed adjusting screw against the index mark (second step) of the fast idle cam. Adjust by turning the fast idle speed adjusting screw until 1600 rpm is attained.

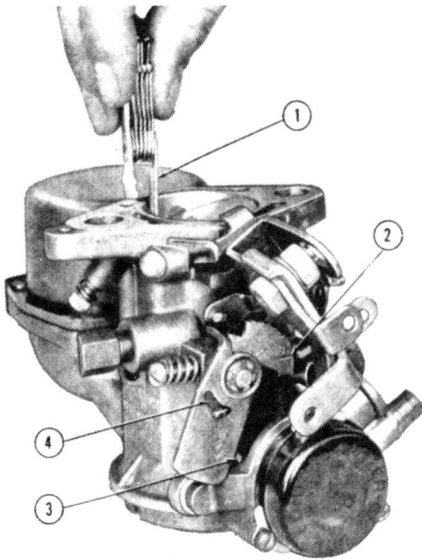

1. Feeler gauge
2. Index mark
3. Offset portion of the connector link
4. Slot

Fast idle linkage adjustment: Carter RBS, 1966–69 6-232

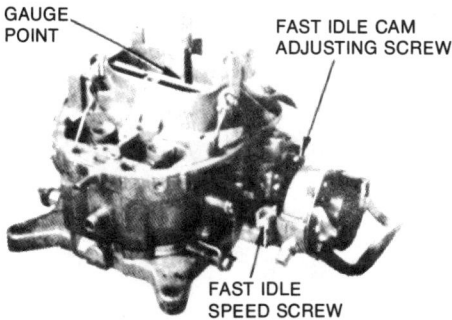

GAUGE POINT

FAST IDLE CAM ADJUSTING SCREW

FAST IDLE SPEED SCREW

Fast idle adjustment: Autolite 4300, 8-360, 401

Fast idle linkage adjustment: Holley 2209, 8-327

8-360, 401 4bbl—Motorcraft 4350

1. Run the engine to normal operating temperature. Connect an accurate tachometer to the engine.

2. Disconnect and plug the EGR and TCS vacuum lines.

3. Position the fast idle screw against the first step of the fast idle cam. Adjust the fast idle screw to give a reading of 1600 rpm.

4. Return the linkage to its normal position, unplug and reconnect the vacuum lines.

327 V8 (2 bbl)

With the throttle stopped on the high step of the fast idle cam, the fast idle speed should be 1800 rpm with the engine at normal operating temperature. Adjust the fast idle speed by bending the tab on the throttle lever.

327 V8 (4 bbl)

The preliminary fast idle setting with the carburetor removed from the engine is made by turning the fast idle adjusting screw to obtain 0.025 in. clearance between the throttle valve and carburetor bore opposite the idle port, with the fast idle adjusting screw on the high step of the cam.

The fast idle setting can be altered with the carburetor on the car to suit individual requirements. However, it has been found that the aforementioned bench setting is most desirable for the best overall performance. The fast idle setting with the adjusting screw on the high step of the cam and the engine at normal operating temperature is about 1700 rpm.

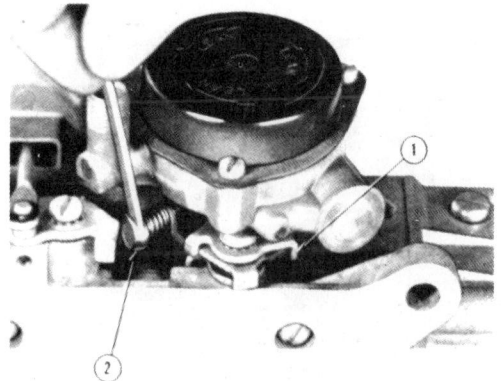

1. Unloader adjustment
2. Fast idle adjustment

Fast idle linkage adjustment: Holley 4160, 8-327

CHOKE LINKAGE ADJUSTMENT
F-Head

The choke is manually operated by a cable that runs from the dash mounted control pull knob to the set screw on the choke actuating arm. To adjust the choke, loosen the set screw at the choke actuating lever and push in the dash knob as far as it will go. Open the choke plate as far as it will go and hold it with your finger while the set screw is tightened.

All Engines with Automatic Chokes

The automatic choke setting is made by loosening the choke cover retaining screws and rotating the cover to align the notch on the cap with the proper index mark on the casting. See the carburetor specifications chart at the end of this section for the proper setting.

UNLOADER ADJUSTMENT
All Engines with Automatic Choke Mechanisms

With the throttle held fully open, apply light pressure on the choke valve toward the closed position and measure the clearance between the edge of the choke valve and the air horn wall. On all 1966–69 engines, the measurement is taken at the upper edge of the choke plate. On 1971 and later engines, the measurement is taken at the lower edge of the choke plate. Adjust the distance by bending the tang on the throttle lever which contacts the fast idle cam. Bend toward the cam to decrease the clearance. See the Carburetor Specifications chart at the end of this section for the proper gap.

NOTE: *Do not bend the unloader down-*

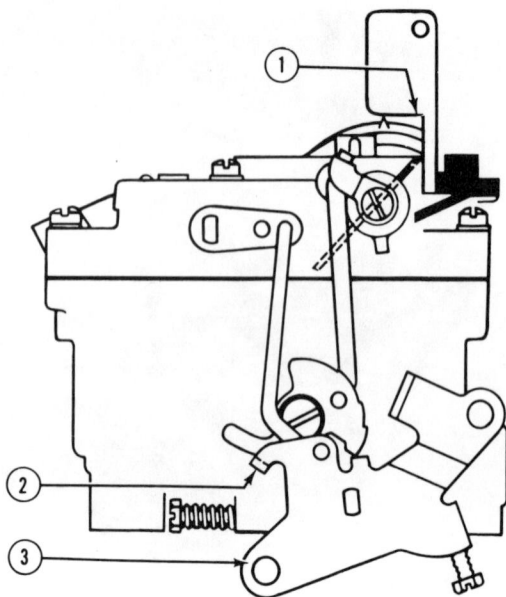

1. Gauge
2. Unloader tang
3. Throttle lever

Choke unloader adjustment: Rochester 2GV, 6-225 and 8-350

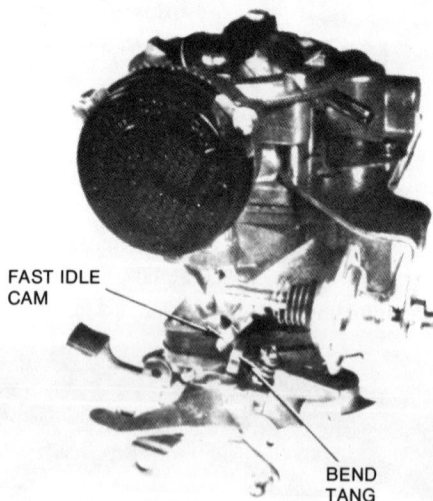

FAST IDLE CAM

BEND TANG

Choke unloader adjustment: Carter YF, 1971 and later 6-232, 258

ward from a horizontal position (1971 and later engines). After making the adjustment, make sure that the unloader tang does not contact the main body flange when the throttle is fully open. Final unloader adjustment must always be done on the vehicle. The throttle should be fully opened by depressing the accelerator pedal to the floor. This is to assure that full-throttle is obtained.

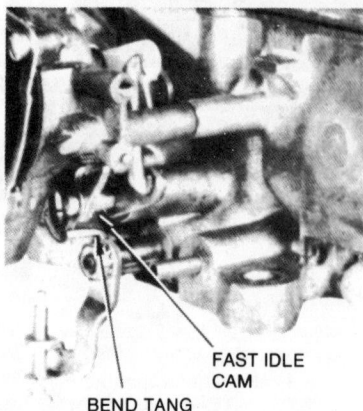

FAST IDLE CAM

BEND TANG

Choke unloader adjustment: Autolite 2100, 4300 and Motorcraft 4350

DASHPOT ADJUSTMENT

All Engines So Equipped, Except the F-Head and V6

With the throttle set at curb idle position, fully depress the dashpot stem and measure the clearance between the stem and the throttle lever. Adjust the clearance by loosening the locknut and turning the dashpot.

F-Head and V6 Engines

The dashpot adjustment is made with the throttle set at curb idle. Loosen the dashpot locknut and turn the dashpot assembly until the dashpot plunger contacts the throttle lever without the plunger being depressed. Then turn the dashpot assembly 2½ turns against the throttle lever, depressing the dashpot plunger. Tighten the locknut securely.

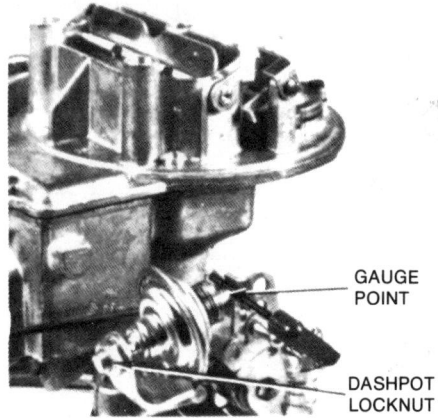

GAUGE POINT

DASHPOT LOCKNUT

Dashpot adjustment: Autolite 2100

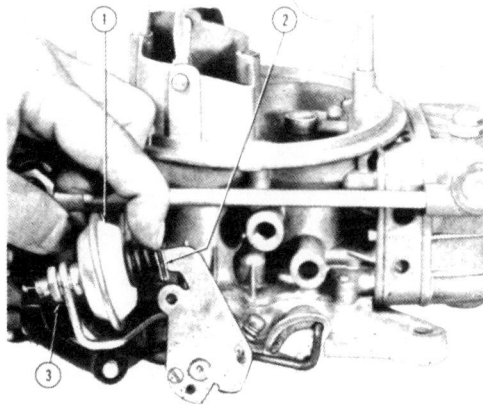

GAUGE POINT

DASHPOT LOCKNUT

Dashpot adjustment: Carter YF

1. Dashpot
2. 3/32 in. gap
3. Locknut

Dashpot adjustment: Holley 4160

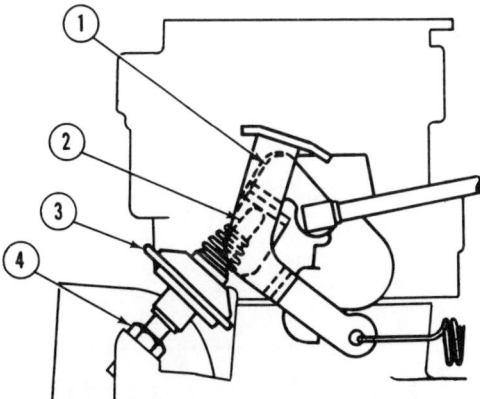

1. Throttle lever
2. Plunger
3. Dashpot
4. Locknut

Dashpot adjustment: Rochester 2GV

1. Throttle lever
2. Plunger
3. Dashpot
4. Locknut

Dashpot adjustment: Carter YF, 4-134

Carburetor Specifications Chart

Model	Year	Float Level	Float Drop	Dashpot	Fast Idle Linkage	Fast Idle Speed, rpm	Choke Unloader	Initial Choke Valve Clear.	Choke Cover Setting Index
Carter YF	1966–69	$17/64$ $29/64$	— $1^1/4$	— $3/32$	See text mark	— 1600	— $19/64$	— $15/64$	— Notch
Carter RBS	1971–74	$21/32$	—	—	$1/50$	1800	$7/64$	$1/4$	1 Lean
Carter BBD	1975–78	$1/4$	—	—	$3/32$	1700	$9/32$	$1/8$	2R
Rochester 2G	1966–69	$15/32$	$1^7/8$	—	—	1800	$1/32$ ①	$1/4$	—
Rochester 2GV	1966–69	$15/32$	$1^3/4$	—	—	1800	$9/64$	$1/4$	—
Autolite 2100	1971–74 1975–76	$3/8$ $1/2$	— —	$9/64$ ② —	$3/32$ ③ $3/32$	1600 1600	$1/4$ ④ $1/4$	$1/8$ $5/32$	2R 2R
Motorcraft 2100	1977–78	$15/16$	—	—	$1/8$	1600	$1/4$	$9/64$	2R
Autolite 4300	1971–75	$13/16$	—	$9/64$	$1/8$	1600	$9/32$	$3/16$	2R
Motorcraft 4350	1976–78	$15/16$	—	—	$1/8$	1600	$1/3$	$5/32$	2NR
Holley 2209	1966–69	$9/16$	—	$5/32$	$1/50$	1800	$3/16$	$11/64$	1 Lean
Holley 4160	1966–69	⑤	—	$7/64$	$1/40$	1800	$11/64$	$1/4$	1 Lean

① $3/64$ in 1969.
② 1972 8-304 with manual trans.: $7/64$.
③ Automatic trans.: $1/10$.
④ 1972: $13/64$.
⑤ Parallel with bowl surface. See text.

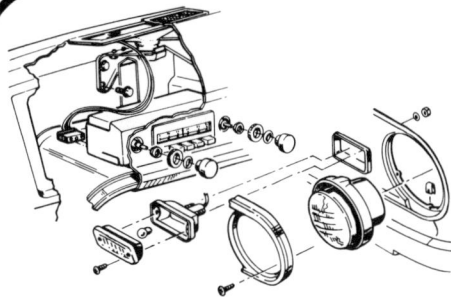

Chassis Electrical

UNDERSTANDING AND TROUBLESHOOTING ELECTRICAL SYSTEMS

Electrical problems generally fall into one of three areas:

1. The component that is not functioning is not receiving current.
2. The component itself is not functioning.
3. The component is not properly grounded.

Problems that fall into the first category are by far the most complicated. It is the current supply system to the component which contains all the switches, relays, fuses, etc.

The electrical system can be checked with a test light and a jumper wire. A test light is a device that looks like a pointed screwdriver with a wire attached to it. It has a light bulb in its handle. A jumper wire is a piece of insulated wire with an alligator clip attached to each end. To check the system you must follow the wiring diagram of the vehicle being worked on. A wiring diagram is a road map of the car's electrical system.

If a light bulb is not working, you must follow a systematic plan to determine which of the three causes is the villain.

1. Turn on the switch that controls the inoperable bulb.

2. Disconnect the power supply wire from the bulb.

3. Attach the ground wire on the test light to a good metal ground.

4. Touch the probe end of the test light to the end of the power supply wire that was disconnected from the bulb. If the bulb is receiving current, the test light will glow.

NOTE: *If the bulb is one which works only when the ignition key is turned on (turn signal), make sure the key is turned on.*

If the test light does not go on, then the problem is in the circuit between the battery and the bulb. As mentioned before, this includes all the switches, fuses, and relays in the system. Turn to the wiring diagram and

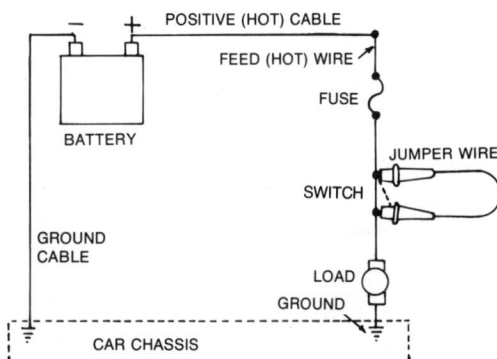

Bypassing a switch with a jumper wire

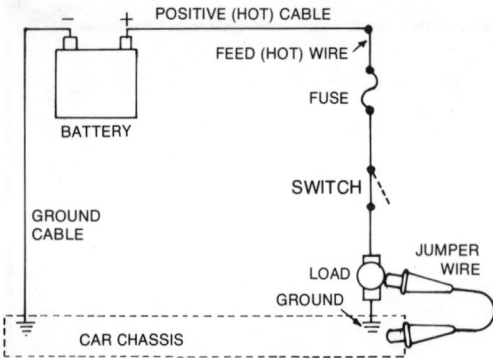

Checking for a bad ground with a jumper wire

find the bulb on the diagram. Follow the wire that runs back to the battery. The problem is an open circuit between the battery and the bulb. If the fuse is blown and, when replaced, immediately blows again, there is a short circuit in the system which must be located and repaired. If there is a switch in the system, bypass it with a jumper wire. This is done by connecting one end of the jumper wire to the power supply wire into the switch and the other end of the jumper wire to the wire coming out of the switch. Again, consult the wiring diagram. If the test light lights with the jumper wire installed, the switch or whatever was bypassed is defective.

NOTE: *Never substitute the jumper wire for the bulb, as the bulb is the component required to use the power from the source.*

5. If the bulb in the test light goes on, then the current is getting to the bulb that is not working in the car. This eliminates the first of the three possible causes. Connect the power supply wire and connect a jumper wire from the bulb to a good metal ground. Do this with the switch which controls the bulb turned on, and also the ignition switch turned on if it is required for the light to work. If the bulb works with the jumper wire installed, then it has a bad ground. This is usually caused by the metal area on which the bulb mounts to the car being coated with some type of foreign matter.

6. If neither test located the source of the trouble, then the light bulb itself is defective.

The above test procedure can be applied to any of the components of the chassis electrical system by substituting the component that is not working for the light bulb. Remember that for any electrical system to work, all connections must be clean and tight.

HEATER

Blower Motor

REMOVAL AND INSTALLATION

1. Disconnect the electrical connection.
2. Remove the screws that hold the motor in place.

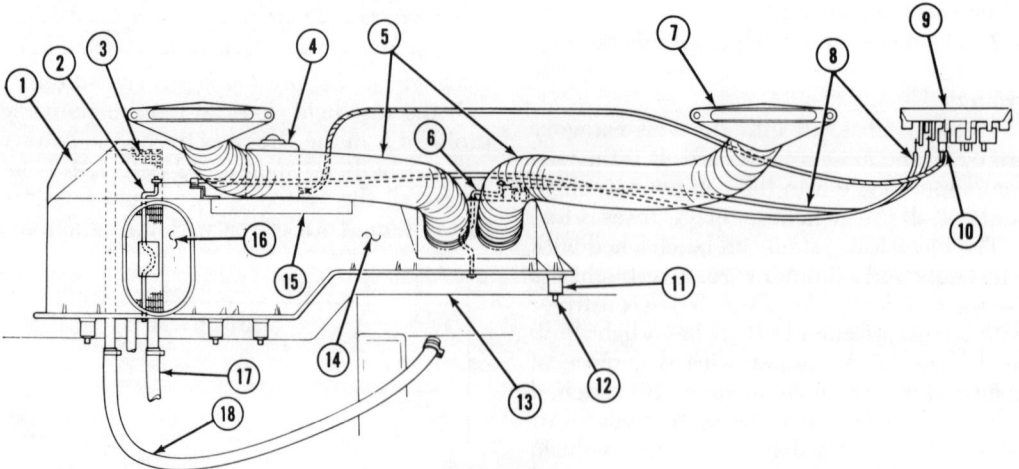

1. Intake air duct	7. Defroster nozzle	13. Firewall
2. Air blender door	8. Heater control cable	14. Heat distributor
3. Heater core	9. Switch panel	15. Fan housing
4. Fan motor	10. Heater switch	16. Transition duct
5. Defroster hose	11. Spacer	17. Outlet heater hose
6. Air distribution door	12. Attaching bolt	18. Inlet heater hose

Commando heater assembly

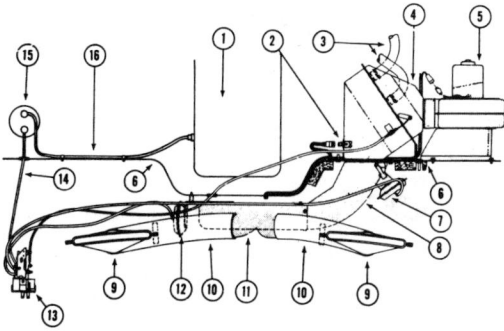

1. Engine
2. Heat resistor
3. Water hose
4. Heater core and duct
5. Blower
6. Firewall
7. Vacuum actuator
8. Transition duct
9. Defroster nozzle
10. Defroster hose
11. Heat distributor
12. Defrost vacuum actuator
13. Heater control
14. Vacuum line—storage tank-to-control
15. Vacuum storage tank
16. Vacuum line—engine-to-storage tank

Wagoneer and Cherokee heater assembly

3. Remove the blower motor.

4. Install the blower motor in the reverse order of removal.

Core

REMOVAL AND INSTALLATION

1. Drain the cooling system.

2. Disconnect the temperature control cable at the heater.

3. Disconnect the heater hoses at the inlet and outlet of the heater.

4. Disconnect the blower motor resistor wires at the plug-type connector on the heater resistor.

5. Remove the nuts that secure the heater core and duct to the firewall. Two of the nuts are on the inside of the vehicle, just to the right of the transition duct on the Wagoneer and Cherokee.

6. Remove the heater core and duct.

7. Mark the duct halves to be sure that they are properly assembled.

8. Remove the screws that fasten the two halves of the duct together.

9. Remove the screws that secure the heater core to the duct.

10. Remove the heater core.

11. Install the heater core in the reverse order of removal. Replace any damaged sealer.

WINDSHIELD WIPERS

Wiper Blade Replacement

1. Insert a screwdriver into the spring release opening of the blade saddle and depress the spring clip. Pull the blade from the arm.

2. Push the blade saddle onto the mounting clip so that the spring clip engages the pin.

Wiper Arm Replacement

1. Raise the blade end of the arm away from the windshield and move the spring tab away from the pivot shaft.

2. Disengage the auxiliary arm retainer clip from the pivot pin and pull the wiper arm from the pivot shaft.

2. Pivot the auxiliary over the pivot pin and engage the retainer clip. Push the wiper arm over the pivot shaft. Be sure that the shaft is in park and the wiper arm is positioned in the down mode.

Motor

REMOVAL AND INSTALLATION

Wagoneer and Cherokee

1. Disconnect the wiper drive link from the crank under the instrument panel.

2. Mark the locations of the wires at the motor to facilitate proper assembly under the hood.

3. Disconnect the motor and washer pump wires at the motor under the hood.

4. Remove the motor-to-dash mounting screws and remove the motor from the vehicle.

5. Install the windshield wiper motor in the reverse order of removal.

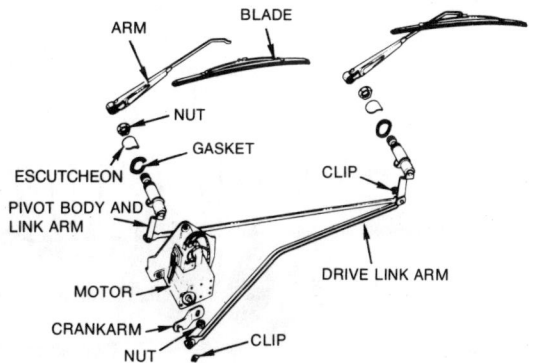

Commando wiper mechanism

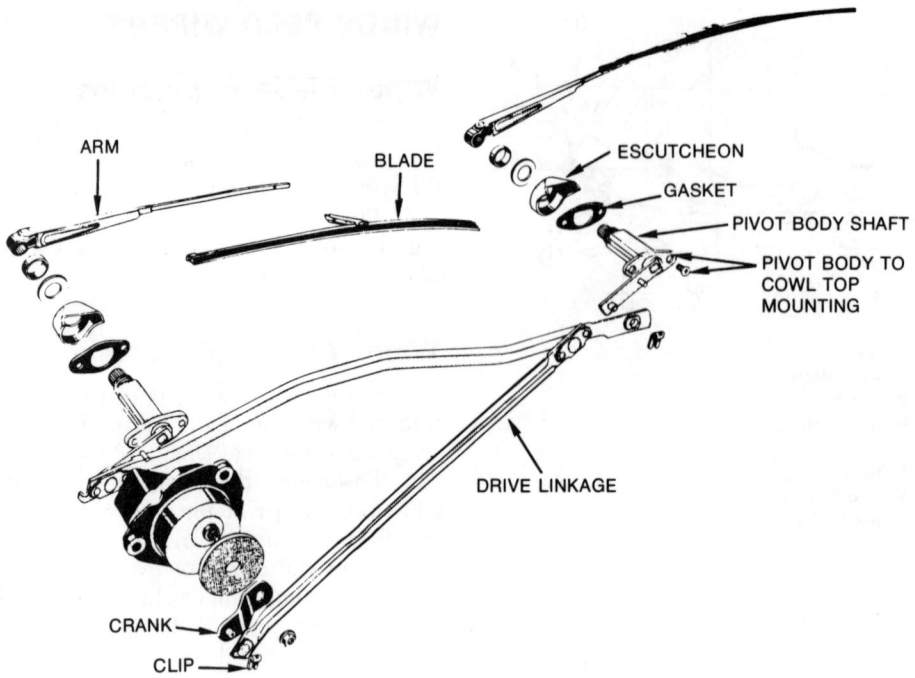

ARM

BLADE

ESCUTCHEON

GASKET

PIVOT BODY SHAFT

PIVOT BODY TO COWL TOP MOUNTING

DRIVE LINKAGE

CRANK

CLIP

Wagoneer and Cherokee wiper mechanism

Commando

1. Disconnect the wire harness plug and speedometer cable from the instrument cluster.

2. Remove the instrument cluster from the instrument panel by depressing the retainer springs at each corner.

3. Remove the 3 motor-to-brake and clutch pedal mounting bracket screws.

4. Disconnect the wiper drive link from the motor crank.

5. Disconnect the washer hoses from the washer pump. Pivot the motor assembly to the right and drop it below the instrument panel.

6. Mark the wire locations for proper assembly.

7. Disconnect the wire harness from the motor and washer pump and remove the motor.

8. Install the windshield wiper motor in the reverse order of removal.

Linkage
REMOVAL AND INSTALLATION

1. Remove the wiper arms and pivot shaft nuts, washers, escutcheons and gaskets.

2. Disconnect the drive arm from the motor crank.

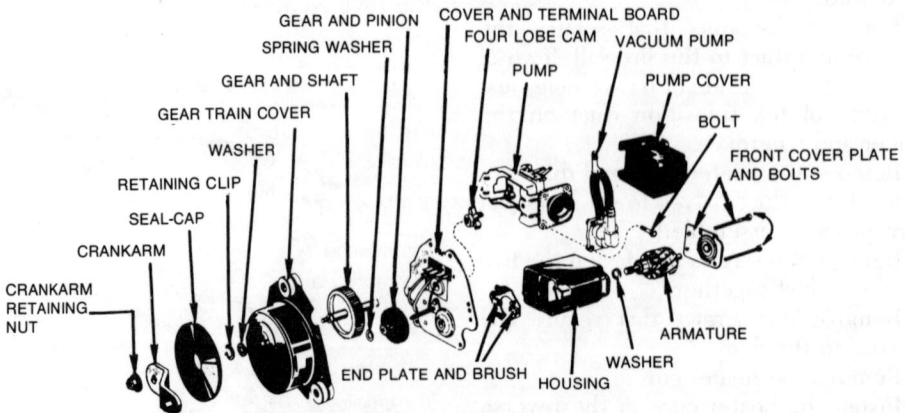

GEAR AND PINION

SPRING WASHER

GEAR AND SHAFT

GEAR TRAIN COVER

WASHER

RETAINING CLIP

SEAL-CAP

CRANKARM

CRANKARM RETAINING NUT

COVER AND TERMINAL BOARD

FOUR LOBE CAM

VACUUM PUMP

PUMP

PUMP COVER

BOLT

FRONT COVER PLATE AND BOLTS

ARMATURE

WASHER

HOUSING

END PLATE AND BRUSH

Two speed wiper motor

3. Remove individual links where necessary, to remove the pivot shaft bodies without excessive interference.

4. Install in the reverse order of removal.

INSTRUMENT CLUSTER

REMOVAL AND INSTALLATION

1966–70 Wagoneers and Commandos

1. Disconnect the speedometer cable and all electrical leads from the rear of the instrument cluster.

2. Compress the retainer springs and slide the instrument cluster forward from the dash panel.

3. Use care in removing the glass frame which is held to the instrument cluster case by clips. Pry up the clips and carefully pry the frame off the case.

4. To remove the fuel gauge, temperature gauge or speedometer, remove the lock-nuts and remove these units from the front.

5. Install the instrument cluster in the reverse order of removal.

1971–74 Wagoneers and Cherokees

The instrument panel is bolted to the surrounding body sheet metal and to the brake and clutch support brackets. When removing the instrument panel, the windshield must be removed to obtain access to the bolts, under the windshield weatherstrip, that attach the instrument panel to the cowl. Other instrument panel attaching bolts must also be removed.

1. Remove the windshield and the windshield weatherstrip to expose the crash pad retaining screws at the base of the windshield.

2. Remove the exposed crash pad retaining screws.

3. Remove the instrument cluster and ash tray.

4. Install the instrument cluster in the reverse order of removal.

1971–73 Commandos

The instrument panel is an integral unit welded across its top flange to the cowl. The panel is bolted to the sides of the cowl and to the steering column, brake and clutch pedal supports and hand brake bracket. It can only be removed by breaking the welds from

under the windshield frame, that secure it to the cowl.

1975–79 Wagoneer and Cherokee

1. Disconnect the battery ground.

2. Disconnect the speedometer cable.

3. Cover the steering column.

4. Remove the cluster attaching screws and tilt the top toward the interior of the vehicle.

5. Mark the electrical connectors and hoses, disconnect them and the blend door air cable.

6. Remove the cluster.

7. Installation is the reverse of removal.

SPEEDOMETER CABLE REPLACEMENT

1. Reach up behind the speedometer head. On some older models, the speedometer cable is connected by a threaded ring. Unscrew the ring and pull the cable sheath from the head. The core may then be pulled from the sheath.

2. On later models, depress the spring tab at the point where the cable attaches to the head and pull the cable straight back away from the head. Pull the core from the sheath.

3. If the cable is broken, raise and support the vehicle and remove the cable from the transmission. Pull the broken end from the sheath.

4. When installing the cable, coat it with speedometer cable lubricant before installation.

RADIO

REMOVAL AND INSTALLATION

Commando

1. Pull off the inner and outer radio control knobs.

2. Remove the two attaching nuts.

3. Remove the antenna lead.

4. Disconnect the rear support strap from the radio, push the radio back to clear the instrument panel and lower the radio far enough to disconnect the speaker leads.

5. Disconnect the fused wire and remove the radio.

6. Install the radio in the reverse order of removal.

Wagoneer and Cherokee

1. Open the glove box door and remove the glove box liner and the lock striker.

2. Remove the antenna lead.

3. Disconnect the fused wire from the fuse panel.

4. Disconnect the rear support strap from the radio.

5. Pull off the radio control knobs and remove the radio attaching nuts.

6. Push the radio back to clear the dash panel and remove the radio through the glove box.

7. Install the radio in the reverse order of removal.

HEADLIGHTS

REMOVAL AND INSTALLATION

1. Remove the screws retaining the headlight door and remove the door.

2. Remove the 3 screws retaining the retaining ring and remove the ring.

3. Pull the headlight out, disconnect the wire harness and remove the headlight from the vehicle.

4. Install the headlight in the reverse order of removal.

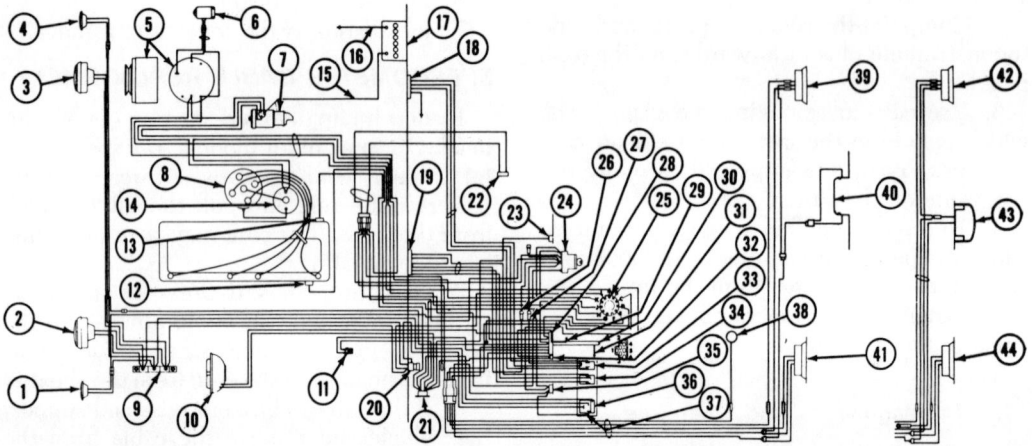

1. Left parking and dir. signal lamp
2. Left head lamp
3. Right head lamp
4. Right parking and dir. signal lamp
5. Alternator
6. Voltage regulator
7. Starting motor
8. Distributor
9. Junction block
10. Horn
11. Stop light switch
12. Temperature sending unit
13. Oil pressure sending unit
14. Ignition coil
15. Positive cable to starting motor
16. Negative cable to ground
17. Battery
18. Heater fan motor
19. Windshield wiper motor
20. Headlight dimmer switch
21. Circuit breaker
22. Backup light switch
23. Cigar lighter
24. Ignition switch
25. Instrument cluster
   A. Temperature gauge
   B. Cluster lights
   C. Fuel gauge
   D. Ground
   E. Left turn indicator
   F. Auxiliary
   G. Oil pressure indicator

   H. Charging indicator
   J. Right turn indicator
   K. Ignition feed
   L. High-Beam indicator
26. Fuse (windshield wiper motor)
27. Fuse (heater)
28. Directional signal switch
   1. Horn
   2. Left front turn
   2. Left turn indicator
   3. Right front turn
   3. Right turn indicator
   4. Left rear turn
   5. Stop light switch
   6. Right rear turn
29. Hazard warning light switch
30. Horn button
31. Directional signal lever
32. Flasher
33. Heater fan switch
34. Windshield wiper and washer switch
35. Instrument cluster dimmer switch
36. Circuit breaker
37. Main light switch
38. Fuel gauge tank unit
39. Right tail and stop lamp (Commando)
40. License plate lamp (Commando)
41. Left tail and stop lamp (Commando)
42. Right tail and stop lamp (convertible)
43. License plate lamp (convertible)
44. Left tail and stop lamp (convertible)

Wiring diagram, 4-134 Commando

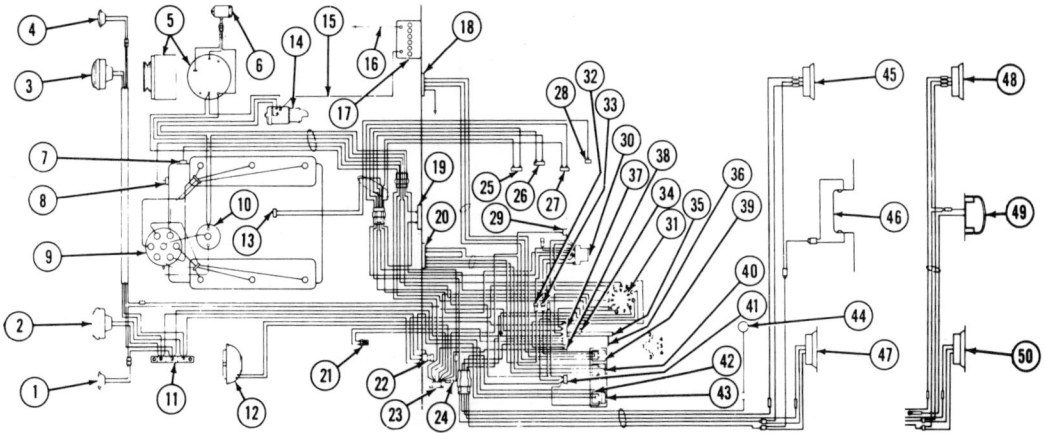

1. Left parking and dir. signal lamp
2. Left head lamp
3. Right head lamp
4. Right parking and dir. signal lamp
5. Alternator
6. Voltage regulator
7. Oil pressure sending unit
8. Temperature sending unit
9. Distributor
10. Ignition coil
11. Junction block
12. Horn
13. Kickdown switch (*)
14. Starting motor
15. Positive cable to starting motor
16. Negative cable to ground
17. Battery
18. Heater fan motor
19. Ballast
20. Windshield wiper motor
21. Stop light switch
22. Headlight dimmer switch
23. Circuit breaker
24. Fuse (clock)
25. Backup light switch
26. Electric clock
27. Neutral safety switch (*)
28. Trans. solenoid (*)
29. Cigar lighter
30. Ignition and starter switch
31. Instrument cluster

A. Temperature gauge
B. Cluster lights
C. Fuel gauge
D. Ground
E. Left turn indicator
F. Auxiliary
G. Oil pressure indicator
H. Charging indicator
J. Right turn indicator
K. Ignition feed
L. High-Beam indicator
32. Fuse (windshield wiper motor)
33. Fuse (heater)
34. Hazard warning light switch
35. Horn button
36. Directional signal lever
37. Directional signal switch
38. Flasher
39. Heater fan switch
40. Windshield wiper and washer switch
41. Instrument cluster dimmer switch
42. Main light switch
43. Circuit breaker
44. Fuel gauge tank unit
45. Right tail and stop lamp (Commando)
46. License plate lamp (Commando)
47. Left tail and stop lamp (Commando)
48. Right tail and stop lamp (convertible)
49. License plate lamp (convertible)
50. Left tail and stop lamp (convertible)
*Automatic Transmission

6-225 Commando wiring diagram

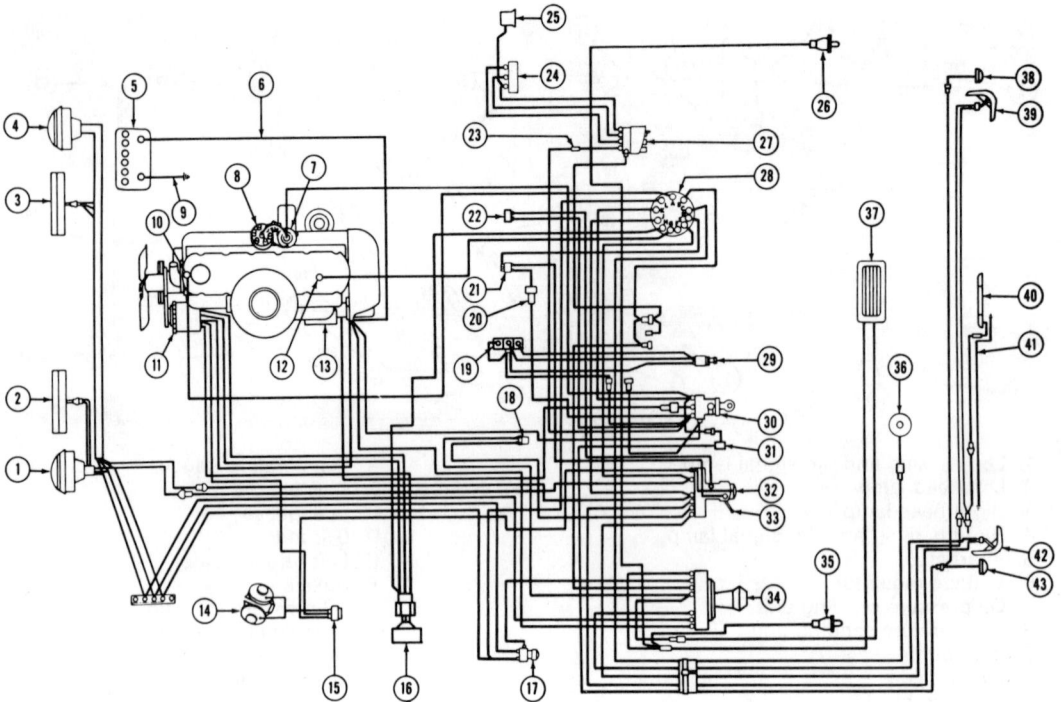

1. Left headlamp
2. Left parking and signal lamp
3. Right parking and signal lamp
4. Right headlamp
5. Battery
6. Positive cable
7. Ignition coil
8. Ignition distributor
9. Negative cable
10. Temperature sending unit
11. Alternator
12. Oil pressure sending unit
13. Starting motor
14. Dual horn (Wagoneer)
15. Horn relay
16. Voltage regulator
17. Foot dimmer switch
18. Stop light switch
19. Wiper motor
20. Fuse
21. Flasher (dir. sig.)
22. Hazard warning flasher
23. Fuse
24. Heater
25. Blower motor
26. Right courtesy and dome lamp switch
27. Heater switch
28. Instrument cluster
    A. Temperature gauge
    B. Cluster lights
    C. Fuel gauge
    D. Ground
    E. Left turn indicator
    F. Auxiliary
    G. Oil pressure indicator
    H. Charging indicator
    J. Right turn indicator
    K. Ignition feed
    L. Left turn indicator
29. Windshield wiper switch
30. Ignition and starter switch
31. Back-Up light switch
32. Horn button
33. Directional signal lever
34. Main light switch
35. Left courtesy and dome lamp switch
36. Fuel gauge tank unit
37. Dome lamp
38. Back-Up lamp
39. Right tail and stop lamp
40. License lamp (Wagoneer)
41. Body cable
42. Left tail and stop lamp
43. Left back-up lamp

1966–69 6-232 Wagoneer wiring diagram

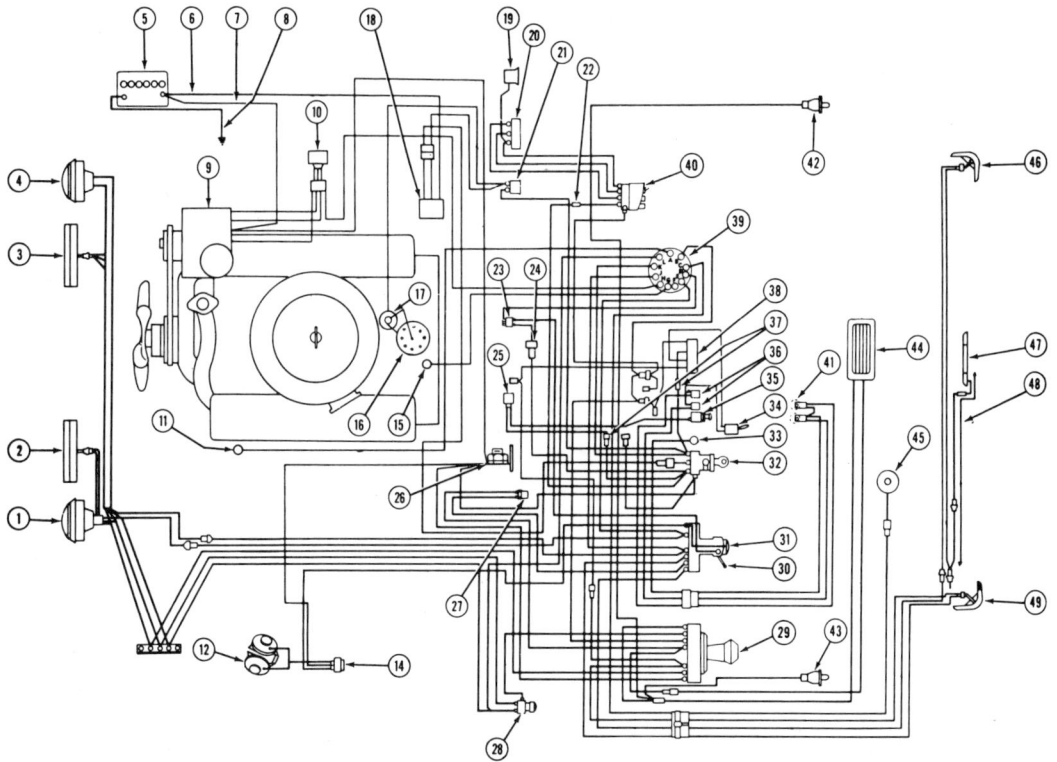

1. Left headlamp
2. Left parking and signal lamp
3. Right parking and signal lamp
4. Right headlamp
5. Battery
6. Positive cable to starting motor
7. Positive cable from alternator
8. Negative cable to ground
9. Alternator
10. Voltage regulator
11. Temperature sending unit
12. Dual horn (Wagoneer)
14. Horn relay
15. Oil pressure sending unit
16. Distributor
17. Ignition coil
18. Starting motor
19. Blower motor
20. Heater
21. Ballast
22. Fuse
23. Flasher
24. Fuse
25. Wiper motor
26. Terminal block
27. Stop light switch
28. Foot dimmer switch
29. Main light switch
30. Directional signal lever
31. Horn button

32. Ignition and starter switch
33. Ground
34. Toggle switch
35. Windshield wiper switch
36. 4WD indicator lights
37. Fuses
38. Relay
39. Instrument cluster
    A. Temperature gauge
    B. Cluster lights
    C. Fuel gauge
    D. Ground
    E. Left turn indicator
F. Auxiliary
    G. Oil pressure indicator
    H. Charging indicator
    J. Right turn indicator
    K. Ignition feed
    L. Left turn indicator
40. Heater switch
41. 4WD indicator switches
42. Right courtesy and dome lamp switch
43. Left courtesy and dome lamp switch
44. Dome lamp
45. Fuel gauge tank unit
46. Right tail and stop lamp
47. License lamp (Wagoneer)
48. License lamp harness
49. Left tail and stop lamp

8-327 Wagoneer wiring diagram

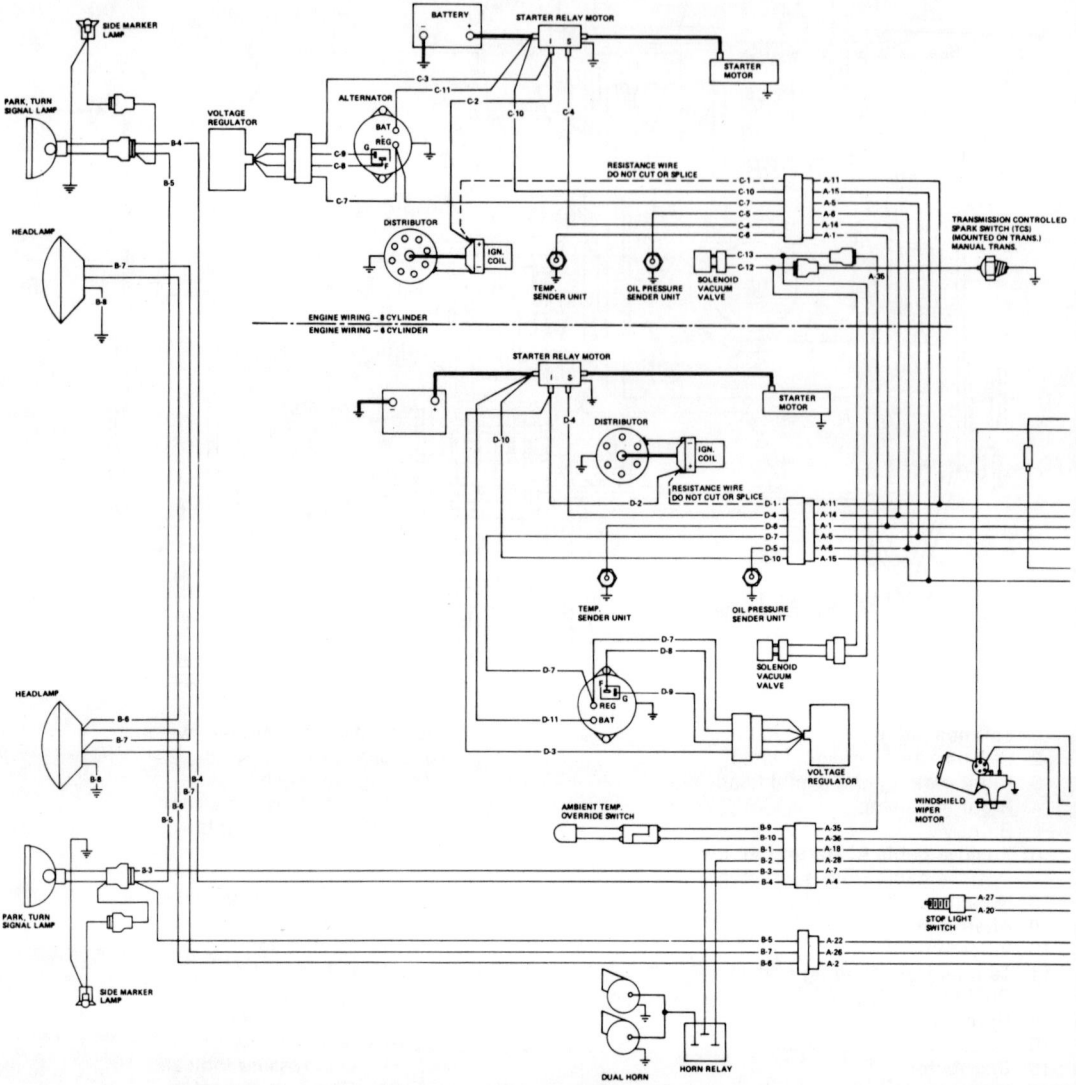

1971 and later commando

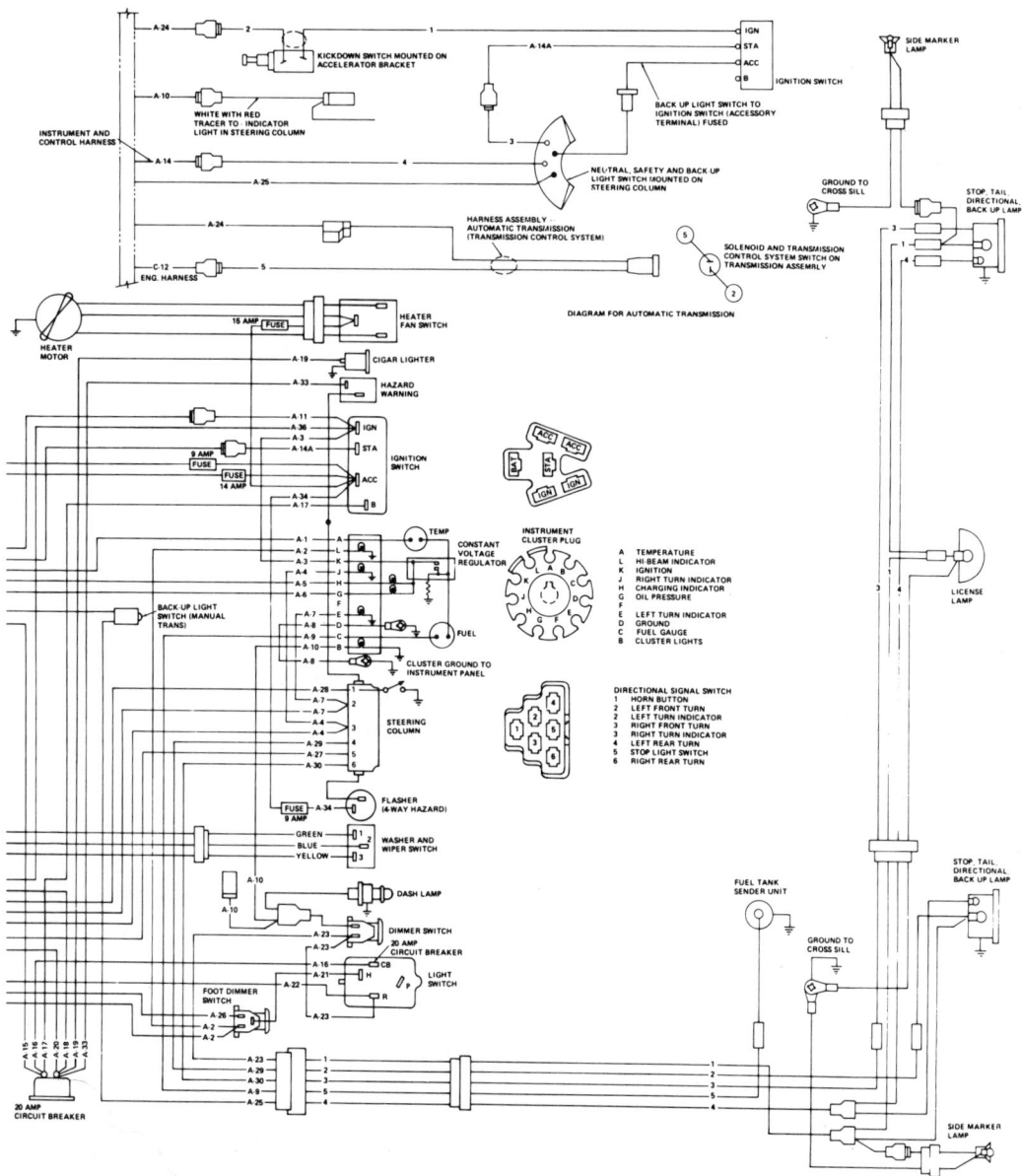

1971 and later commando

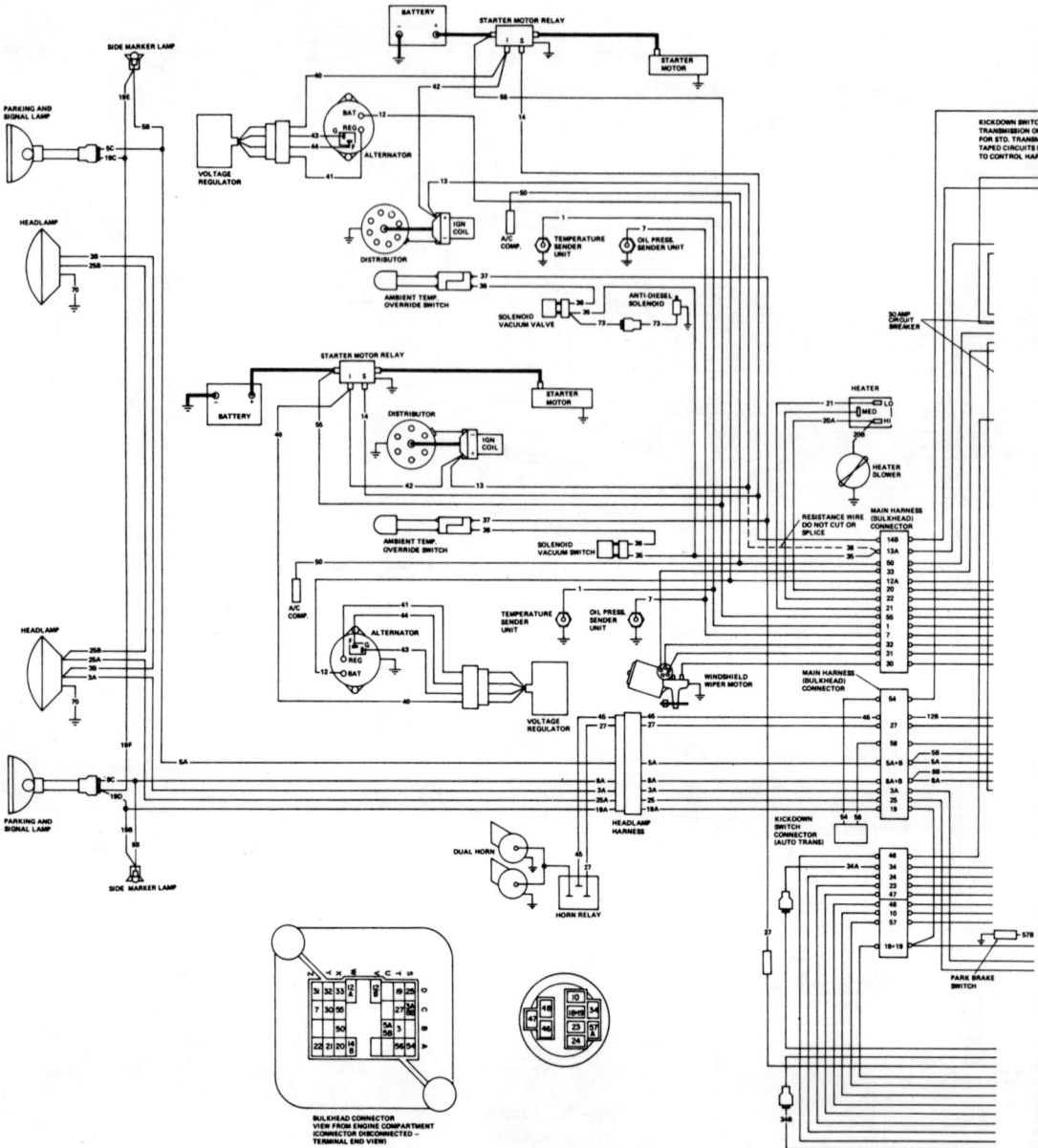

1971–74 Wagoneer Cherokee

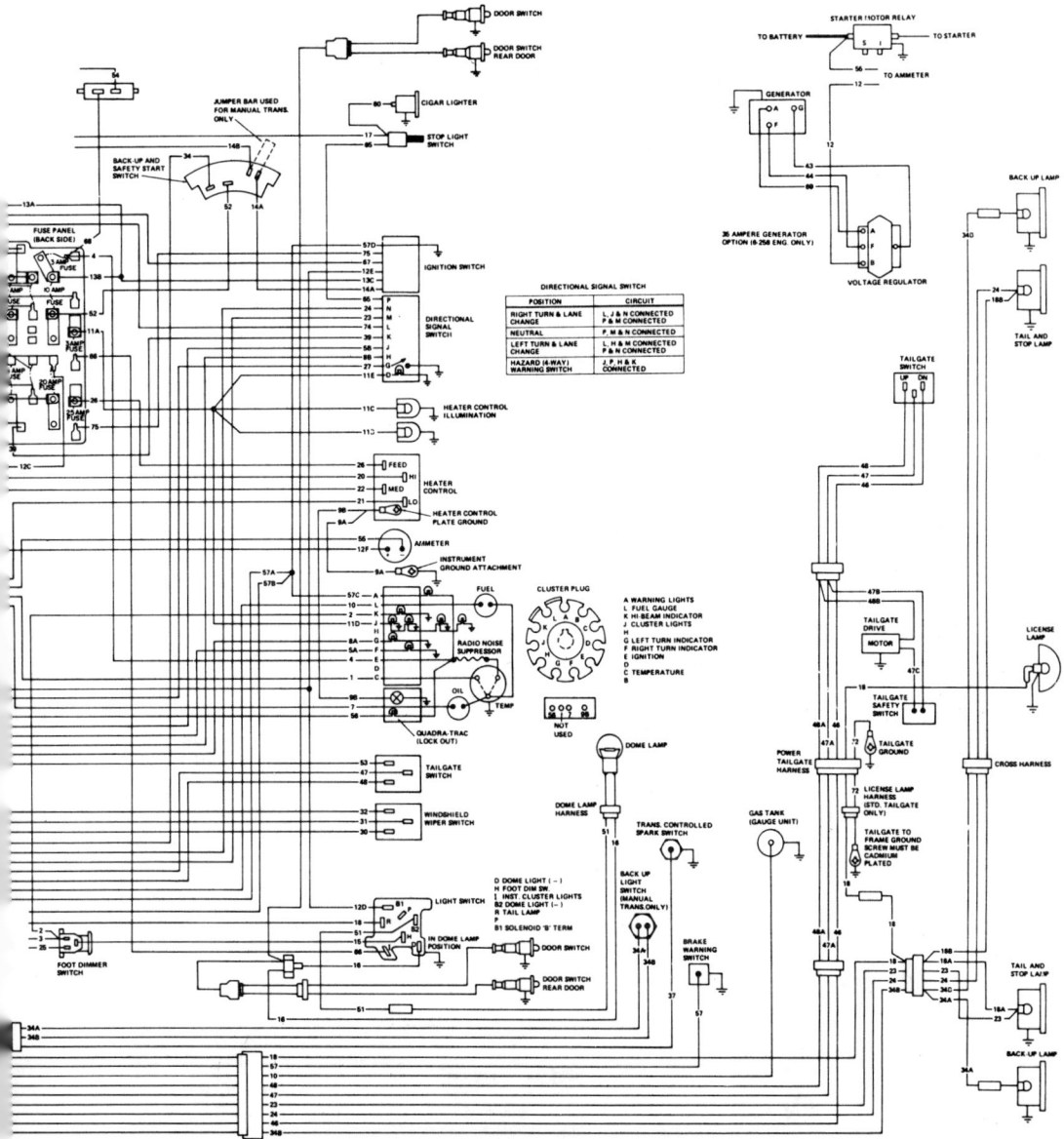

1971-74 Wagoneer Cherokee

5457847* (S.O.5457845-C)
5350456* (S.O.919337-B) ──┐ OPT.

MARKER & REFLECTOR LAMP

BATTERY

ELECTRONIC CONTROL UNIT

VOLTAGE REGULATOR

55B YELLOW 14 (FUSIBLE LINK)

55A YELLOW 10

SPLICE "K"

STARTER SOLENOID

5B GREEN 16

40 YELLOW 16

14 LT BLUE 16

OIL PRESSURE SENDING UNIT

STARTING MOTOR

5C GREEN 16

19E WHITE 16

5457862-C
5350447-C ── OPT.

CHASSIS

BODY

43 BLACK 16

12 RED 10

BAT

FLD

7 PURPLE 18 (8)

ALTERNATOR

PARK & SIGNAL LAMP

19C WHITE 16

SPLICE "I"

44 GREEN 18

THROTTLE CLOSING SOLENOID

HEADLAMP

SPLICE "I"

CAPACITOR JUMPER

CARBURETOR SOLENOID

79 GREEN 16

81 WHITE 16

80 BLUE 16

2

4

6

8

70 BLACK 16

5450589-B

25B GRAY 16

38 GRAY W/TR 14

58B BLACK 16

5A GREEN 16

19F WHITE 16

13C RED W/TR-14

43 BLACK 16

73 RED W/TR-16

82 RED W/TR 16

DISTRIBUTOR

1

3

5

7

55A YELLOW 10 (8)

13B RED W/TR 14

TEMPERATURE SENDER

AIR CONDITIONER COMPRESSOR

13A RED W/TR 14 (8)

SPLICE "L"

ENGINE WIRING 8 CYL.

1 PURPLE W/TR 18 (8)

50 BROWN 16 (8)

55A YELLOW 10 (8)

55B YELLOW 14 (FUSIBLE LINK)

95 RED 14

SPLICE "I"

50 BROWN 16 (6)

14 LT BLUE 16 (8)

ELECTRONIC CONTROL UNIT

AIR CONDITIONER COMPRESSOR

7 PURPLE 18 (6)

BATTERY

BODY

STARTER SOLENOID

OIL PRESSURE SENDER

14 LT BLUE 16 (6)

CHASSIS

DISTRIBUTOR

SPLICE "N"

82 RED W/TR 16

80 BLUE 16

81 WHITE 16

CAPACITOR JUMPER

79 GREEN 16

12 RED 10 (8)

STARTING MOTOR

13B RED W/TR -14

1

2

3

4

5

6

14 LT BLUE 16 (6 & 8)

12 RED 10 (6 & 8)

12 RED 10 (6)

TEMPERATURE SENDER

1 PURPLE W/TR 18 (6)

50 BROWN 16 (6 & 8)

13A RED W/TR 14 (6)

58B BLACK 16

58A BLACK 16

12 RED 10

95 RED 14

RESISTOR ASSEMBLY

1 PURPLE W/TR 18 (6 & 8)

12
14
50

BODY GROUND

ALTERNATOR & INTEGRAL VOLTAGE REGULATOR (DELCO)

HEATER BLOWER MOTOR

7 PURPLE 18 (6 & 8)

208 TAN 14 (6 & 8)

78 BLACK W/TR 24 (10 OHMS)

ENGINE WIRING 6 CYL.

13 RED W/TR 14 (6 & 8)

13

DASH CONNECTOR AS VIEWED FROM INSTRUMENT PANEL SIDE

25B GRAY 16

38 GRAY W/TR 14

19F WHITE 16

21 BROWN 16 (6 & 8)

22 BROWN W/TR 16 (6 & 8)

20A TAN 14 (6 & 8)

21
22
20A

W/S WIPER MOTOR

GRAY

SERIES BLACK

SHUNT

RED

PARK SWITCH

NO. 3 (32)

NO. 1 (31)

NO. 2 (33)

32 BLUE W/TR 14 (6 & 8)

W/S WIPER MOTOR

31 BLUE 14 (6 & 8)

33 RED W/TR 14 (6 & 8)

WINDSHIELD WASHER BOTTLE

55A YELLOW 10 (6 & 8)

32
31
33
55A

HEADLAMP

3A GRAY W/TR 14

92 BLACK W/TR 18

91 PINK 14

3A GRAY W/TR 14

92
91
3A

70 BLACK 16

19A WHITE 16

25A GRAY 16

8A GREEN W/TR 16

5A GREEN 16

19A WHITE 16

25A GRAY 16

5A
19A
25A

PARK & SIGNAL LAMP

8C GREEN W/TR 16

19D WHITE 16

58A BLACK 16

SPLICE "H"

SPLICE "G"

5459221-D

8A GREEN W/TR 16

8A

19B WHITE 16

8B GREEN W/TR 16

HORNS

45 RED W/TR 14

54 YELLOW 16

56 ORANGE 16

10 PINK 16

18 WHITE 16

23 LT GREEN W/TR 16

45
54
56
10
18
23

MARKER & REFLECTOR LAMP

45A RED W/TR 14

KICKDOWN & QUADRATRAC SWITCH

54 YELLOW 16

56 ORANGE 16

5350253 C

24 LT GREEN 16

34 WHITE W/TR-16

57 BLACK 18

34
57

10
18
23
24
34
57

FRAME HARNESS CONNECTOR

1975–79 Wagoneer and Cherokee

17A RED W/TR 14
24 LT GREEN 16
23 LT GREEN W/TR 16
74 RED W/TR 18
39 PINK 16
5B GREEN 18
8B GREEN W/TR 18

STOP LAMP SWITCH & CRUISE CONTROL

67 YELLOW 12
68 YELLOW 16
4 RED 16
60 RED 16
CIGAR LIGHTER
65A RED 16
17B RED W/TR 14
65B PINK 16

13B RED W/TR 14
33 RED W/TR 14
52 RED 16
75 RED W/TR 14
67 YELLOW 12
TURN SIGNAL SWITCH
11E ORANGE 18
27 BLK W/TR 18
57D BLACK 18

STOP LAMP SWITCH

11A ORANGE 18
52 & 213
53
50 BROWN
66 RED W/TR 18
IGNITION SWITCH

5450049 B (MAN TRANS)
IGNITION SWITCH
12E RED 12

34 GRN W. WHT TR 18
14A LT BLUE 16
3186890-B (MAN. TRANS.)
BACK UP SWITCH
NEUTRAL SAFETY SWITCH
13C RED W/TR 18
9B BLACK 18

A WARNING LTS.
L GAS GAUGE
K HI-BEAM INDICATOR
J CLUSTER LIGHTS
G LEFT TURN INDICATOR
F RIGHT TURN INDICATOR
E IGNITION
C TEMPERATURE

12C RED 12
26 RED W/TR 14
14B LT BLUE 16
TRANSMISSION KICKDOWN SWITCH
12F RED 12
AMMETER

RADIO NOISE SUPPRESSOR
OIL

34 WHITE W/TR 18
SPLICE "C"
57A BLACK 18
68 YELLOW 18
7 PURPLE 18
56 ORANGE 16
CLUSTER CONNECTION
57B BLACK 18
SPLICE "D"
57D BLACK 18

TEMP

5A GREEN 18
8A GREEN W/TR 18
4 RED 18
11D ORANGE 18
CLUSTER CONNECTION
10 PINK 18
1 PURPLE W/TR 18
57C BLACK 18
57B BLACK 18
12E RED 12
9B BLACK 18

FUEL

SF 213 RED W/TR 14
13A RED W/TR 14
50 BROWN 16
54 YELLOW
56 ORANGE 16
5A GREEN 16
58 GREEN 18
8B GREEN W/TR 18

33 RED W/TR 14
14B LT BLUE 16
55 YELLOW 10

12C RED 10
2 GRAY W/TR 18
12A RED 10
27 BLACK W/TR 18
SPLICE "B"

HEATER GROUND
9C
26 RED W/TR 14
21 BROWN 16
FROM I.P.

FUSE PANEL

11B ORANGE 18
SPLICE "A"
11A ORANGE 18
11C ORANGE 18
HEATER LAMPS
11E ORANGE 18
SF 213 RED W/TR-14
HEATER CONTROL SWITCH

9B BLACK 18

23 LT GREEN W/TR 16
24 LT GREEN 16
57A BLACK 18
34 WHITE W/TR 18
12A RED 10
50 BROWN 16
1 PURPLE W/TR 18
13A RED W/TR 14
22 BROWN W/TR 16

20 TAN 14
21 BROWN 16
22 BROWN W/TR 16
22 BROWN W/TR 16
SPLICE "M"
DEFOGGER & HEATER GROUND
9E BLACK-18
9A BLACK-18
9D BLACK-18

'2A
14B
50
1
2 GRAY W/TR 18
HORN RELAY
12B RED 12
66 RED W/TR 18
213
SEAT BELT BUZZER

13A
21
22
20
32
31
33
55
91 RED W/TR 18
92
91
3
DIMMER SWITCH
45 RED W/TR 16
15 RED W/TR 14
10 PINK 18
12D RED 12
SEAT BELT TIMER
218A YELLOW-18
220 DK BLUE-18
220
SEAT BELT GND

21 BROWN 16
20 TAN 14
32 BLUE W/TR 14
31 BLUE 14
33 RED W/TR 14
7 PURPLE 18
55 YELLOW 10
92 BLACK 18
3 GRAY W/TR 14
5A GREEN 18
5B GREEN 18

92 BLACK 18
31 BLUE 14
3 (32)
W/S WIPER & WASHER SWITCH
LOW
(31)
LIGHT SWITCH
16 BLACK 18
51 ORANGE 16
218B YELLOW-18

5
19
25
19 WHITE 16
WASHER 3 (92)
OFF HIGH
18 WHITE 16

PARKING BRAKE WARNING SWITCH

25 GRAY 16
57B BLACK 18

8
45
54
56
10
18
24
34
57A
8B GREEN W/TR 18
45 RED W/TR 16
54 YELLOW 16
56 ORANGE 16
10 PINK 16
18 WHITE 16

BRAKE WARNING
57

18A WHITE 16
10 PINK 18
24 LT GREEN 16
23 LT GREEN W/TR 16
34A WHITE W/TR 18
57 BLACK 18

48 TAN 12
4/ BROWN 12
46 RED W/TR12
86 BROWN 18
85 DK BLUE-18
154 BLACK W/WHITE TR 18
89 PURPLE 16
90 ORANGE 16

FOR OPTIONAL EQUIPMENT ONLY

10
18
23
24
34A
57
10 PINK 18
18A WHITE 16
23 LT GREEN W/TR 16
24 LT GREEN 16
34A WHITE W/TR 18
57 BLACK 18

AUXILIARY HARNESS CONNECTOR

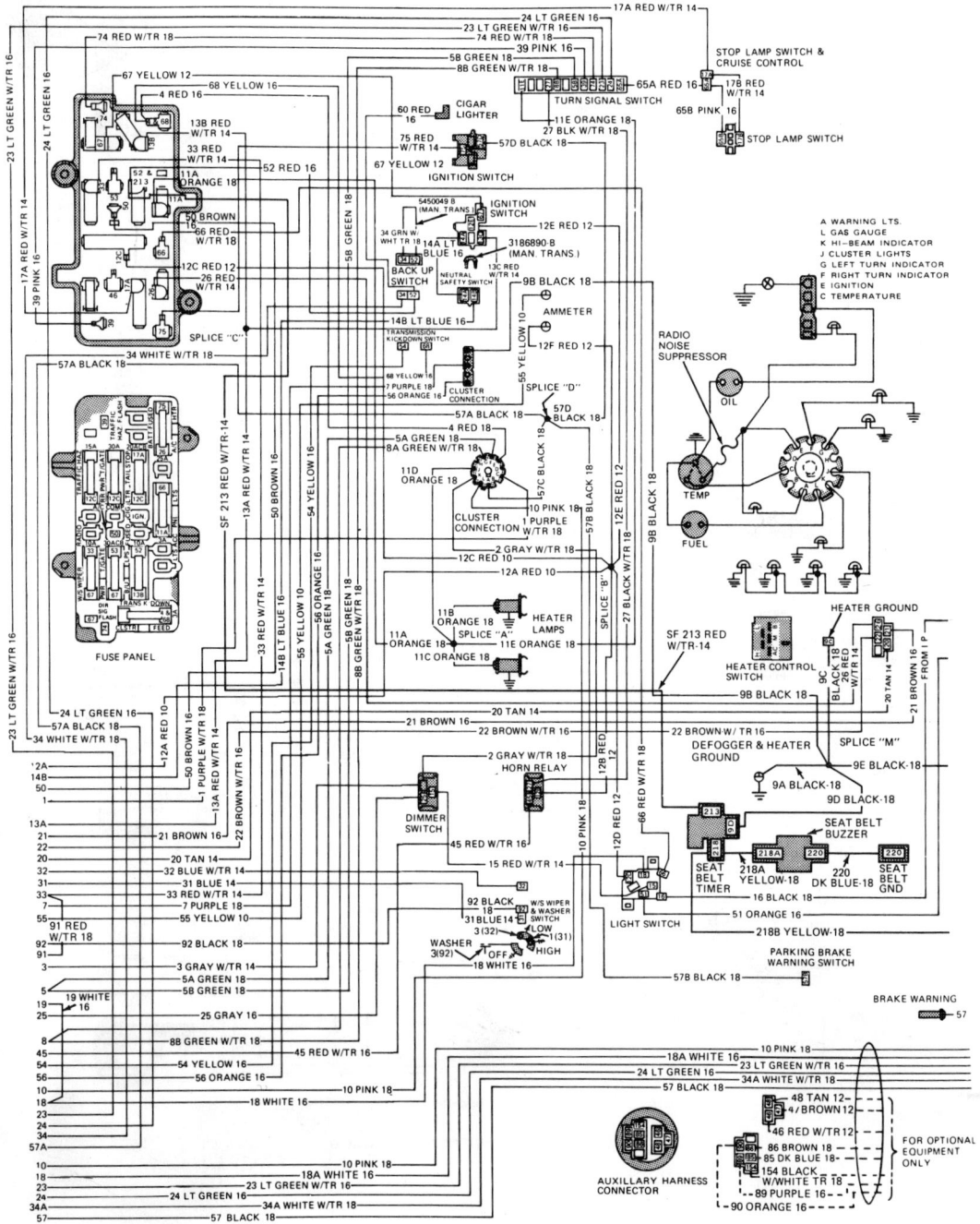

1975-79 Wagoneer and Cherokee

COURTESY LAMP

(4 DOOR 15 & 18)
5456014C

BLACK W/TR 18

BLACK W/TR 18

BLACK W/TR 18

(2 DOOR 16 & 17)
(OPT 25, 45 & 46)

MARKER & REFLECTOR
LAMP (25, 45 & 46)

16B BLK W/YEL TR 18

18B BLK W/YEL TR 18

51B BROWN 18

18B WHITE 16
24 LT GREEN 16

TAIL, STOP & MARKER LAMP
(15)

34B WHITE
W/TR 18

BACK UP LAMP
(15)

MARKER & REFLECTOR
LAMP (16, 17 & 18)

18E WHITE 16

TAIL & STOP LAMP
(25, 45 & 46)

BACK UP LAMP
(25, 45 & 46)

34B WHITE
W/TR 18

TAIL, STOP & BACK UP
LAMP (16, 17 & 18)

DOME LAMP
(SEE ACCESSORY DRAWING)

(SEE ACCESSORY DRAWING)

16 BLK

51 BROWN 18

16B BLK W/YEL TR 18

5450120 C

LICENSE LAMP
(25, 45 & 46)

BACK UP LAMP
(25, 45 & 46)

24 LT GREEN 16

34C WHITE W/TR 18

18D WHITE 16

34B WHITE W/TR 18
72 BLACK 18
18B WHITE 16

LICENSE LAMP
(15, 16, 17 & 18)

DOME & COURTESY
LAMP GROUND

16B BLACK W/YEL TR 18

16 BLACK
W/YEL TR 18

TAIL & STOP LAMP
(25, 45 & 46)

16A BLK W/YEL TR 18

FROM I P

DOME & COURTESY
LAMP FEED

5450122 C

34B WHITE W/TR 18

18C WHITE 16

18B WHITE 16

18A WHITE 16

23 LT GREEN W/TR 16

BACK UP
LIGHT SWITCH

34B WHT W/TR 18

5450036 D

18B WHITE 16

24 LT GREEN 16

23 LT GREEN W/TR 16

34A WHITE W/TR 18

TAIL, STOP & BACK UP
LAMP (16, 17 & 18)

994034 (S O 994033 B)

(4 DOOR 15)
(OPT. FOR 16, 17, 18, 25, 45 & 46)

51A
BROWN
-18

COURTESY LAMP

MARKER & REFLECTOR
LAMP (25, 45 & 46)
5350431 (S O 535252 C)

24 LT GREEN 16

5459062-D

MARKER & REFLECTOR
LAMP (16, 17 & 18)

34A WHITE W/TR 18

BACK UP LAMP
(15)

(2 DOOR 16 & 17)
(OPT 25, 45 & 46)

BLACK W/TR 18

SEAT
BELT
LAMP

BLACK W/TR 18

DOOR SWITCH
(4 DOOR 15 & 18)

34B WHT
W/TR 18

34B WHT W/TR 18
(15, 16, 17 & 18)

34A WHITE
W/TR 18

TAIL, STOP & MARKER LAMP
(15)

GAS TANK
GAUGE UNIT

18A WHITE 16

23 LT GREEN W/TR 16

24 LT GREEN 16

18A WHITE 18

24 LT GREEN 16

18B WHITE 16

23 LT GREEN 16

LIGHT
BLACK 18

10 PINK 18
FUEL SENDER

25, 45 & 46

(15, 16, 17 & 18)

18 23 24
JUNCTIONS

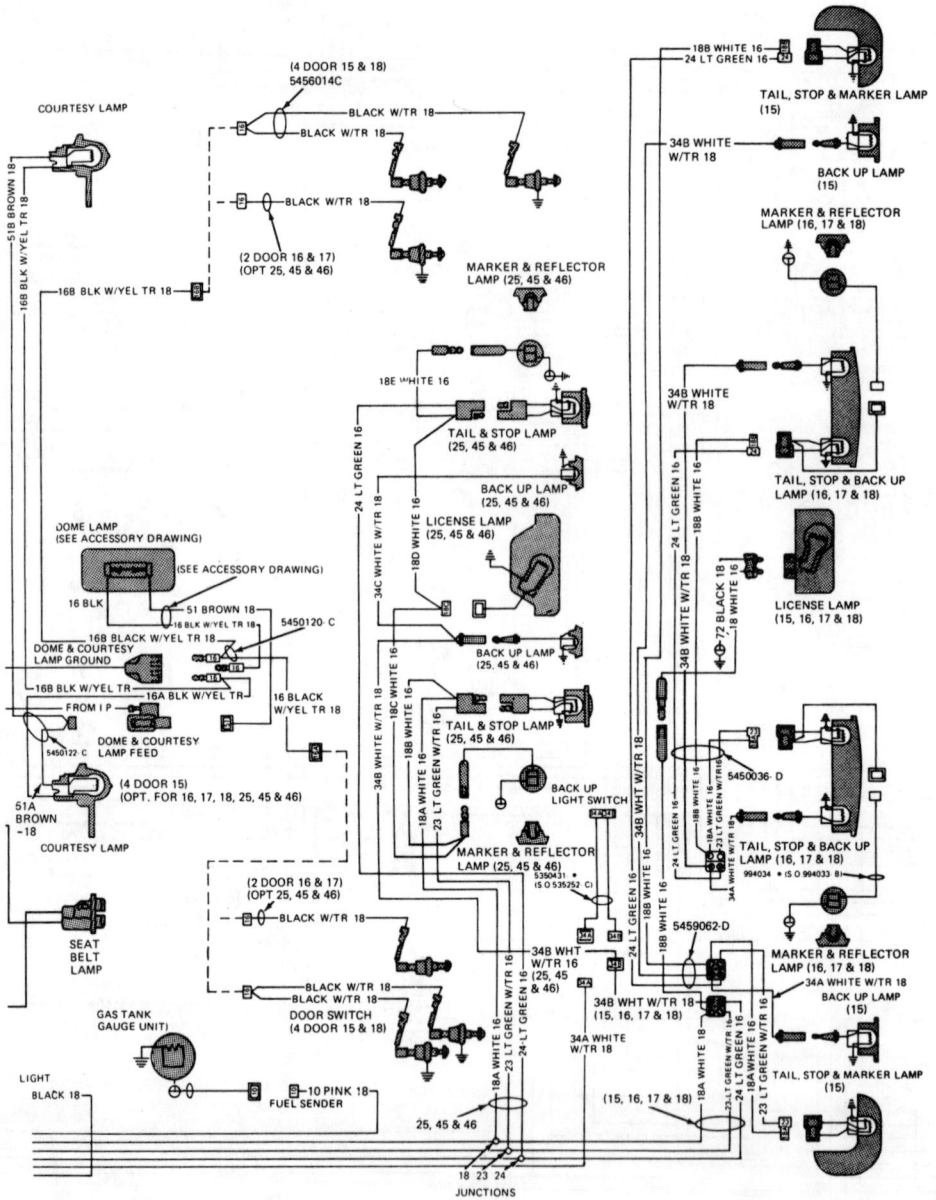

1975–79 Wagoneer and Cherokee

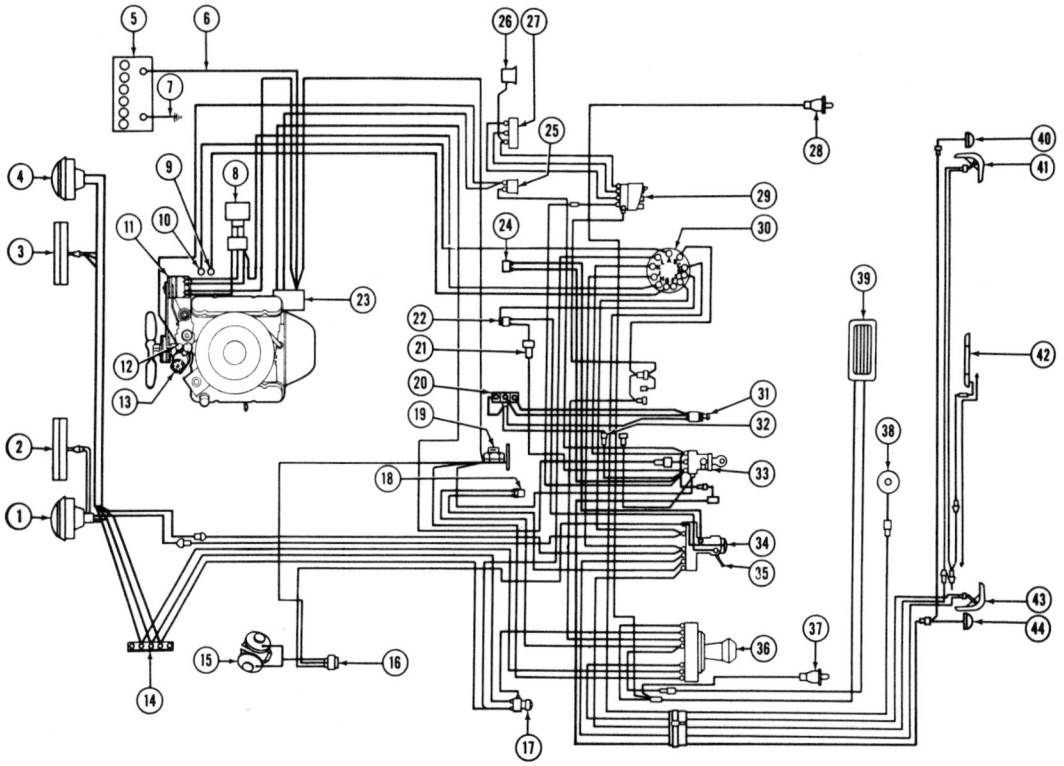

1. Left headlamp
2. Left parking and signal lamp
3. Right parking and signal lamp
4. Right headlamp
5. Battery
6. Positive cable
7. Negative cable
8. Voltage regulator
9. Oil Pressure sending unit
10. Temperature sending unit
11. Alternator
12. Ignition coil
13. Ignition distributor
14. Junction block
15. Horns
16. Horn relay
17. Foot dimmer switch
18. Stop light switch
19. Terminal block
20. Wiper motor
21. Fuse
22. Flasher (dir. sig.)
23. Starting motor
24. Hazard warning flasher
25. Ballast
26. Blower motor
27. Heater
28. Right courtesy and dome lamp switch

29. Heater switch
30. Instrument cluster
   A. Temperature gauge
   B. Cluster lights
   C. Fuel gauge
   D. Ground
   E. Left turn indicator
   F. Auxiliary
   G. Oil pressure indicator
   H. Charging indicator
   J. Right turn indicator
   K. Ignition feed
   L. Left turn indicator
31. Windshield wiper switch
32. Fuse
33. Ignition and starter switch
34. Horn button
35. Directional signal lever
36. Main light switch
37. Left courtesy and dome lamp switch
38. Fuel gauge tank unit
39. Dome lamp
40. Right back-up lamp
41. Right tail and stop lamp
42. License lamp (Wagoneer)
43. Left tail and stop lamp
44. Left back-up lamp

8-350 Wagoneer

Wagoneer and Cherokee Light Bulb Specifications

	Trade Number	
Light Description	*1966–70 Models*	*1971 and Later Models*
Headlights	6012	6014
Front Parking Light	1034 or 1157	1157NA
Stop, Tail and Directional Signal	1034 or 1157	1157
License Plate Light	67 or 1155	1155
INDICATOR LIGHTS		
Headlight Beam	158	158
Directional Signals	158	158
Charge Indicator	158	158
Oil Pressure	158	158
Instrument Cluster	158	158
Dome Lamp	212	212
Heater Control	1816	1815
Clock	1816	1816
Ignition Switch	1445	—
Shift Selector	1445	1816
4-Wheel Drive Indicator	—	158
Parking Brake Warning	57	158
Radio	1892	1892
Courtesy Light	90	89
Back-Up	1156	1156
Glove Box	1891	1891
4-Way Flasher Warning	53	552
Side Marker	—	194

Wagoneer and Cherokee Fuse Application Chart

Fuse Application	*Fuse*
Air Conditioner (1971 and Later)	25 Amp
Back-Up Lights (includes cigar lighter 1975–79)	9 Amp (1966–70) 15 Amp (1971–79)
Cigarette Lighter (1971–74)	20 Amp
Clock (1966–70)	2 Amp
Cluster Feed (1971 and Later)	3 Amp
Control Panel (1971 and Later)	20 Amp Circuit Breaker
Directional Signal	10 Amp
Electric Tailgate Window (1971 and Later)	30 Amp and 3 Amp Circuit Breaker
4-Way Flasher Warning	9 Amp (1966–70 350 V8) 14 Amp (1966–70 327 V8 and 232 Six) 15 Amp (1971 and Later)
Headlights (1971 and Later)	25 Amp Circuit Breaker
Heater	14 Amp (1966–70) 25 Amp Circuit Breaker (1971 and Later)
Horn (1971 and Later)	20 Amp Circuit Breaker
Parking Brake Warning and Brake Failure (1971 and Later)	3 Amp
Parking Brake (1966–70)	9 Amp
Radio	2 Amp (1966–70) 5 Amp (1971 and Later)
Windshield Wiper/Washer	14 Amp (1966–70) 10 Amp (1971 and Later)

Commando Light Bulb Specifications

Light Description	Trade Number	
	Pre-1971 Models	1971 and Later Models
Headlights	6012	6012
Parking and Directional Signals	1157NA	1157NA
License Plate Light	1155 or 67	1155
Side Marker	194	194
INDICATOR LIGHTS		
Headlight Beam	158	158
Directional Signals	158	158
Charge Indicator	158	158
Temperature	158	—
Oil Pressure	158	158
Clock	57	—
Radio	1892	1892
Courtesy	90	90
Back-Up	1156	1156
Console Light (Auto. Trans.)	1816	1445 (Column)
Control Panel	53	—
Instrument Panel	158	158
4-Way Flasher Warning	57	170

Commando Fuse Application Chart

Fuse Application	Fuse
Air Conditioner Heater (1971 and Later)	25 Amp
Back-Up Lights	25 Amp
Cigarette Lighter (1971 and Later)	20 Amp Circuit Breaker
Clock (1966–70)	1 Amp
Control Panel (1971 and Later)	20 Amp Circuit Breaker
Directional Signal	9 Amp (1966–70) 10 Amp (1971–73)
Hazard 4-Way Flasher	9 Amp (1966–70)
	20 Amp Circuit Breaker (1971 and Later)
Headlights (1971 and Later)	25 Amp Circuit Breaker
Horn (1971 and Later)	20 Amp Circuit Breaker
Radio	2 Amp (Pre-1971)
	5 Amp (1971 and Later)
Windshield Wiper/Washer	14 Amp

6

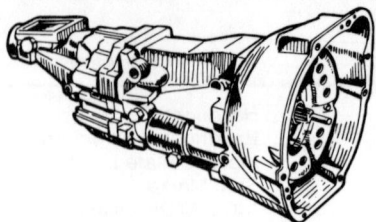

Clutch and
Transmission

MANUAL TRANSMISSION

REMOVAL AND INSTALLATION
Wagoneer Through 1970

NOTE: *The transfer case and the transmission can be removed as an assembly. However, it is recommended that the transfer case be removed separately for ease in handling the transmission.*

1. Drain the oil from the transmission and transfer case. Replace the drain plugs.

2. Disconnect the front and rear driveshafts from the transfer case.

3. Remove the transfer case.

4. Remove the transmission access cover from the floor and disconnect the shift rods from the transmission on the 3-speed transmission.

5. On the 4-speed transmission, remove the control housing assembly. Place a clean cloth over the hole to prevent dirt from entering.

6. Disconnect the speedometer cable.

7. Place a jack under the engine. Protect the engine oil pan with a block of wood.

8. Place a transmission jack under the transmission and remove the nuts holding the rear mounting to the frame crossmember.

9. Remove the bolts holding the trans-mission to the flywheel housing or adapter plate.

10. Slide the transmission assembly toward the rear of the vehicle until the main drive gear shaft (transmission input shaft) clears the flywheel housing.

11. Lower the transmission on the jack and remove the transmission from under the vehicle.

To install the manual transmission:

12. Install the manual transmission in the reverse order of removal, taking note of the following concerning the installation of the 4-speed transmission:

a. When installing the transmission adapter plate to the clutch housing, always check the alignment of the transmission adapter plate bore with the crankshaft. Attach a special tool and dial indicator gauge over the inside circumference of the clutch pressure plate and rotate the crankshaft slowly, noting the dial gauge reading. The maximum runout is 0.005 in. If the dial gauge reading is more than this, loosen and reposition the adapter plate as necessary until the correct alignment is obtained.

b. When installing the rear adapter plate, be sure that the capscrew heads do not protrude beyond the adapter plate face or interfere with the transfer case fitting tightly against the rear of the adapter plate.

Commandos Through 1970

1. Drain the lubricant from the transmission and transfer case. Replace the drain plugs.

2. Remove the right front seat, floor mat and floor board center section.

3. Disconnect the wires from the back-up light switch.

4. Remove the capscrews securing the cane shift lever housing, then remove the shift lever and shift housing assembly and its gasket from the transmission. Cover the transmission opening with a clean cloth to keep out dirt.

5. On vehicles equipped with the remote control console shift lever, disconnect the remote control rods from the transmission shift shaft arm levers.

6. Remove the cotter pins and clevis pins securing the transfer shift rods to the control rods of the transfer case.

7. Remove the capscrews securing the transfer shift mechanism tube to the support shaft and remove the transfer shift mechanism assembly.

8. Remove the skid plate.

9. Disconnect the speedometer cable from the transfer case connection.

10. Disconnect the parking brake linkage at the bellcrank arm located directly under the transfer case.

11. Disconnect the rear driveshaft from the transfer case drive flange, slide the propeller shaft yoke rearward and tie the shaft to one side.

12. Disconnect the front driveshaft at the axle.

13. On vehicles equipped with the V6 engine, remove the intermediate shaft support bearing cap after the transfer case is out of the vehicle.

14. Disconnect the clutch release cable from the clutch cross-shaft assembly.

15. Place a jack under the engine, protecting the oil pan with a block of wood.

16. Remove the capscrews and insulators securing the transmission stabilizer bracket to the frame. Note the sequence of removal to facilitate installation.

18. Remove the nuts and washer securing the frame crossmember to the transmission mount. Lower the frame crossmember and slide the crossmember with the parking brake mechanism attached, forward and out of the way.

19. Place a transmission jack under the transmission and transfer case.

20. Remove the capscrew and lockwasher securing the stabilizer bracket to the transfer case and remove the stabilizer bracket.

21. Remove the bolts and lockwashers securing the transmission case to the flywheel bellhousing adapter.

22. Carefully move the transmission and transfer case assembly slightly to the right and remove the clutch control crossshaft assembly from the frame and transfer case.

23. Slide the transmission and transfer case assembly rearward until the transmission main driveshaft is clear of the clutch and bellhousing.

24. Lower the transmission jack and remove the transmission case as an assembly from under the vehicle.

25. Remove the intermediate shaft support bearing cap from the transfer case.

26. Install the transmission in the reverse order of the removal procedure.

1971 and Later Commando, Wagoneer and Cherokee

1. Remove the floor lever knobs, trim rings and boots.

2. On 3-speed models, remove the floor covering. Remove the floor pan section from above the transmission. Remove the transmission shift control and lever assembly from the transmission.

3. On 4-speed models, remove the shift control housing cap, washer, spring, shift lever and pin.

4. Remove the transfer case shift lever and bracket assembly.

5. Raise the vehicle.

6. Mark the driveshafts and flanges for assembly alignment. Remove the front driveshaft. Disconnect the front end of the rear driveshaft from the transfer case.

7. Disconnect the clutch cable and remove the cable mounting bracket from the transfer case.

8. Disconnect the speedometer cable, back-up light switch wires, transmission control spark advance (TCS), if so equipped, and the parking brake cable if it is connected to the crossmember.

9. On 1979 Models, lower the exhaust system to clear the catalytic converter from the work area.

10. Support the transmission with a transmission jack.

11. Disconnect the support crossmember from the side sill.

12. Remove the bolts which attach the transmission to the clutch housing.

13. Lower the transmission slightly. Move the transmission, transfer case and crossmember backward far enough for the transmission clutch shaft to clear the clutch housing.

14. Remove the assembly from the vehicle.

15. Install the transmission in the reverse order of the removal procedure.

LINKAGE ADJUSTMENT

Wagoneer

1. If shifting action is not smooth and positive, first disconnect the transmission shift rods from the remote control levers at the lower end of the steering column.

2. Check and correct any binding of the remote control shaft in the steering column.

3. Reconnect the rods to the levers at the lower end of the steering column.

4. Shift the transmission into Neutral.

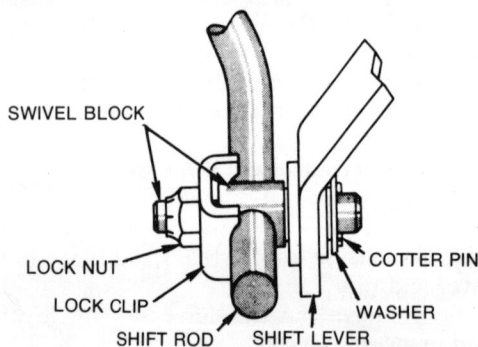

SWIVEL BLOCK

LOCK NUT

LOCK CLIP

SHIFT ROD   SHIFT LEVER

COTTER PIN

WASHER

Transmission linkage junction, Wagoneer and Cherokee 3-speed

5. Free the shift rods in the swivel blocks on the transmission shift levers by loosening the locknuts and lock clips. Insert a piece of snug-fitting $3/16$ in. rod through the remote control shift levers and housing at the lower end of the steering column. Make sure that the transmission gears are in Neutral. The gears are in Neutral when the two shift levers on the side of the transmission are in such a position that either of the levers can be moved freely while the other remains stationary.

6. With the $3/16$ in. rod in place in the remote control shift levers and housing at the lower end of the steering column, tighten the lock clips with the locknuts at the shift levers on the transmission.

7. Remove the $3/16$ in. rod.

Commando

1. Remove the button plug from the left side of the console.

2. Reach through the hole in the console, lift the shift tower rubber cover and remove the button plug in the shift tower.

1. Low and Reverse shift arm and link
2. Second and High shift arm and link
3. Transmission-to-bellhousing adapter
4. Transmission
5. Transfer case
6. Second and High shift rod
7. Low and Reverse shift rod
8. Speedometer cable and housing
9. Clutch cross-shaft

Commando shift linkage

3. Move the gearshift lever to Neutral.

4. Insert a $5/16$ in. diameter rod through the holes in the console and shift tower and through the aligning holes in the two shift levers.

5. Loosen the nut on each of the shift lever adjustments at the transmission. This will free the outer shift levers to move in the elongated holes on the inner shift levers. Make sure the transmission shift levers are in Neutral. Then tighten the nuts to 15–20 ft lbs torque.

6. Remove the $5/16$ in. diameter aligning rod. Operate the gearshift lever to be sure the transmission detents are engaging in their respective positions.

7. Install the button plugs.

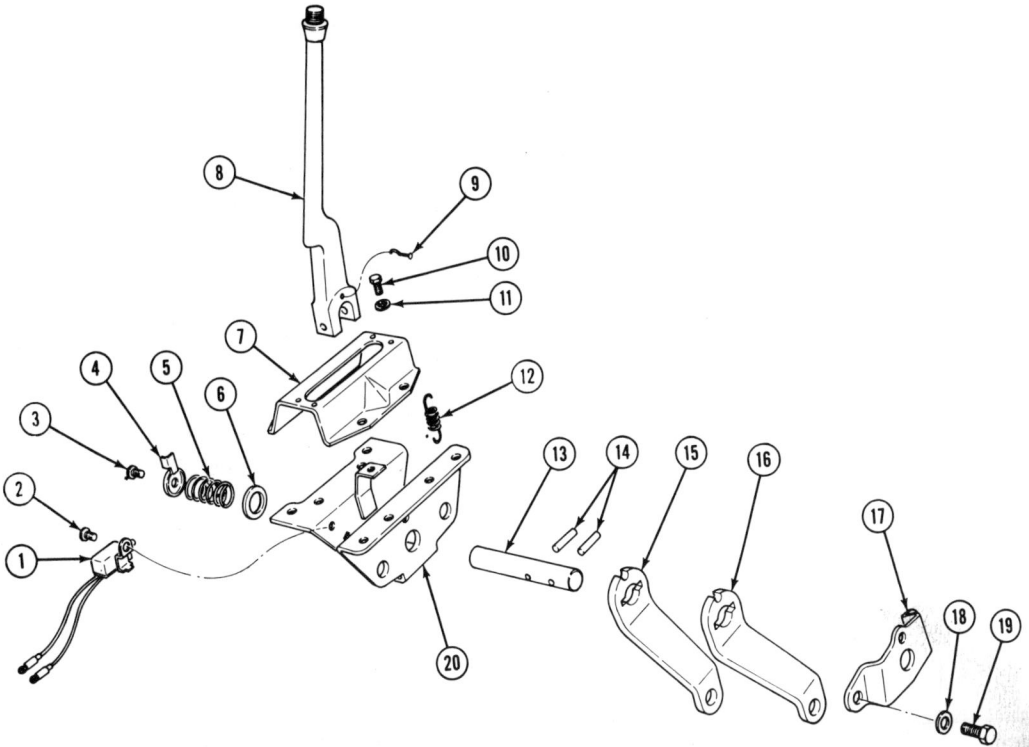

1. Back-up light switch
2. Screw and washer
3. Screw and lockwasher
4. Actuator
5. Shaft compression spring
6. Spring washer
7. Lever bracket
8. Shift lever
9. Spring pin
10. Screw
11. Lockwasher
12. Return spring
13. Lever shaft
14. Pin
15. 1st and Reverse arm
16. 2nd and 3rd arm
17. Bracket
18. Washer
19. Screw
20. Base bracket

Commando shift lever assembly. The ⁵/₁₆ in. diameter adjustment rod is inserted through the base of the bracket and through the notches in the top of the two shifter arms

CLUTCH

The clutch installed in 1966–70 Commandos is either a 3-pressure-springed, 3-fingered, 9¼ in. diameter Auburn clutch installed behind the F-Head engine with 1150 lbs of plate pressure or, a diaphragm type, 10.4 in. diameter G.M. clutch installed behind the V6 engine with 1600 lbs of plate pressure.

The clutch installed in the 1966–70 Wagoneers are as follows:

Behind the 232 Six, there is a Borg & Beck, 10½ in. diameter, 9-pressure-springed, 3-fingered clutch with 1350 lbs of plate pressure.

Behind the 327 V8, there is also a Borg & Beck 10½ in. diameter, 3-fingered clutch. This clutch, however, has 12 pressure

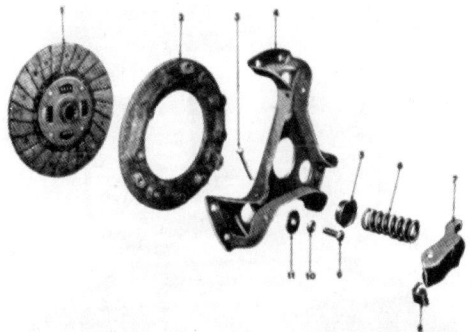

1. Clutch disc and hub
2. Pressure plate
3. Pivot pin
4. Bracket
5. Spring cup
6. Pressure spring
7. Release lever
8. Return spring
9. Adjusting screw
10. Jam nut
11. Washer

Auburn clutch assembly

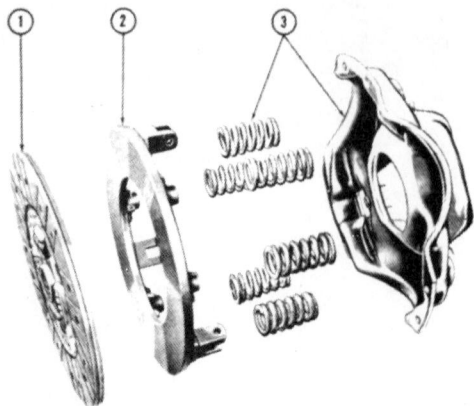

1. Clutch disc and hub
2. Pressure plate
3. Backing plate and pressure spring

Rockford clutch assembly

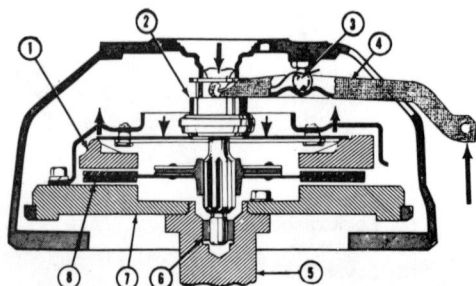

1. Pressure plate
2. Throwout bearing
3. Pivot point
4. Clutch fork
5. Engine crankshaft
6. Pilot bearing
7. Flywheel
8. Clutch disc

GM diaphragm clutch assembly

Borg & Beck clutch assembly

springs, providing a plate pressure of 1858 lbs.

The clutch installed behind the 350 V8 is a G.M. diaphragm type, 11 in. in diameter with 2450–2750 lbs of plate pressure.

All of the six cylinder engines and the 304 V8 in vehicles made in 1971 and after, have a 10½ in. diameter clutch installed behind them. The 360 and 401 V8 have an 11 in. clutch.

When the clutch pedal is depressed, linkage moves the clutch fork, which pivots on a ball stud and activates the throwout bearing. The throwout bearing then depresses the fingers, or in the case of the V6 and 350 V8, the prongs of the diaphragm spring. The fingers or prongs are mounted on pivot pins, which reverse the direction of force. Force is then applied directly to the retracting springs or fingers, thus lifting the pressure plate from the clutch disc and disengaging the clutch.

REMOVAL AND INSTALLATION

1966–70 Commando—F-Head

1. Remove the transmission and transfer case from the vehicle.

2. Remove the flywheel housing.

3. Mark the clutch pressure plate and engine flywheel with a center punch so that the clutch assembly can be reinstalled in the same position after adjustments or replacement are completed.

4. Remove the clutch pressure plate bracket bolts equally, a little at a time, to prevent distortion and to relieve the clutch springs evenly.

5. Remove the pressure plate assembly and clutch disc from the flywheel.

Before the clutch is reinstalled, it should be inspected for warpage. If grease or oil is evident on the disc friction facings, the facings should be replaced and the cause of the oil accumulation corrected.

6. Put a small amount of cup grease in the flywheel pilot bushing and install the clutch disc with the short end of the hub toward the flywheel. Next, place the pressure plate assembly in position.

7. With a clutch disc aligning arbor or a spare transmission mainshaft, align the clutch disc splines leaving the arbor in position while tightening the pressure plate screws evenly.

8. Assemble the flywheel housing to the engine and reinstall the transmission and transfer case. Make sure that the clutch re-

lease bearing carrier return spring is hooked in place.

9. Assemble the rest of the components in the reverse order of removal. Adjust the clutch control cable so that there is 1 in. of free pedal travel.

1966–70 Commando—V6

1. Remove the transmission.
2. Remove the clutch throwout bearing and pedal return spring from the clutch fork.
3. Remove the flywheel housing from the engine.
4. Disconnect the clutch fork from the ball stud by forcing it toward the center of the vehicle.
5. Mark the clutch cover and flywheel with a center punch so that the cover can later be installed in the same position on the flywheel. This is necessary to maintain engine balance.
6. Loosen the clutch attaching bolts alternately, one turn at a time, to avoid distorting the clutch cover flange, until the diaphragm spring is released.
7. Support the pressure plate and cover assembly while removing the last bolts.
8. Remove the pressure plate and clutch disc from the flywheel.

NOTE: *Use extreme care to keep the clutch disc clean.*

Before installing the clutch, lubricate the inner surface of the pilot bushing in the crankshaft; the splines of the transmission drive gear shaft, sliding the clutch disc over the shaft and wiping off excess lubricant; the front bearing retainer of the transmission, sliding the throwout bearing along the retainer several times and removing any excess grease; the groove in the throwout bearing collar; and the ball stud in the flywheel housing together with the seat in the clutch fork.

NOTE: *Use wheel bearing grease SPARINGLY to lubricate all of the above. If excessive grease is applied, operating heat will cause it to run out onto the clutch disc, ruining it.*

9. Install and reassemble the clutch in the reverse order of removal, taking note of the following:

a. Be certain that the mark on the clutch cover is aligned with the mark made on the flywheel during clutch removal.

b. Tighten the clutch attaching bolts alternately, so that the clutch is drawn into position squarely on the flywheel. Turn each bolt one turn at a time to avoid bend-

ing the clutch cover flange. Tighten the attaching bolts to 30–40 ft lbs.

c. Be certain the dowel pins on the flywheel housing are installed in the cylinder block when the housing is installed.

1966–70 Wagoneer—350 V8

Remove and install the clutch in the same manner as outlined under 1966–70 "Commando—V6."

1966–70 Wagoneer—232 Six and 327 V8

Remove and install the clutch in the same manner as outlined under "1966–70 Commando—F-Head."

1971 and Later Commando, Wagoneer and Cherokee

1. Remove the transmission.
2. Remove the starter, throwout bearing and sleeve assembly and clutch housing.
3. Mark the clutch cover, pressure plate and flywheel with a center punch to insure correct alignment during assembly.
4. Loosen the attaching screws in sequence, one or two turns at a time until the spring tension on the cover is released. This is to prevent the clutch cover from becoming warped, which could result in clutch chatter when reinstalled.
5. Install the clutch in the reverse order of removal, tighten the attaching bolts of the cover in sequence, one or two turns at a time to 40 ft lbs.

NOTE: *The clutch pedal is not to be depressed until the transmission has been installed.*

PEDAL HEIGHT, FREE-PLAY AND LINKAGE ADJUSTMENT

1966–70 Commando

There are two types of clutch linkage used on 1966–70 Commandos: the crossshaft tube and lever type and the clutch control cable type. Both use a cable to actuate the release and engagement of the clutch.

To adjust the cross-shaft lever and tube type linkage:

1. With the clutch cross-shaft tube and lever assembly positioned at a 52° angle for the F-Head engine and 58° angle for the V6 engine, adjust the clutch control cable assembly from the clutch pedal to the cross-shaft lever, making sure that the clutch pedal is against the pedal support bracket stop.

2. With the cross-shaft assembly posi-

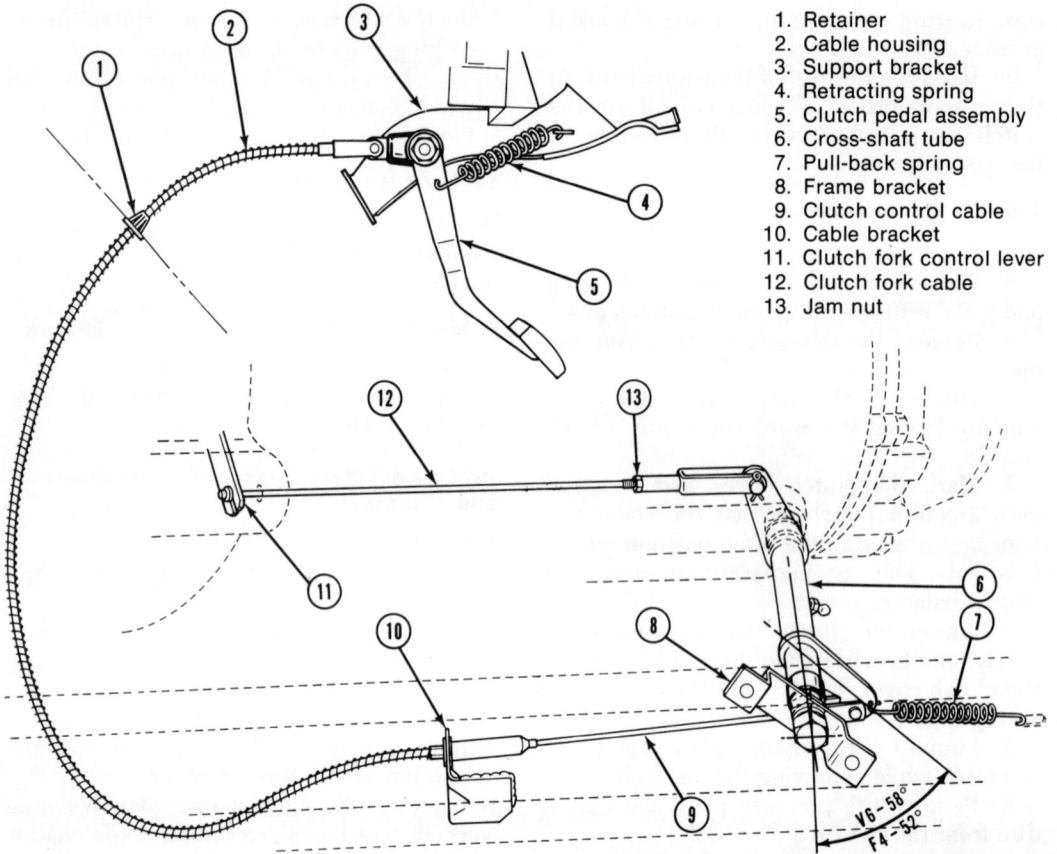

1. Retainer
2. Cable housing
3. Support bracket
4. Retracting spring
5. Clutch pedal assembly
6. Cross-shaft tube
7. Pull-back spring
8. Frame bracket
9. Clutch control cable
10. Cable bracket
11. Clutch fork control lever
12. Clutch fork cable
13. Jam nut

Clutch linkage adjustment: 1966–70 Commando with cross-shaft tube and lever type linkage

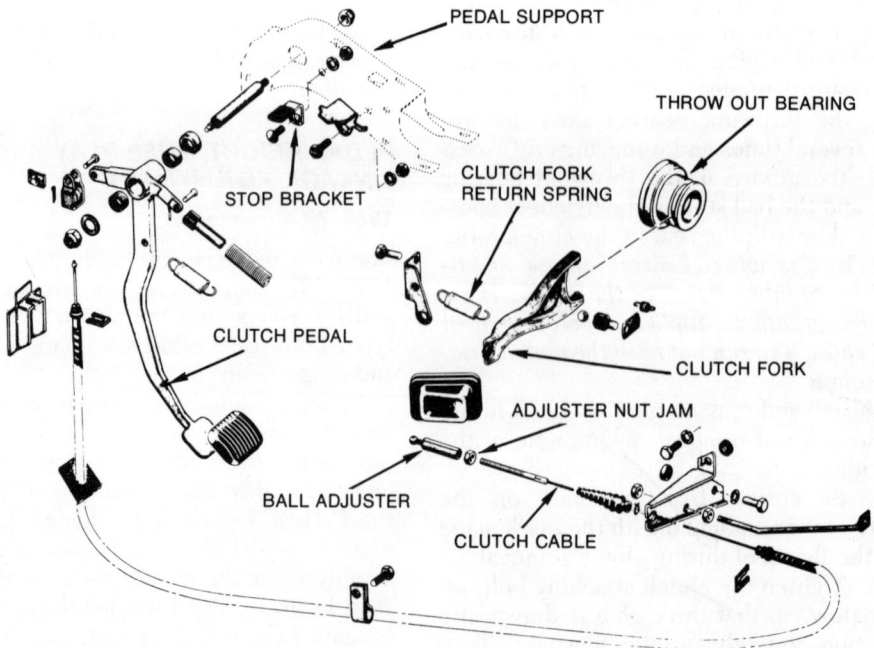

PEDAL SUPPORT

THROW OUT BEARING

CLUTCH FORK
RETURN SPRING

STOP BRACKET

CLUTCH FORK

CLUTCH PEDAL

ADJUSTER NUT JAM

BALL ADJUSTER

CLUTCH CABLE

Clutch linkage: 1971–72 Commando

tioned as above, push the clutch housing release lever rearward, until the clutch throwout bearing contacts the clutch pressure plate release levers or diaphragm plate.

3. Adjust the clutch fork cable from the release lever to the clutch cross-shaft lever, until the cable clevis just enters the lever hold.

4. Remove the clevis pin and lengthen the cable 2½ turns of the clevis for the V6 engine and 3 turns for the F-Head engine, to obtain about ¾ in. of free-play.

5. Release the clutch lever and insert the cable clevis pin and lock the clevis jam nut.

To adjust the clutch control cable type linkage:

1. With the clutch pedal against the pedal support bracket stop, adjust the ball adjusting nut until the slack is removed from the cable and the clutch throwout bearing con-

tacts the clutch pressure plate, release levers or diaphragm plate.

2. Loosen the ball adjusting nut 2½ turns to obtain about ¼ in. of free-play.

3. Lock the hex nut.

1966–70 Wagoneer

1. Disconnect the adjustable rod from the clutch pedal.

2. Adjust the clutch pedal stop bolt for a positive overcenter position. This is done by turning the bolt in or out of the crossmember to obtain ¼–½ in. of clutch pedal travel before the overcenter spring assists the pedal to the floorboard. The pedal may be ⅜–⅝ in. higher than the pad on the brake pedal. This is a normal condition.

3. With the clutch pedal resting against the stop bolt, adjust the length of the adjustable rod by turning the ball joint in or out on

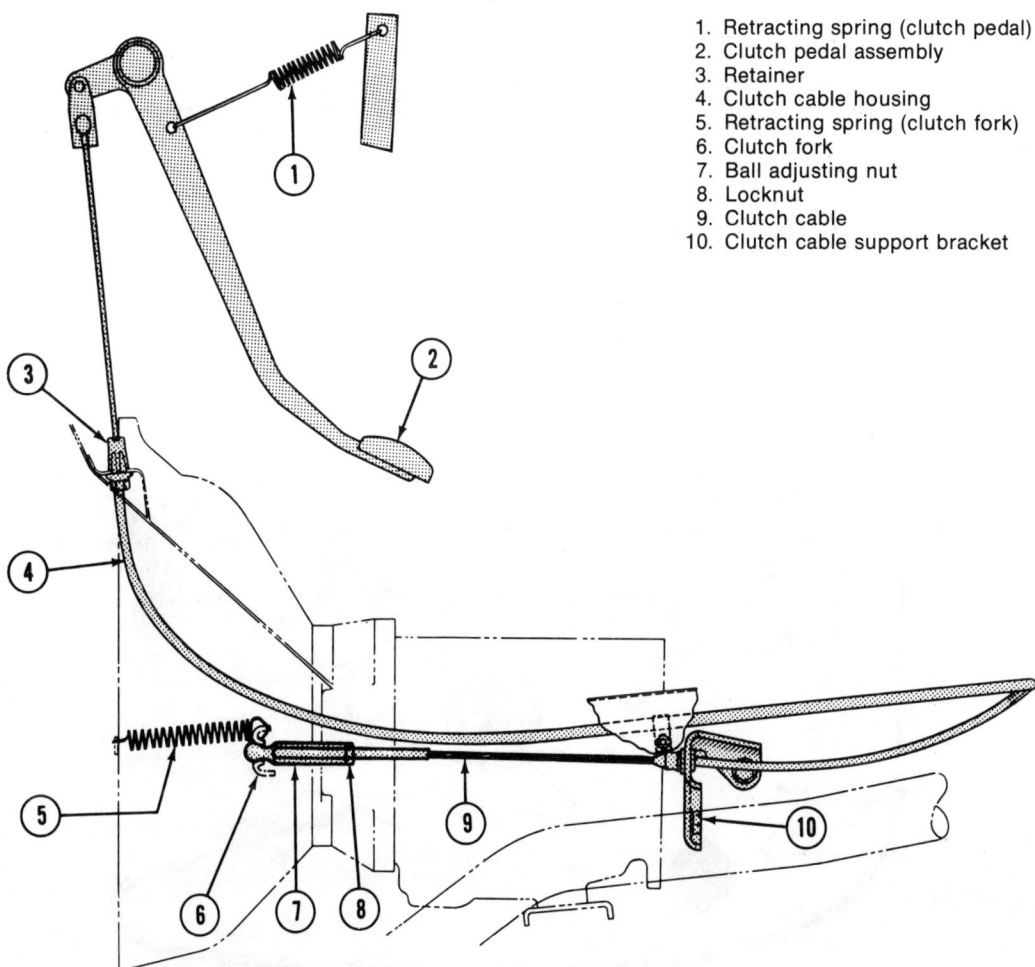

1. Retracting spring (clutch pedal)
2. Clutch pedal assembly
3. Retainer
4. Clutch cable housing
5. Retracting spring (clutch fork)
6. Clutch fork
7. Ball adjusting nut
8. Locknut
9. Clutch cable
10. Clutch cable support bracket

Clutch linkage adjustment: 1966–70 Commando with the control cable type linkage

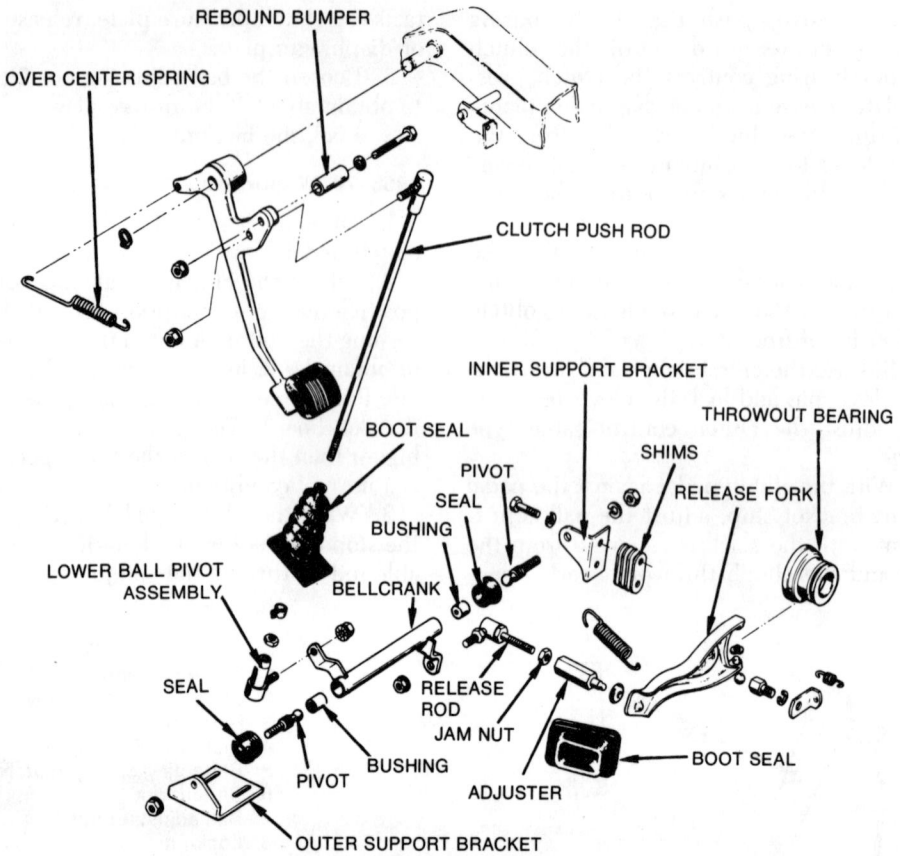

REBOUND BUMPER

OVER CENTER SPRING

CLUTCH PUSH ROD

INNER SUPPORT BRACKET

THROWOUT BEARING

BOOT SEAL

SHIMS

RELEASE FORK

PIVOT

SEAL

BUSHING

LOWER BALL PIVOT ASSEMBLY

BELLCRANK

SEAL

RELEASE ROD

JAM NUT

BOOT SEAL

PIVOT

BUSHING

ADJUSTER

OUTER SUPPORT BRACKET

Clutch linkage: 1973 Commando

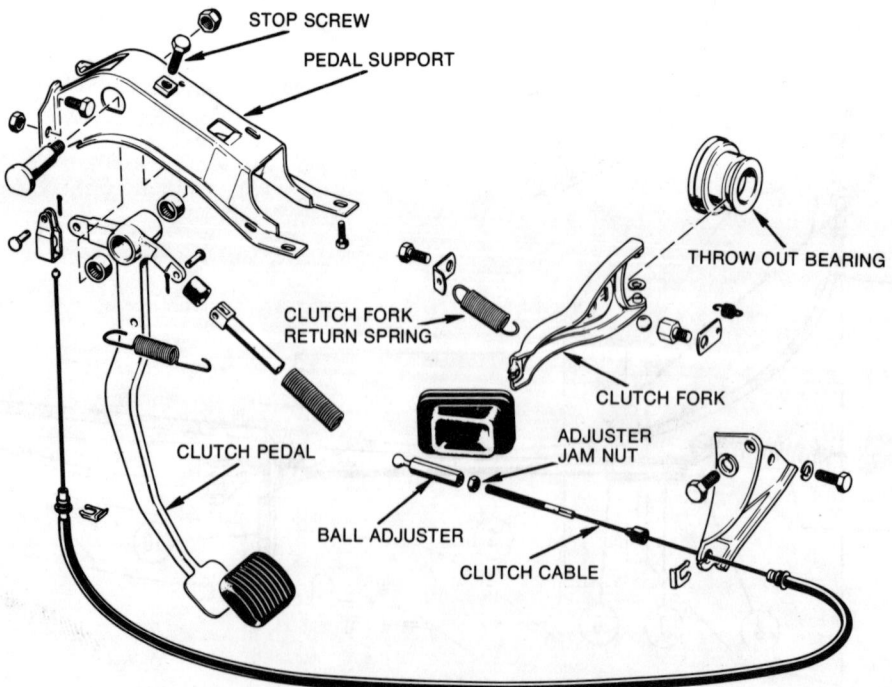

STOP SCREW

PEDAL SUPPORT

THROW OUT BEARING

CLUTCH FORK RETURN SPRING

CLUTCH FORK

CLUTCH PEDAL

ADJUSTER JAM NUT

BALL ADJUSTER

CLUTCH CABLE

Clutch pedal height adjustment: 1971–72 models

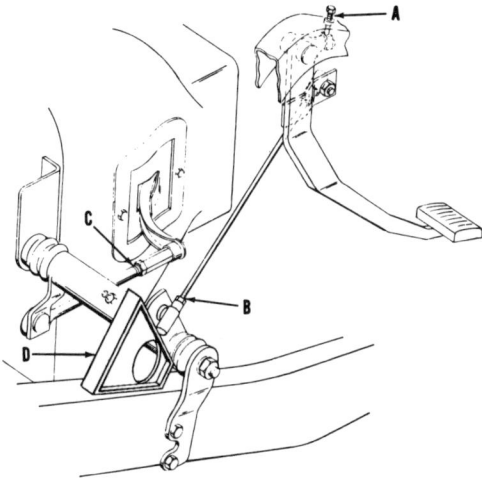

A. Pedal height adjustment point
B. Cross-shaft arm position adjustment
C. Clutch pedal free-play adjustment
D. Tool W-341 or W-317

Clutch linkage adjustment: 1966–70 Wagoneer

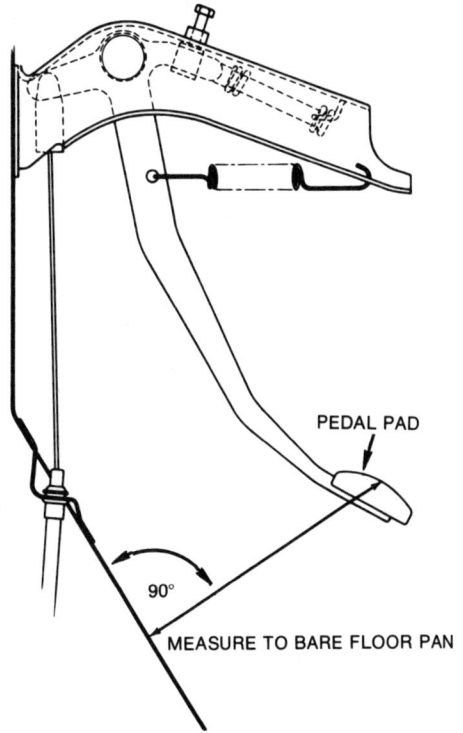

PEDAL PAD

90°

MEASURE TO BARE FLOOR PAN

Clutch linkage: 1971–72 Wagoneer

the rod, so that when the adjustable rod is connected, the arm of the clutch pedal cross-shaft is at a 30° angle with a 350 V8 and 49° with a 232 Six and 327 V8, in relation to the top of the left frame side rail.

4. To adjust clutch pedal free-play, adjust the cross-shaft-to-throwout lever link, so that the clutch pedal can be depressed 1 in. before the clutch starts to disengage.

NOTE: *There is a special tool available for setting the angle of the arm of the clutch pedal cross-shaft as outlined in Step 3 of the above procedure. Tool W-341 is for the vehicle equipped with a 350 V8 and gives the 30° angle and tool W-317 is for the vehicle equipped with a 232 Six or 327 V8 and gives the 49° angle.*

1971 and 1972 Commando, Wagoneer and Cherokee

1. The clutch pedal has an adjustable stop located on the pedal support bracket, directly behind the instrument cluster. Adjust the stop to provide 8⅜ in. on the Wagoneer and 8 in. on the Commando, between the top of the pedal pad and the closest point, 90° on the bare floor pan.

2. Lift the clutch pedal up against the pedal support bracket stop.

3. Unhook the clutch fork retrun spring.

4. Loosen the ball adjusting nut until some cable slack exists.

5. Adjust the ball adjusting nut until the slack is removed from the cable and the clutch throwout bearing contacts the pressure fingers.

6. Back off the ball adjusting nut ¾ of turn to provide free-play.

7. Tighten the jam nut.

8. Hook the clutch fork return spring.

1973 and 1974 Commando, Wagoneer and Cherokee

1. Adjust the bellcrank outer support bracket to provide approximately ⅛ in. bellcrank end-play.

2. Lift the clutch pedal up against the pedal stop.

3. On the clutch pushrod (pedal-to-bellcrank) adjust the lower ball pivot assembly onto or off the rod as required to position the bellcrank inner lever parallel to the front face of the clutch housing (slightly forward from vertical).

4. Adjust the clutch fork release rod (bellcrank-to-release fork) to obtain the maximum specified clutch pedal freeplay. For the Commando, the free-play should be ¾–½ in.; for the Wagoneer and Cherokee, the free-play should be ⅝–⅜ in.

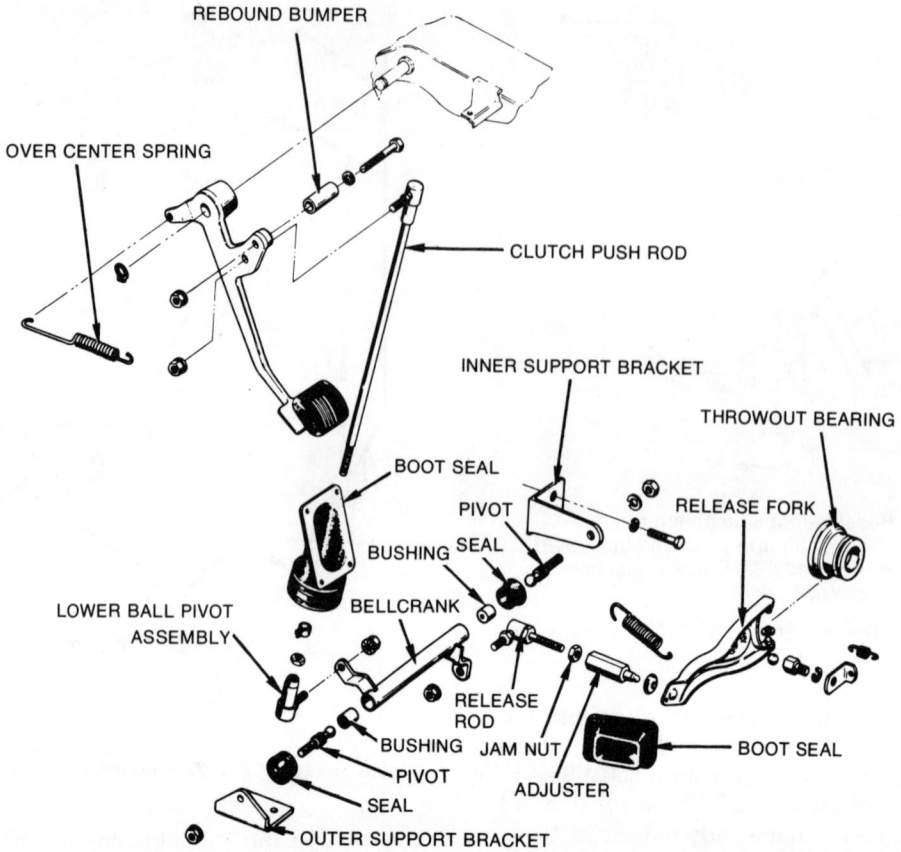

Clutch linkage: 1973 and later Wagoneer and Cherokee

1975–79 Models

1. Lift the pedal up against the pedal stop.
2. Adjust the clutch pushrod lower ball pivot assembly in or out to position the bellcrank inner lever parallel to the front face of the clutch housing. Proper positioning should be slightly forward from vertical.
3. Loosen the jamnut and turn the throwout fork adjuster in or out to obtain ⅓ to ⅝ inch freeplay. Tighten the jamnut.

AUTOMATIC TRANSMISSION

The automatic transmission used in all Jeep vehicles is the Turbo Hydra-Matic 400, produced by the General Motors Corporation.

PAN REMOVAL

Since the Turbo Hydra-Matic transmission doesn't have a drain plug, the fluid is drained by loosening the pan and allowing the fluid to run out over the top of the pan.

To avoid making a really big mess, place a drain pan under one corner of the transmission pan and remove the two attaching screws nearest to either side of that particular corner. One by one, and in a progressive manner, loosen all of the other attaching screws holding the transmission pan, leaving the ones farthest away from the "drain" corner tighter than the rest. When the majority of the fluid has drained, hold the pan up with one hand, remove the remaining at-

Removing or installing the transmission pan

taching screws and carefully lower the pan. There will be some automatic transmission fluid left in the pan, so be careful not to spill any.

Clean the pan in a suitable solvent and wipe it dry with a clean, lint-free cloth.

Install the pan in the reverse order of removal using a new pan-to-transmission gasket and torqueing the bolts in an alternating pattern to 10–13 ft lbs.

FILTER SERVICE

The filter is located directly under the oil pan.

There are filter replacement kits available for changing the transmission fluid filter. The kit includes a new filter, pan gasket and, in most cases, a new rubber O-ring to seal the intake pipe. If a new O-ring is not provided, leave the old one in place. If you can see that the old O-ring is cracked or damaged in any way, it is necessary to replace it with a new one, which can be obtained at a Jeep or G.M. dealer.

1. Remove the bottom pan and gasket.

2. Remove the oil filter retainer bolt and remove the oil filter assembly from the transmission.

3. Remove the intake pipe from the filter and the intake pipe-to-case O-ring if it is to be replaced.

4. Coat the new rubber O-ring with transmission fluid and position it in the groove at the inlet opening.

5. Slide the inlet pipe onto the new filter and position the filter on the transmission, guiding the inlet pipe in place.

6. Install the filter retaining bolt and tighten securely.

7. Install the bottom pan.

Removing or installing the transmission filter

TRANSMISSION SHIFT LINKAGE ADJUSTMENT
1966–73

1. Remove the cotter pin, flat washer and spring washer from the adjusting block at the transmission end of the shift rod and remove the block from the shift lever.

2. Make sure that the transmission shift lever is in the Neutral detent position.

3. Place the selector lever in the Neutral position and hold it firmly forward against the stop.

4. Loosen the locknuts at either end of the adjusting block and position the block on the shift rod, so that it may be freely inserted on the transmission shift lever without moving the lever. Tighten the nuts to 6–12 ft lbs torque.

5. Operate the selector lever to be sure that the transmission detents are engaging in their respective positions.

1974–79

1. Place the shift lever in Neutral.

2. Raise and support the vehicle.

3. Loosen the locknut on the gearshift rod trunnion just enough to permit movement of the rod in the trunnion.

4. Place the transmission outer range selector lever into neutral and tighten the trunnion locknut to 9 ft lb torque.

5. Lower the vehicle and check lever operation in all ranges.

TRANSFER CASE

All 4WD Commandos and Wagonners came equipped with a Spicer Model 20 two-speed transfer case until 1973, when the Warner Gear Quadra-Trac transfer case became available as an option on the Wagoneer. On 1974 models, Quadra-Trac became standard equipment on the Wagoneer and an option on the Cherokee with 2-bbl 360 V8 and standard on all other V8 engines. Quadra-Trac was not available in the Cherokee equipped with the 258 six until 1976. No Commando was ever equipped with Quadra-Trac.

REMOVAL AND INSTALLATION
Spicer Model 20

1. Remove the transfer case shift lever knob and trim boot.

2. Remove the transfer case shift lever.

3. Drain the transfer case lubricant.

NOTE: *It is not necessary to drain the automatic transmission when removing the transfer case.*

4. Disconnect the torque reaction bracket from the crossmember, if so equipped.

5. Disconnect the front half of the parking brake from the equalizer clevis.

6. Disconnect the front driveshaft from the transfer case.

7. Disconnect the rear driveshaft at the transfer case. Slide the driveshaft yoke back and move the shaft to one side.

8. Disconnect the speedometer cable.

9. On 1966–70 models, remove the center cotter pin of the shaft linkage at the transfer case to disconnect the shift linkage.

10. On 1966–70 models, remove the exhaust pipe bracket bolts.

11. Remove the bolts attaching the transfer case to the transmission and remove the transfer case.

12. Remove the gasket which seals between the transmission and the transfer case.

13. Install the transfer case in the reverse order of removal.

Quadra-Trac

NOTE: *Complete assembly removal is normally not required except when the front output shaft, front annular bearing, transmission output shaft seals or the transfer case (front housing) require service. To service the chain, drive sprocket, differential unit, diaphragm control system, needle bearing, thrust washer or rear output shaft, the rear half of the Quadra-Trac transfer case can be removed, giving access to these components without removing the unit from the vehicle.*

1. Lift and support the vehicle.

2. Mark the front and rear output shaft yokes and universal joints to provide alignment references to be used during assembly.

3. Disconnect the front driveshaft rear universal joint from the transfer case front yoke.

4. Disconnect the rear driveshaft front universal joint from the transfer case rear yoke.

5. Remove the bolts that attach the exhaust pipe support bracket to the transfer case.

6. Mark the diaphragm control vacuum hoses, lock-out indicator switch wire and speedometer cable.

7. Disconnect the parking brake cable guide from the pivot on the right frame side.

8. Remove the two transfer case-to-transmission bolts which enter from the front side and the two that enter from the rear side.

9. Move the transfer case assembly backward until the unit is free of the transmission output shaft and lower the assembly from the vehicle.

10. Remove all gasket material from the rear of the transmission and install the Quadra-Trac unit in the reverse order of removal. Tighten the attaching bolt to 15–25 ft lbs.

TRANSFER CASE LINKAGE ADJUSTMENT

Spicer Model 20

The shifter rails of the transfer case lever assembly connect to the shifter rails of the transfer case either directly or through non-adjustable links. The linkage should be lubricated periodically.

Warner Quadra-Trac

NOTE: *This procedure only applies to early production Quadra-Trac transfer case equipped with a reduction unit. Later production units are equipped with a lever controlled unit for which there is no adjustment.*

There are two features which can be operated manually concerning the transfer case: the "Lock-Out" feature and the engagement of the optional "Low Range Reduction Unit."

Since the "Lock-Out" feature is a vacuum actuated unit, there are no external adjustments that can be made other than making sure that all vacuum lines are in place, connected and not damaged in any way.

The reduction on early models unit is actuated by a shift cable and can be adjusted in the following manner:

1. Loosen the nut which clamps the cable to the shift lever pivot. Be sure that the cable can move freely in the pivot.

2. Move the reduction shift lever to the most rearward detent position (Hi-Range position).

3. Push the Low Range lever inward until it stops. Pull the Low Range lever out slightly, no more than $1/16$ in.

4. Tighten the cable clamp nut at the reduction unit shift lever.

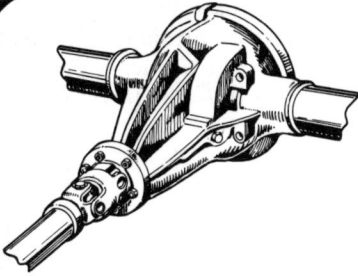

Drive Train

DRIVELINE

Power from the transfer case to the front and
rear axles is accomplished by means of tubu-
lar driveshafts. Each shaft is equipped with a
universal joint at each end to allow the driveshaft angles to change while the vehicle is in
motion.

Each shaft is equipped with a splined slip-
joint at one end to allow for variations in
length caused by the spring action of the
vehicle's suspension. The yokes at the front
and rear of each shaft must be aligned in the
same horizontal plane. This is necessary to

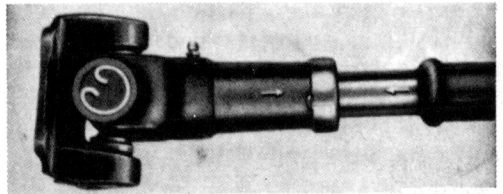

Driveshaft alignment markings

avoid vibration. Conventional Cardan cross
universal joints have needle bearings surrounding the 4 center cross bearing journals.
The center cross bearing journal assembly is
a simple design; correct assembly is just a

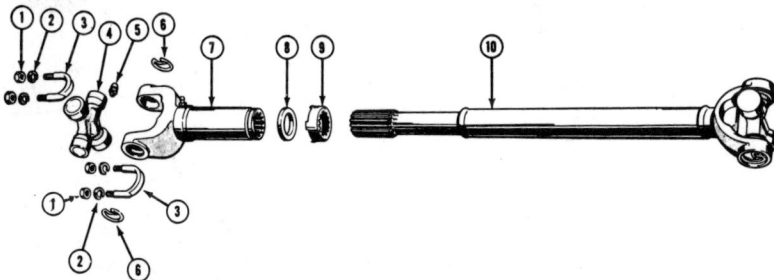

1. U-bolt nut
2. U-bolt washer
3. U-bolt
4. Universal joint journal

5. Lubrication fitting
6. Snap-ring
7. Universal joint sleeve yoke

8. Rubber washer
9. Dust cap
10. Driveshaft tube

Typical 1966–70 driveshaft

matter of arranging the needles around the inside of the bearing cups and placing the cups on the center cross journals.

The driveshafts on the Wagoneer/Cherokee 4WD vehicles equipped with automatic transmissions are installed in production with the slip-joint end at the axle rather than at the transfer case. Whenever the front driveshaft is removed, it must be installed with the slip-joint end at the axle.

Because of the variety of engine and transmission combinations, numerous driveshafts are used. Make sure that the correct driveshaft is installed, if one is being replaced.

Both driveshafts and universal joints should be checked regularly for foreign matter around the shafts, dented or bent shafts and loose attaching bolts. Should the vehicle be undercoated, care must be taken not to allow any undercoating to remain on the driveshafts, as this can cause vibration.

Front and Rear Driveshaft
REMOVAL AND INSTALLATION

In order to remove the front and rear driveshafts, unscrew the attaching nuts from the universal joint's U-bolts, remove the U-bolts and slide the shaft forward or backward toward the slip-joint. The shaft can then be removed from the end yokes and removed from under the vehicle. Install the driveshaft in the reverse order.

NOTE: *Some driveshafts are marked at the slip-joints with arrows on the spline and sleeve yoke. When installing the driveshaft, align the arrows to have the yokes at the front and rear of the shaft in the same parallel plane.*

U-Joints

There are two types of universal joints used on Wagoneers, Commandos and Cherokees: the Cardan cross-type with U-bolts and snap-rings, Cardan cross-type with just snap-rings, and a double Cardan cross-type which uses only snap-rings to hold it together; and the Ball and Trunnion-type universal joint which serves as a combination slip-joint and universal joint.

Universal joints fail for numerous reasons, the primary one being lack of lubrication. Others could be structural damage incurred

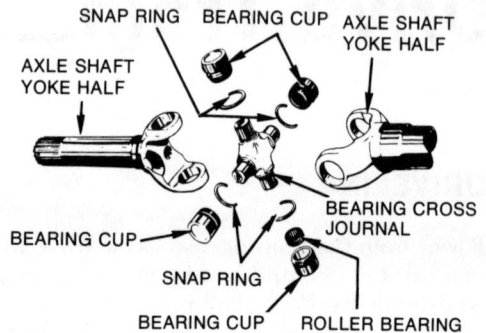

1971 and later Cardan cross-type U-joint

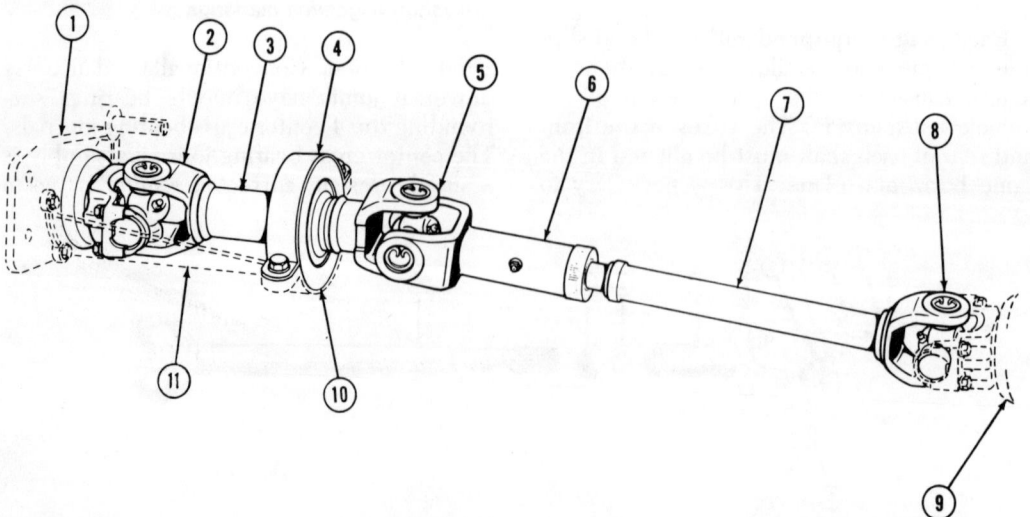

1. Transfer case
2. Rear universal joint
3. Intermediate shaft
4. Bearing support clamp
5. Intermediate universal joint
6. Slip-joint
7. Front shaft
8. Front universal joint
9. Front axle housing
10. Intermediate shaft bearing
11. Bearing support bracket

Front driveshaft assembly: Commando 6-225

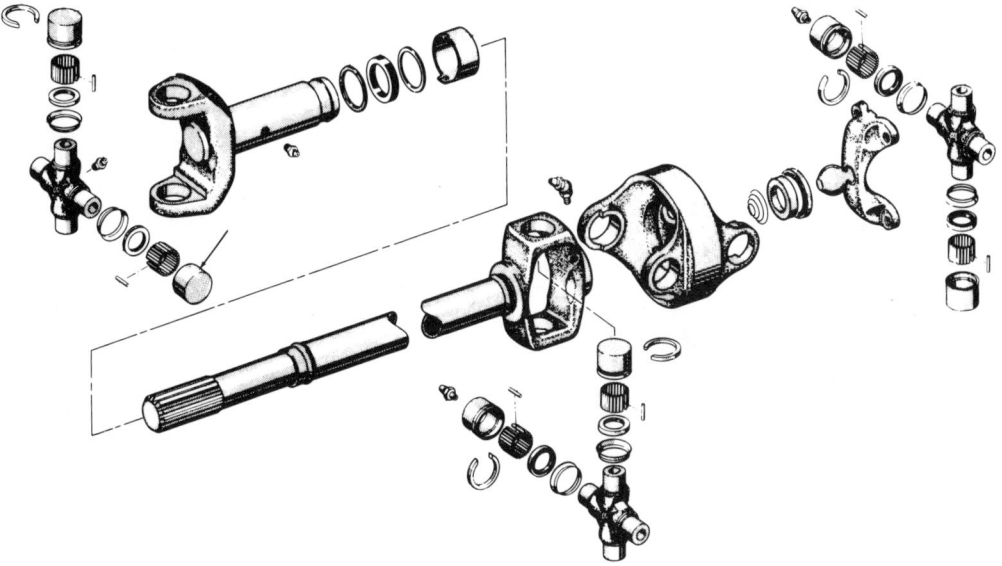

Cardan double, cross-type U-joint

from hitting something, an unbalanced drive-shaft, the entrance of dirt or water due to a rotted rubber seal or just plain wear from excessive mileage.

Rebuilding kits are available for both types of universal joints. The kit for a Cardan cross-type universal joint includes the entire cross assembly, the cross bearing journal, roller bearings, bearing cap with new rubber seal and new snap-rings.

The rebuilding kit for the Ball and Trunnion-type universal joint includes a new grease cover and gasket, universal joint body, two centering buttons and spring washers, two ball and roller bearing assemblies, two thrust washers, dust cover and two dust cover clamps.

OVERHAUL

Cardan Cross-Type—Snap-Rings

1. Remove the driveshaft from the vehicle.

2. Remove the snap-rings by pinching the ends together with a pair of pliers. If the rings do not readily snap out of the groove tap the end of the bearing lightly to relieve pressure against the rings.

3. After removing the snap-rings, press on the end of one bearing until the opposite bearing is pushed from the yoke arm. Turn the joint over and press the first bearing back out of that arm by pressing on the exposed end of the journal shaft. To drive it out, use a soft ground drift with a flat face, about $1/32$ in.

smaller in diameter than the hole in the yoke; otherwise there is danger of damaging the bearing.

4. Repeat the procedure for the other two bearings, then lift out the journal assembly by sliding it to one side.

5. Wash all parts in cleaning solvent and inspect the parts after cleaning. Replace the journal assembly if it is worn extensively. Make sure that the grease channel in each journal trunnion is open.

6. Pack all of the bearing caps $1/3$ full of grease and install the rollers (bearings).

7. Press one of the cap/bearing assemblies into one of the yoke arms just far enough so that the cap will remain in position.

8. Place the journal in position in the installed cap, with a cap/bearing assembly placed on the opposite end.

9. Position the free cap so that when it is driven from the opposite end it will be inserted into the opening of the yoke. Repeat this operation for the other two bearings.

10. Install the retaining clips. If the U-joint binds when it is assembled, tap the arms of the yoke slightly to relieve any pressure on the bearings at the end of the journal.

Cardan Cross-Type—U-Bolts

1. Removal of the attaching U-bolt releases one set of bearing races. Slide the driveshaft into the yoke flange to remove that set of bearing races being careful not to lose the rollers (bearings).

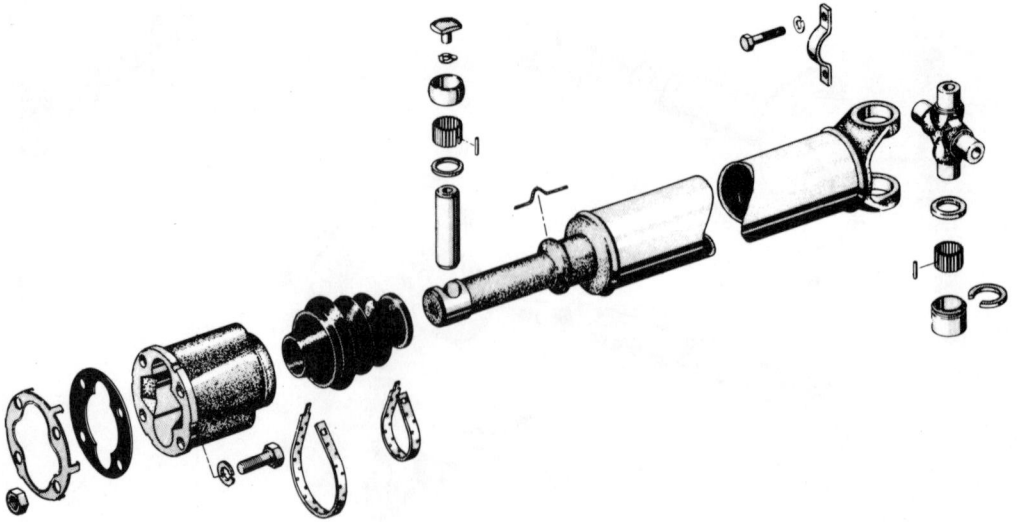

Ball and trunnion-type U-joint

2. After removal of the first set of bearings, release the other set by pinching the ends of the snap-rings with pliers and removing them from the sleeve yoke. Should the rings fail to snap readily from the groove, tap the end of the bearing lightly, to relieve the pressure against them.

3. Press on the end of one bearing, until the opposite bearing is pushed out of the yoke arm.

4. Turn the universal joint over and press the first bearing out by pressing on the exposed end of the journal assembly. Use a soft round drift with a flat face about $1/32$ in. smaller in diameter than the hole in the yoke arm. Then drive out the bearing.

5. Lift the journal out by sliding it to one side.

6. Install in the reverse order of removal, using the procedures for the snap-ring U-joints from Step 4 on as a guide.

Ball and Trunnion-Type

1. Remove the driveshaft from the vehicle.

2. Position the tube of the driveshaft near the ball-type universal joint, in a bench vise; clamp lightly.

3. Bend the lugs of the grease cover away from the universal joint body and remove the cover and the gasket.

4. Remove the two clamps from the dust cover. Push the joint body toward the driveshaft tube. Remove the two centering buttons and spring washers, the two ball and roller bearings and the two thrust washers from the trunnion pin.

5. Press the trunnion pin from the ball head with an arbor press.

NOTE: *It is strongly recommended that the trunnion pin be pressed out and in by an arbor press because of the possible damage that could result from other methods such as hammer and drift. This is a very critical component in the assembly and any damage could result in a premature failure or even necessitating replacement of the entire driveshaft.*

6. Clean the ball head of the driveshaft with a suitable solvent and dry thoroughly.

To assemble the ball and trunnion-type universal joint:

7. Secure the larger end of the dust cover to the universal joint body with the larger of the two clamps. Install the smaller clamp at the smaller end of the dust cover, then fit the cover over the ball head shaft.

8. Push the universal joint cover toward the driveshaft tube.

9. With an arbor press, press the trunnion pin into a centered position in the ball head.

NOTE: *The trunnion pin must project an equal distance from each side of the ball head. If it is not centered within 0.006 in., driveshaft vibration may result.*

10. Hold the universal joint body toward the tube of the driveshaft to gain access to the trunnion pin. Install one thrust washer, one ball and roller bearing and one spring washer

at one side of the ball head. Compress the centering buttons into the trunnion pin, then move the joint body away from the driveshaft tube, into position to surround the buttons and to hold them in place.

11. Insert the breather between the dust cover and ball head shaft, along the length of the shaft. The breather must extend no more than ½ in. beyond the dust cover, along the shaft. Tighten the clamp screw to secure the cover to the shaft. Cut away any portion of the dust cover which protrudes from beneath either clamp.

12. Pack the raceways of the universal joint body (inner surfaces which surround the ball and roller bearings) with about 2 oz. of universal joint grease. Divide the grease equally between the raceways. Position the gasket and grease cover on the body of the universal joint and bend the lugs of the cover in place. Move the body inward and outward, toward and away from the driveshaft tube to distribute the grease in the raceways.

13. Install the driveshaft on the vehicle.

REAR AXLE

Axle Shaft

REMOVAL AND INSTALLATION

Tapered Shaft

1. Jack up the vehicle and remove the hub cap.

2. Remove the wheel.

3. Remove the axle nut dust cap.

4. Remove the axle shaft cotter pin, castle nut and flat washer.

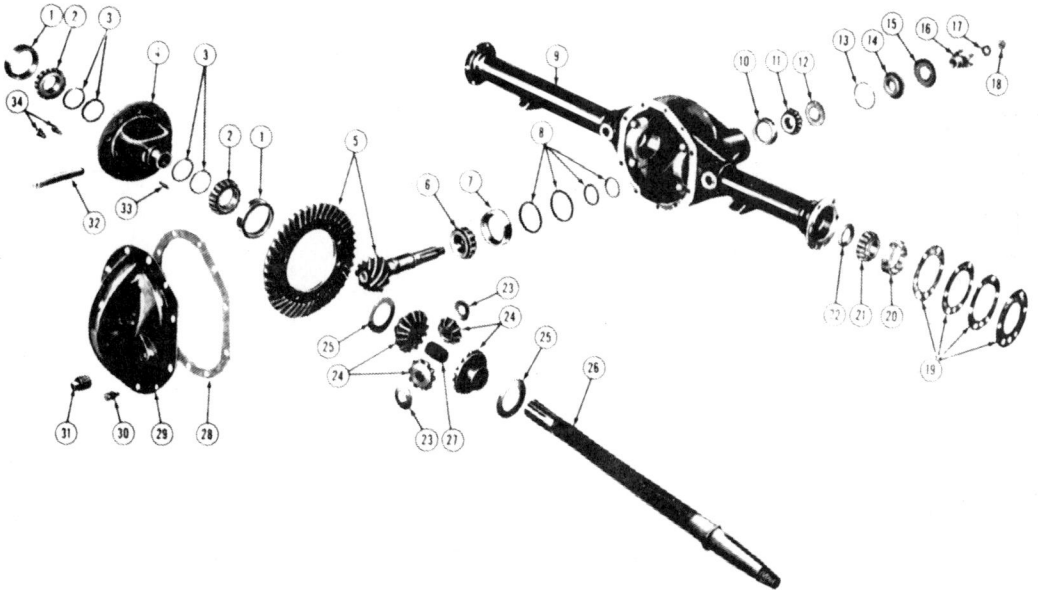

1. Differential bearing cup	18. Pinion nut
2. Differential bearing cone and rollers	19. Wheel bearing shims
3. Shims	20. Bearing cup
4. Differential case	21. Bearing cone and rollers
5. Ring gear and pinion	22. Oil seal
6. Pinion inner bearing cone and rollers	23. Thrust washer
7. Pinion inner bearing cup	24. Differential pinion gears
8. Pinion shims	25. Thrust washer
9. Axle housing	26. Axle shaft
10. Pinion outer bearing cup	27. Spacer
11. Pinion outer bearing cone and rollers	28. Gasket
12. Oil slinger	29. Housing cover
13. Gasket	30. Screw and lockwasher
14. Pinion oil seal	31. Filler plug
15. Dust shield	32. Differential shaft
16. Yoke	33. Lockpin
17. Flat washer	34. Ring gear screw

Exploded view of the rear axle assembly with a tapered shaft

5. Back-off the brake adjustment.

6. Use a puller to remove the wheel hub.

7. Remove the screws attaching the brake dust protector, grease and bearing retainers, brake assembly and shim to the housing.

8. Remove the hydraulic line from the brake assembly.

9. Remove the dust shield and oil seal.

NOTE: *If both shafts are being removed, keep the shims separated. Axle shaft end play is adjusted at the left side only.*

10. Use a puller to remove the axle shaft.

11. Install the axle shaft in the reverse order of removal, using a new grease seal and installing the hub assembly before the woodruff key.

NOTE: *Should the axle shaft be broken, the inner end can usually be drawn out of the housing with a wire loop after the outer oil seal is removed. However, if the broken end is less than 8 in. long, it usually is necessary to remove the differential assembly.*

Removing the rear wheel hub with a puller

Flanged Shaft

1. Jack up the vehicle and remove the wheels.

2. Remove the brake drum spring lock nuts and remove the drum.

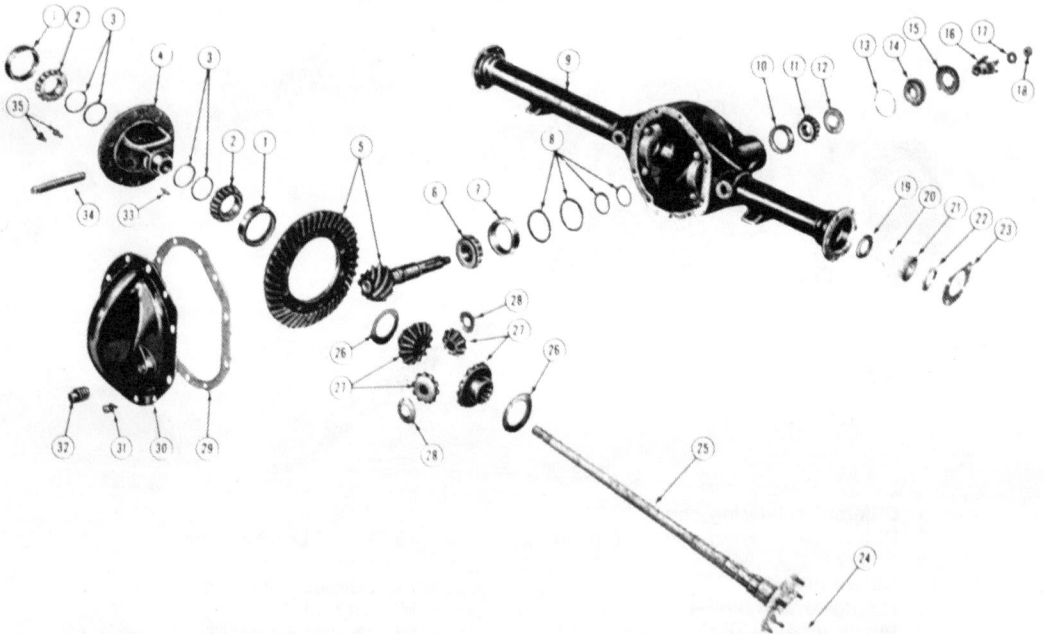

1. Differential bearing cup	13. Gasket	25. Axle shaft
2. Differential bearing	14. Pinion oil seal	26. Thrust washer
3. Shims	15. Dust shield	27. Differential pinion gears
4. Differential	16. Yoke	28. Thrust washer
5. Ring gear and pinion	17. Flat washer	29. Gasket
6. Pinion inner bearing	18. Pinion nut	30. Housing cover
7. Pinion inner bearing cup	19. Axle housing oil seal	31. Screw and lockwasher
8. Pinion shims	20. Axle shaft retainer ring	32. Filler plug
9. Axle housing	21. Axle shaft bearing	33. Lockpin
10. Pinion outer bearing cup	22. Axle shaft oil seal	34. Differential shaft
11. Pinion outer bearing	23. Axle shaft retainer plate	35. Ring gear screw
12. Oil slinger	24. Axle shaft cup plug	

Exploded view of the rear axle assembly with a flanged axle shaft

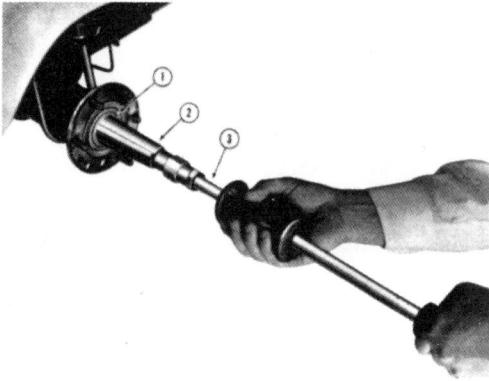

1. Bearing cone and roller
2. Axle shaft
3. Tapered axle puller

Removing the tapered axle shaft with a puller

3. Remove the axle shaft flange cup plug by piercing the center with a sharp tool and prying it out.

4. Using the access hole in the axle shaft flange, remove the nuts which attach the backing plate and retainer to the axle tube flange.

FLANGE ADAPTER

AXLE FLANGE   PULLER

Removing a flanged axle shaft with a puller

5. Remove the axle shaft from the housing with an axle puller.

6. Install in reverse order of removal.

FRONT AXLE

Front Hub

REMOVAL AND INSTALLATION

1. Remove the hub cap. On models with disc brakes, remove the caliper.

2. Remove the drive flange snap-ring.

3. On models with disc brakes, remove the rotor hub bolts, cover and gasket. On models with drum brakes, remove the axle flange bolts, lockwashers, and flatwashers.

4. If the axle is on the vehicle, apply the

Removing the hub cover with a puller

foot brakes. Remove the axle flange with a puller.

5. Release the locking lip of the lockwasher, and remove the outer nut, lockwasher, adjusting nut, and bearing lockwasher.

6. On models with disc brakes, remove the bearing and rotor. On models with drum brakes, back off on the brake adjusting star wheel adjusters and remove the brake drum assembly with the bearings. Be careful not to damage the oil seal.

7. On models with drum brakes, remove the brake backing plate. If the axle is on the vehicle, it will first be necessary to disconnect the brake hose between the front brake line and the flexible connection. On models with disc brakes, remove the adapter and splash shield.

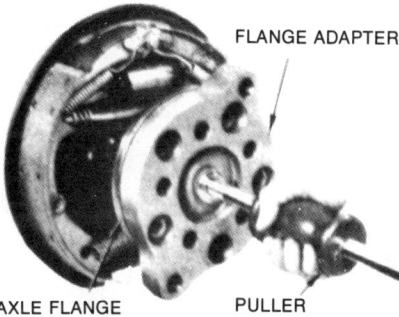

Removing the front axle shaft drive flange with a puller

Removing the wheel bearing nut

Front axle shaft removal

8. Remove the spindle and spindle bushing.

9. Remove the axle shaft and universal joint assembly.

10. Clean all parts.

11. Insert the universal joint and axle shaft assembly into the axle housing, being careful not to knock out the inner seal. Insert the splined end of the axle shaft into the differential and push into place.

12. Install the wheel bearing spindle and bushing.

13. Install the brake backing plate, or adapter and splash shield.

14. Grease and assemble the wheel bearings and oil seal.

15. Install the wheel hub and drum on the wheel bearing spindle. On disc brakes, install the rotor, hub, and caliper. Install the wheel bearing washer and adjusting nut. Tighten the nut to 50 ft lb, and back it off ¼ turn while rotating the hub. Install the lock-

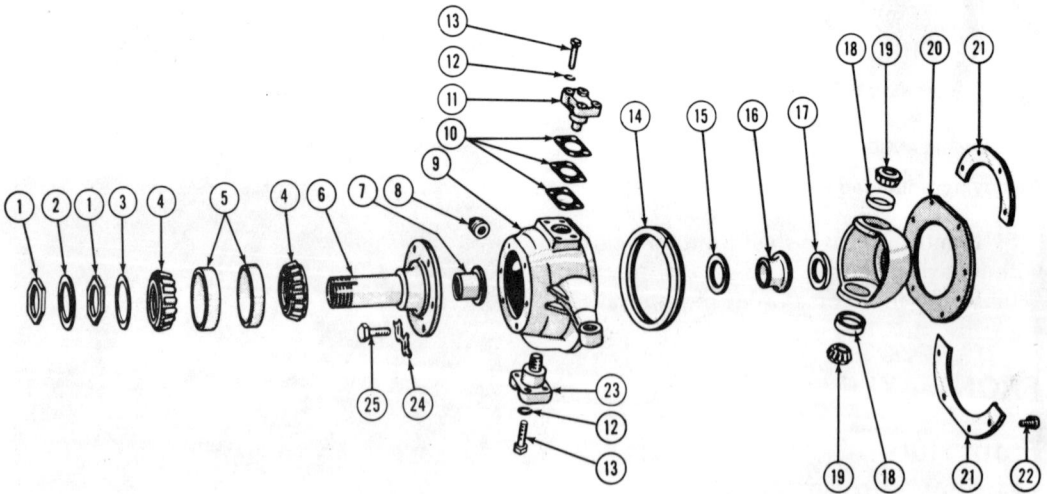

1. Bearing adjusting nut	10. Shims	18. Bearing cup
2. Lockwasher	11. Upper bearing cap	19. Bearing cone and rollers
3. Lockwasher	12. Lockwasher	20. Oil seal
4. Bearing cone and rollers	13. Bolt	21. Retainer
5. Bearing cup	14. Oil seal and backing ring	22. Bolt
6. Spindle	15. Thrust washer	23. Lower bearing cap
7. Bushing	16. Axle pilot	24. Lockstrap
8. Filler plug	17. Oil seal	25. Bolt
9. Left knuckle		

Steering knuckle and wheel bearing assembly

washer and nut, tighten the nut into place, and then bend the lip of the lockwasher over onto the locknut.

16. Install the drive flange and gasket onto the hub and attach with six capscrews and lockwashers. Install the snap-ring onto the outer end of the axle shaft.

17. Install the hub cap.

18. Install the wheel, lug nuts, and wheel disc.

19. If the tube was installed with the axle assembly on the vehicle, check the front wheel alignment, bleed the brakes and lubricate the front axle universal joints.

8

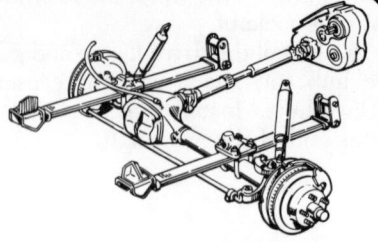

Suspension
and Steering

FRONT AND REAR SUSPENSION

All springs should be examined periodically for broken or shifted leaves, loose or missing clips, angle of the spring shackles, and position of the springs on the saddles. Springs with shifted leaves do not retain their normal strength. Missing clips may permit the spring leaves to fan out or break on rebound. Broken leaves may make the vehicle hard to handle or permit the axle to shift out of line.

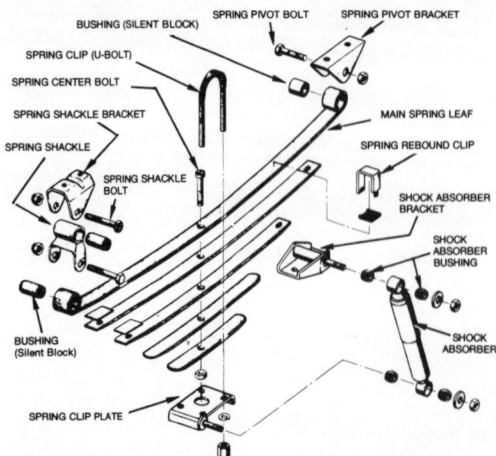

Exploded view of the front spring assembly: Commando

Weakened springs may break causing difficulty in steering. Spring attaching clips or bolts must be tight. It is suggested that they be checked at each vehicle inspection.

The springs are mounted under the front and rear axles on the Commando, and the front axle of the Wagoneer and Cherokee. On the rear axle of the Wagoneer and Cherokee, the springs are mounted on top of the axle.

On 1966–70 Commandos, the rear spring was a single-leaf unit with a six-leaf spring optional.

Springs

REMOVAL AND INSTALLATION

Spring Mounted Below the Axle.

1. Raise the vehicle with a jack under the axle. Place a jackstand under the frame side rail. Then lower the axle jack so the load is relieved from the spring and the wheels rest slightly on the floor.

2. Disconnect the shock absorber from the spring clip plate (except on the Wagoneer and Cherokee).

3. Remove the nuts which secure the spring clips (U-bolts). Remove the spring plate and spring clips. Free the spring from the axle by raising the axle.

4. Remove the pivot bolt nut and drive

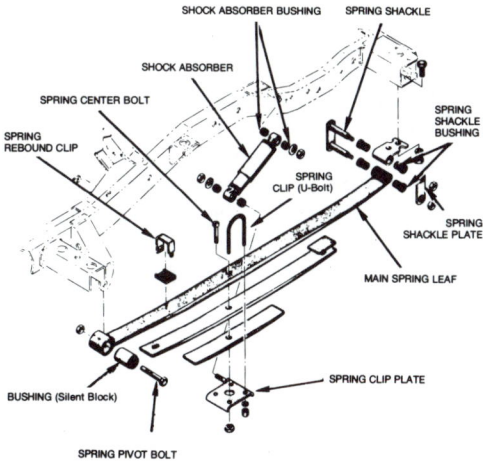

Exploded view of the rear spring assembly: Commando

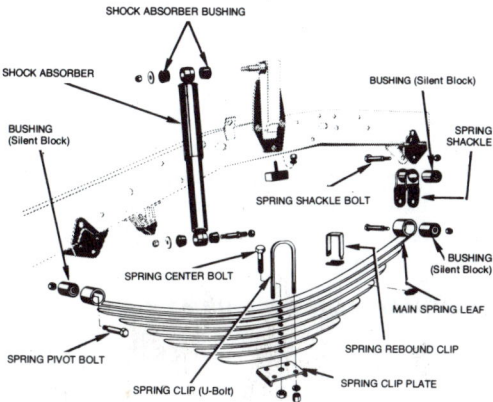

Exploded view of the front spring assembly: Wagoneer and Cherokee

out the pivot bolt. Disconnect the shackle from the shackle bracelet by removing the locknut, locknut and bolt or nut, or lockwasher and bolt.

5. With the spring removed, the spring shackle and/or shackle plate may be removed from the spring by removing the lock nut, or lock nut and shackle bolt or nut, or lockwasher and shackle bolt.

6. Inspect the bushings in the eye of the main spring leaf and the bushings of the spring shackle for excessive wear. Replace if necessary.

7. The spring can be disassembled for replacing an individual spring leaf, by removing the clips an and the center bolts.

8. To install the spring on the vehicle, with the bushings in place and the spring shackle attached to the springs, position the spring in the pivot hanger and install the pivot bolt and lock nut. Only tighten the lock

nut enough to hold the bushings in position until the vehicle is lowered from the jack.

9. Position the spring and install the shackle, shackle bolts, shackle plate if applicable, lockwasher, and nut. Only finger tighten the nuts at this time.

10. Move the axle into position on the spring by lowering the axle jack. Place the spring center bolt in the axle saddle hole. Install the spring clips, spring plate, lockwashers and nuts. Torque the $^7/_{16}$ in. nuts to 36–42 ft lbs and the ½ in. nuts to 45–65 ft lbs.

NOTE: *Be sure that the center bolt is properly centered in the axle saddle.*

11. Connect the shock absorber as necessary.

12. Remove the axle and allow the weight of the vehicle to seat the bushings in their operating positions. Then torque the $^7/_{16}$ in. spring pivot bolt nuts and spring shackle nuts to 25–40 ft lbs. Torque the ⅝ in. shackle nuts 55–75 ft lbs.

Spring Mounted Above the Axle (Rear Spring on the Wagoneer and Cherokee)

1. Remove the spring clip nuts, lockwashers, spring clips, and clip plate.

2. Raise the vehicle and place a stand under the frame side rail. Adjust the stand so that the load is released from the spring with the wheel resting on the floor and the spring center bolt is clear of the spring saddle.

3. Remove the lubrication fittings on models so equipped. Remove the self-locking nuts and flat washers from the spring hangers.

4. Remove the spring evenly off the spring hangers.

5. With the spring removed, the spring shackle can be removed from the spring by

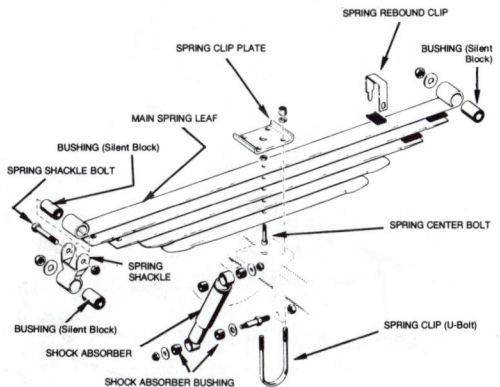

Exploded view of the rear spring assembly: Wagoneer and Cherokee

removing the locknut, flat washer and spring shackle bolt.

6. Inspect the bushings in the eye of the main leaf and the bushings of the spring shackle for excessive wear and replace them if necessary.

7. The spring can be disassembled for the purpose of replacing individual spring leaves by removing the rebound clips and the center bolt.

To install:

8. With the bushings in place, spring shackle attached to the springs and the shackle bolt torqued to 55–75 ft lbs, slide the spring into position on the spring hangers.

9. Raise the vehicle slightly with a jack at the frame side rail. Remove the stand and lower the vehicle so that the spring rests on the spring saddle and the center bolt aligns properly with the guide hole in the spring saddle.

10. Install the spring clips (U-bolts), spring clip plate, lockwashers and spring clip nuts. Torque the 7/16 in. nuts to 36–42 ft lbs and the ½ in. nuts to 45–65 ft lbs.

Shock Absorbers
REMOVAL AND INSTALLATION

NOTE: *Before installing new shocks, they should be purged of air. To do this, hold the shock upright and fully extend it, then invert and compress it. Do this several times.*

1. Remove the locknuts and washers.

2. Pull the shock absorber eyes and rubber bushings from the mounting pins.

3. Install the shocks in the reverse order of the removal procedure.

NOTE: *Squeaking usually occurs when movement takes place between the rubber bushings and the metal parts. The Squeaking may be eliminated by placing the bushings under greater pressure. This is accomplished either be adding additional washers or by tightening the locknuts. Do not use mineral lubricant to stop the squeaking as it will deteriorate the rubber.*

STEERING

Steering Knuckle Pivot-Pins (1966–71 Commandos and 1966–73 Wagoneers)

The steering knuckle pivot pins take the place of ball joints in a conventional vehicle. The pins pivot on tapered roller bearings located in the axle yoke. Replacement of these bearings requires removal of the hub and brake drum assembly, wheel bearings, axle shaft, spindle, steering tie rod, and steering knuckle. Disassemble the steering knuckle as follows:

REMOVAL AND INSTALLATION

1. Remove the eight screws that hold the oil seal retainer in place.

2. Remove the four screws which secure the lower pivot pin bearing cap.

3. Remove the four screws which hold the upper bearing cap in place.

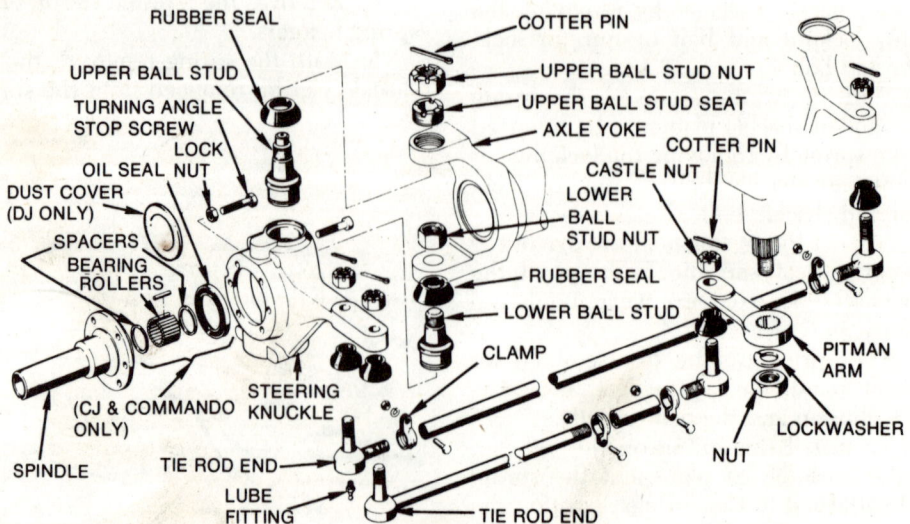

Steering and linkage assembly: 1972–73 Commando

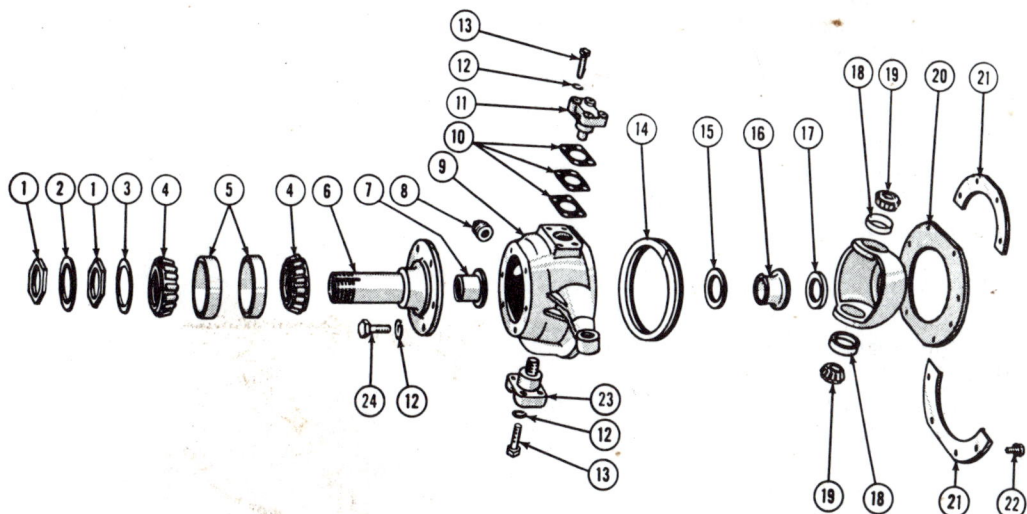

1. Bearing adjusting nut
2. Lockwasher
3. Lockwasher
4. Bearing cone and rollers
5. Bearing cup
6. Spindle
7. Bushing
8. Filler plug
9. Left knuckle and arm
10. Shims
11. Upper bearing cap
12. Bolt
14. Oil seal and backing ring
15. Thrust washer
16. Axle pilot
17. Oil seal
18. Bearing cup
19. Bearing cone and rollers
20. Oil seal
21. Retainer
22. Bolt
23. Lower bearing cap
24. Bolt

Exploded view of the steering knuckle, wheel bearings and spindle: 1966–71

4. Remove the bearing cap.

5. The steering knuckle can now be removed from the axle.

6. Wash all of the parts in cleaning solvent.

7. Replace any worn or damaged parts. Inspect the bearings and races for scores, cracks, or chips. Should the bearing cups be damaged, they may be removed and installed with a driver.

8. To install, reverse the removal procedure. When reinstalling the steering knuckle sufficient shims must be installed under the top bearing cap to obtain the correct preload on the bearing. Shims are available in 0.003, 0.005, 0.010, and 0.030 in. thicknesses. Install only one shim of the above thicknesses at the top only. Install the bearing caps, lockwashers, and screws, and tighten securely.

You can check preload on the bearings by hooking a spring scale in the hole in the knuckle arm for the tie rod socket. Take the scale reading when the knuckle has just started its sweep.

The pivot pin bearing preload should be 12–16 lbs with the oil seal removed. Remove or add shims to obtain a preload within these limits. If all shims are removed and adequate preload is still not obtained, a washer may be used under the top bearing cap to increase preload. When a washer is used, shims may

have to be reinstalled to obtain proper adjustment.

Open Type Steering Knuckle Ball Joints (Studs) on 1972 and Later Commandos and 1974 Wagoneers and Cherokees

REMOVAL AND INSTALLATION

NOTE: *Throughout this procedure, where a ball stud is either removed or installed, a hydraulic press or a two jawed gear puller can be used and, if at all possible, should be used to make the job easier. However it is possible to complete the job using a mallet, drift and a large socket the same size as the ball studs.*

1. Replacement of the ball joints, or ball stud, as they will be called from here on, requires the removal of the steering knuckle. To remove the steering knuckle, first remove the wheel, brake drum, and hub as an assembly. Remove the brake assembly from the spindle. Position the brake assembly on the front axle in a convenient place. Remove the snap ring from the axle shaft.

2. Remove the spindle and bearing assembly. It may be necessary to tap the spindle with a soft mallet to disengage it from the steering knuckle.

3. Slide the axle shaft out through the steering knuckle.

4. Disconnect the steering rods from the knuckle arm.

5. Remove the lower ball stud nut.

6. Remove the cotter pin from the upper stud. Loosen the upper stud until the top edge of the nut is flush with the top end of the stud.

77. Use a lead hammer to unseat the upper and lower studs from the yoke. Remove the upper nut and the knuckle assembly.

8. Remove the ball stud seat from the upper hold in the axle yoke. It is threaded in the hole. There are special wrenches available for removing the seat.

9. Securely clamp the knuckle assembly in a vise with the upper ball stud pointed down.

10. Using a large socket or drift, of approximately the same size as the ball stud, and a

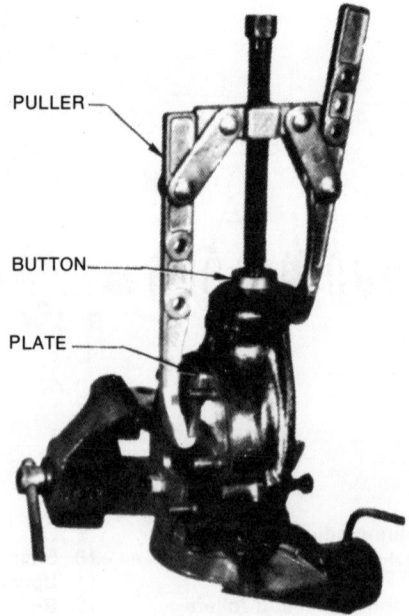

Using a puller to remove the lower ball stud

mallet, drive the lower stud out of the knuckle.

11. Place the socket on the bottom surface of the upper ball stud. Place the drift through the hole where the lower ball stud was and place it on the socket. Drive the upper ball stud out of the knuckle with a mallet.

12. Before installing the lower ball stud, run the lower ball stud nut onto the stud just far enough so the head of the stud is flush with the top edge of the nut.

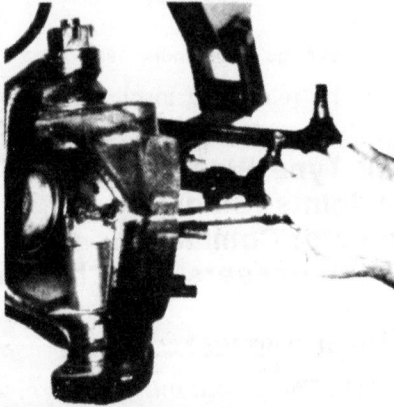

Removing the lower stud

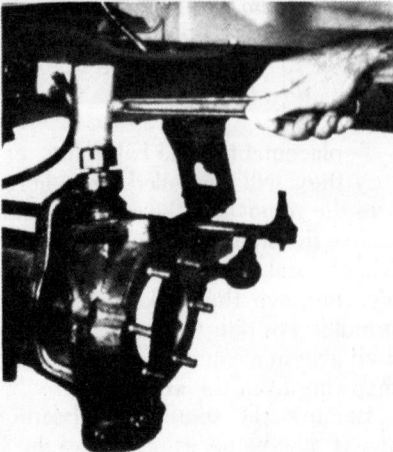

Removing the upper and lower ball studs from the yoke

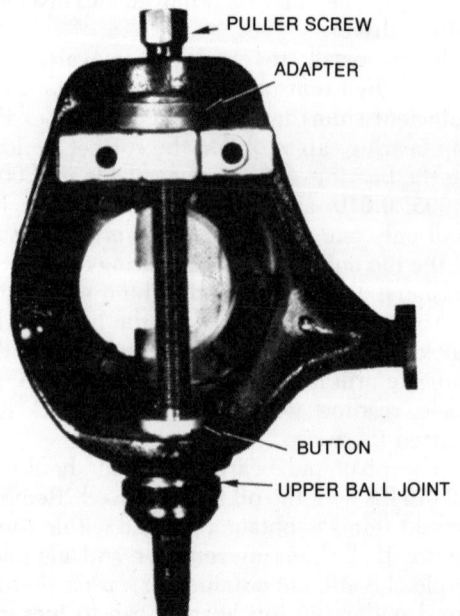

Removing the upper ball stud

13. Invert the knuckle in the vise. Position the lower ball stud in the knuckle with the nut in place. Place the same size socket over the nut and drive the ball stud into place with the drift and mallet.

14. Use the same procedure for installing the upper ball stud. The drift will not be needed to install the upper ball stud.

15. Install the upper ball stud seat into the axle yoke. Use a new one if the old one shows evidence of wear. The top of the seat should be flush with the top of the yoke.

16. Install the knuckle assembly onto the axle yoke. Install the lower stud nut. Tighten it to 70–90 ft lbs.

17. Install the upper stud nut and tighten it to 100 ft lbs. Install the cotter pin. If the cotter pin holes do not align, tighten the nut until the pin can be installed. Do not loosen the nut to align the holes.

18. Install the axle shaft, spindle and bearing assembly, and brake assembly. Connect the steering rods. Install the drum and hub, and wheel assembly. Adjust the wheel bearings. Check the turning angle. Adjust the stop screw to permit the proper turning angle of 31°.

Steering Knuckle Oil Seal
REPLACEMENT
1966–70 Commandos and 1966–73 Wagoneers

1. Remove the old steering knuckle oil seal by removing the 8 screws which hold it in place. Remove the seal retainer plate halves and remove the oil seal felt.

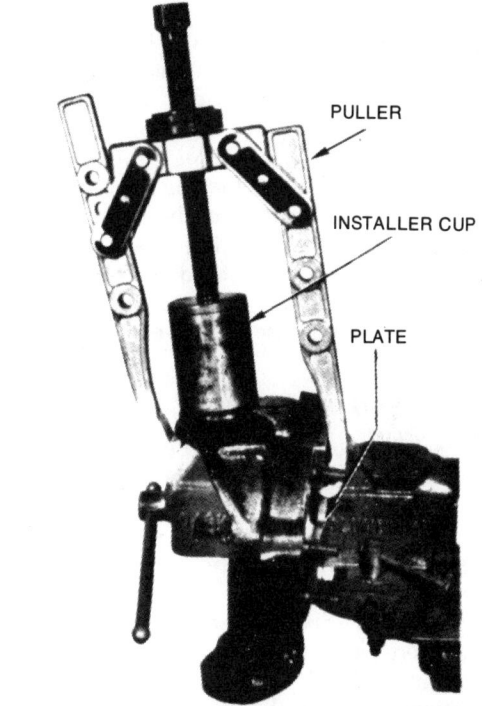

Installing the upper ball stud

2. Examine the spherical surface of the axle for scores or scratches which could damage the seal. Smooth any roughness with emery cloth.

3. Before installing the oil seal felt, make a diagonal cut across the top side of the felt so that it may be slipped over the axle. Install

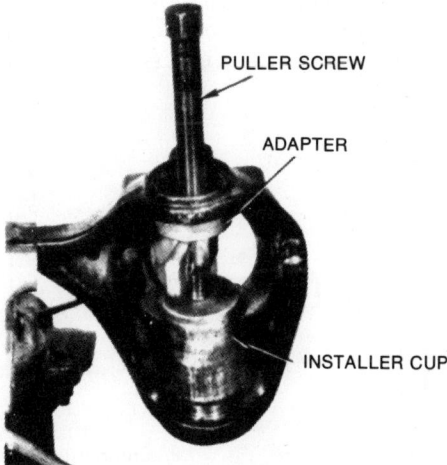

Installing the lower ball stud

1. Stop screw

Turning angle adjustment screw

the oil seal assembly in the reverse sequence mentioned above, making sure that the backing ring (of the backing ring and oil seal assembly) is toward the wheel.

After driving in wet, freezing weather, swing the front wheels from right to left to remove the moisture adhering to the oil seal and the spherical surface of the axle housing. This will prevent freezing with resultant damage to the seals. Should the vehicle be stored for any period of time, coat the surfaces with light grease to prevent rusting.

Front End Alignment

Proper alignment of the front wheels must be maintained in order to ensure ease of steering and satisfactory tire life.

The most important factors of front wheel alignment are wheel camber, axle caster, and wheel toe-in.

Wheel toe-in is the distance by which the wheels are closer together at the front than at the rear.

Wheel camber is the amount the top of the wheels incline outward from the vertical.

Front axle caster is the amount in degrees that the steering pivot pins are tilted toward the rear of the vehicle. Positive caster is inclination of the top of the pivot pin toward the rear of the vehicle.

These points should be checked at regular intervals, particularly when the front axle has been subjected to a heavy impact. When checking wheel alignment, it is important that wheel bearings and knuckle bearings be in proper adjustment. Loose bearings will affect instrument readings when checking the camber, pivot pin inclination, and toe-in.

Front wheel camber is preset. Caster can be altered by use of shims between the axle pad and the springs. Wheel toe-in may be adjusted. To measure wheel toe-in, follow the procedure given later on in this section.

CASTER ADJUSTMENT

Caster angle is established in the axle design by tilting the top of the kingpin toward the rear, and the bottom of the kingpin forward so that an imaginary line through the center of the kingpin would strike the ground at a point ahead of the point of tire contact.

The purpose of caster is to provide steering stability which will keep the front wheels in the straight ahead position and also assist in straightening the wheels when coming out of a turn.

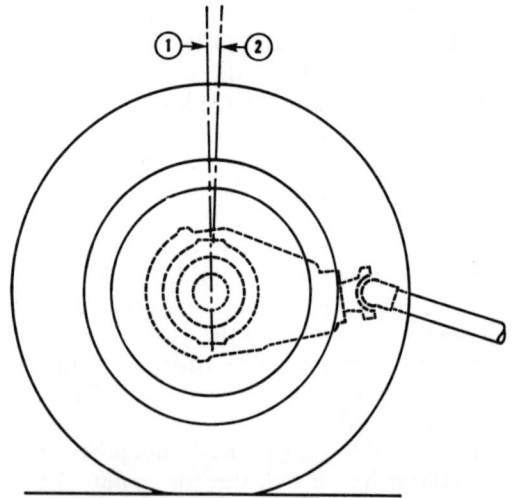

1. Vertical line   2. Caster angle

Axle caster

Caster is corrected by installing shims between the axle pad and the springs.

If the camber and toe-in are correct and it is known that the axle is not twisted, a satisfactory check may be made by testing the vehicle on the road. Before road testing, make sure all tires are properly inflated, being particularly careful that both front tires are inflated to exactly the same pressure.

If the vehicle turns easily to either side but is hard to straighten out, insufficient caster for easy handling of the vehicle is indicated. If correction is necessary, it can usually be accomplished by installing shims between the springs and axle pads to secure the desired result.

CAMBER ADJUSTMENT

The purpose of camber is to more nearly place the weight of the vehicle over the tire contact patch on the road to facilitate ease of steering. The result of excessive camber is ir-

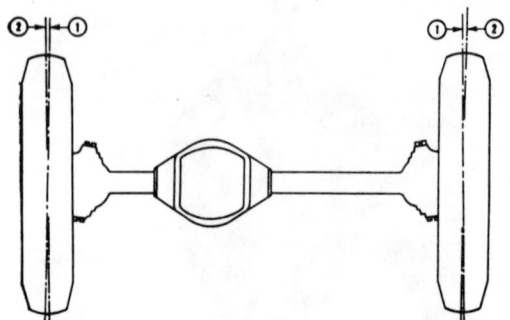

1. Vertical line   2. Camber angle

Wheel camber

regular wear of the tires on the outside shoulders and is usually caused by bent axle parts.

The result of excessive negative or reverse camber will be hard steering and possibly a wandering condition. Tires will also wear on the inside shoulders.

Unequal camber may cause any or a combination of the following conditions: unstable steering, wandering, kick-back or road shock, shimmy or excessive tire wear. The cause of unequal camber is usually a bent steering knucle or axle end.

Correct wheel camber is set in the axle at the time of manufacture and cannot be altered by any adjustment. It is important that the camber be the same on both front wheels. Heating of any parts to facilitate straightening usually destroys the heat treatment given them at the factory. Cold bending may cause a fracture of the steel and is also unsafe. Replacement with new parts is recommended rather than any straightening of damaged parts.

Toe-in Adjustment

First raise the front of the vehicle to free the front wheels. Turn the wheels to the straight ahead position. Use a steady rest to scribe a pencil line in the center of each tire tread as the wheel is turned by hand. A good way to do this is to first coat the wheel with a strip with chalk around the circumference of the tread at the center to form a base for a fine pencil line.

Measure the distance between the scribed lines at the front and rear of the wheels using care that both measurements are made at an equal distance from the floor. The distance between the lines should be greater at the rear than at the front by $3/64$ in. to $3/32$ in. To adjust, loosen the clamp bolts and turn the tie rod with a small pipe wrench. The tie rod is threaded with right and left hand threads to provide equal adjustment at both wheels. Do not overlook retightening the clamp bolts to 15–20 ft lbs.

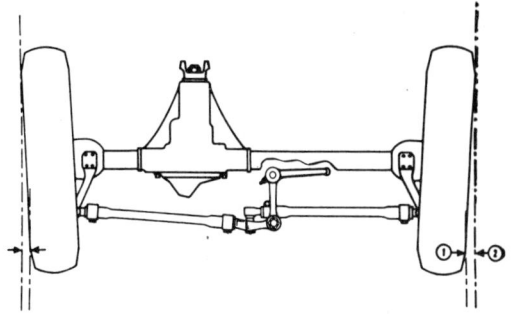

1. Vertical line   2. Toe-in angle

Front wheel toe-in

It is common practice to measure between the wheel rims. This is satisfactory providing the wheels run true. By scribing a line on the tire tread, measurement is taken between the road contact points reducing error by wheel run-out.

STEERING WHEEL

REMOVAL AND INSTALLATION

1. Disconnect the negative battery cable.
2. Set the front tires in a straight ahead position.
3. Pull the horn button from the steering wheel. With sport wheel, remove the button, nut, washer, retainer and horn ring.

On 1972 and later models it is necessary to remove the attaching screws from under the steering wheel spoke to remove the horn cover.

4. Remove the steering wheel nut and horn button contact cup.
5. Scribe a line mark on the steering wheel and steering shaft if there is not one already. Release the turn signal assembly from the steering post and install a puller.
6. Remove the steering wheel and spring.
7. To install, align the scribe marks on the steering shaft with the steering wheel and secure the steering wheel spring, steering wheel, and horn button contact cup with the steering wheel nut.

Wheel Alignment Specifications

Year	Model	Preferred Caster (deg.)	Preferred Camber (deg.)	Toe-in (in.)	Steering Axis Inclination (deg.)
1966–73	All	3P	$1^1/_2$P	$3/_{64}$–$3/_{32}$	$7^1/_2$
1974–79	All	4P	$1^1/_2$P	$3/_{64}$–$3/_{32}$	$8^1/_2$

8. Install the horn button.

9. Connect the battery cable and test the horn.

TURN SIGNAL SWITCH REPLACEMENT
Commando

1. Disconnect the battery and remove the steering wheel.

2. Remove the directional signal lever. A milled flat on the lever will accept a ¼ in. wrench for facilitating removal and installation.

3. Disconnect the directional signal harness from the frame harness. Remove the wires from each plastic connector.

4. Remove the wires by inserting a screwdriver with a narrow blade into the terminal end of the connector. Insert the blade into the narrowest part (the groove) of the opening to depress the retaining tang and pull the wire out of the connector.

5. Remove the harness loom.

6. Remove the two retaining screws from the directional switch and the two retaining screws from the 4-way flasher switch.

7. Pry out the horn contact button.

8. Remove the two bowl attaching screws and allow the bowl to slide down the column.

9. Tape the disconnected wires together. Fold some of the wires back to flatten the bulge.

10. Attach a wire or heavy string to the turn signal harness and pull the harness out from the top of the steering column.

11. To install the wiring harness, use the wire or string to pull the harness down through the steering column.

12. Install the switch and assemble the steering column in the reverse order of removal and disassembly.

Wagoneer and Cherokee

1. Disconnect the battery and remove the steering wheel.

2. Loosen the anti-theft cover retaining screws and lift the cover from the column. It is not necessary to completely remove these screws as they are held on the cover by plastic retainers.

3. Depress the lockplate and pry the round wire snap-ring from the steering shaft groove. Remove the snap-ring, lockplate, directional signal cancelling cam, upper bearing preload spring and thrust washer from the steering shaft.

4. Place the directional signal actuating lever in the right turn position and remove the lever.

5. Depress the hazard warning light switch, located on the right-side of the column adjacent to the key lock on newer models, and remove the button by turning it in a counterclockwise direction.

6. Remove the directional signal wire harness connector block from its mounting bracket on the right-side of the lower column. On vehicles equipped with an automatic transmission, use a stiff wire, such as a paper clip, to depress the locktap which retains the shift quadrant light wire in the connector block.

7. Remove the directional signal switch retaining screws and pull the directional signal switch and wire harness from the column.

8. Install the directional signal switch and assemble the steering column in the reverse order of removal.

Power Steering Pump
REMOVAL AND INSTALLATION

If the power steering pump has to be removed to service another component, it is not necessary to remove the hoses from the pump. Just disconnect the mounting fixtures and lift the pump away from the engine and lay it out of the way. The only time the power steering hoses have to be removed from the pump is when the pump has to be removed from the vehicle for service or replacement.

1. Remove the pump drive belt tension adjusting bolt. Disconnect the belt from the pump.

2. Disconnect the return and pressure hoses from the pump. Cover the hose connector and union on the pump and open ends of the hoses to avoid the entrance of dirt.

3. Remove the front bracket from the engine on 1972 and later model vehicles equipped with V8 engines.

4. Remove the two nuts which secure the rear of the pump to the bracket, and the two bolts which secure the front of the pump to the bracket and remove the pump.

5. To install, position the pump in the bracket and install the rear attaching screws. On 1972 and later model vehicles equipped with V8 engines, install the front bracket.

6. Connect the hydraulic hoses. Adjust the drive belt tension.

7. Fill the pump reservoir to the correct level.

8. Start the engine and wait for at least

three minutes before turning the steering wheel. Check the level frequently during this time and add fluid as necessary.

9. Slowly turn the steering wheel through its entire range a few times with the engine running. Recheck the level and inspect for possible leaks.

NOTE: *If air becomes trapped in the fluid, the pump may become noisy until all of the air is out. This may take some time since trapped air does not bleed out rapidly.*

Tie Rod End Removal and Installation

1. Remove the cotter pins and retaining nuts at both ends of the tie rod and from the end of the connecting rod where it attaches to the tie rod.

2. Remove the nut attaching the steering damper push rod to the tie rod bracket and move the damper aside.

3. Remove the tie rod ends from the steering arms and connecting rod with a puller.

4. Count the number of threads showing on the tie rod before removing the ends, as a guide to installation.

5. Loosen the adjusting tube clamp bolts and unthread the ends.

6. Installation is the reverse of removal. Torque the connecting rod-to-tie rod nut to 40 ft.lb. on 1966–70 models and to 70 ft.lb. on 1971 and later models.

7. Adjust toe-in, if necessary.

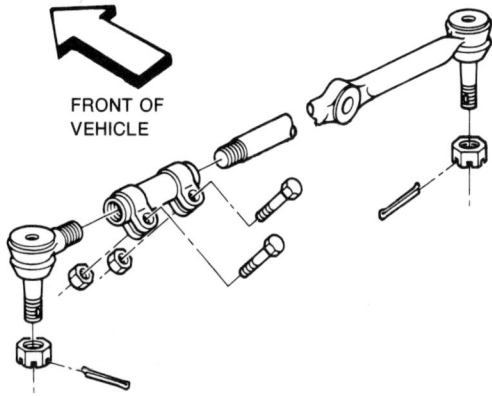

FRONT OF VEHICLE

Tie rod assembly

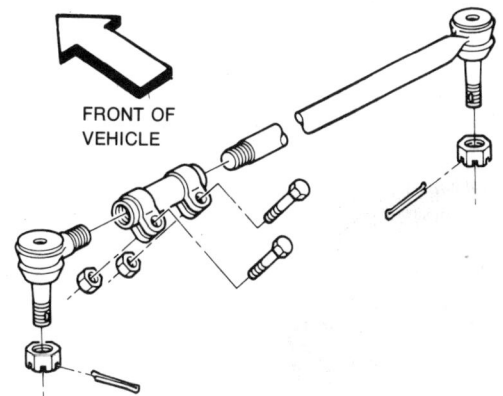

FRONT OF VEHICLE

Connecting rod assembly

9

Brakes

BRAKE SYSTEM

A double hydraulic brake system in conjunction with self-adjusting wheel brake units is used on all vehicles covered in this text. The purpose of two completely segregated systems is that failure in one part of the brake system does not result in disability of the entire hydraulic brake system. Failure in the front brake system will leave the rear brake system still operative and vice versa. The double hydraulic master cylinder has two outlets, two residual check valves, two separate fluid reservoirs and two hydraulic pistons, all of which are operated in tandem by a single hydraulic pushrod.

All of the vehicles covered here have mechanical, hand or foot-actuated parking brakes. When the parking brake is applied either by hand or foot, tension is exerted on the front parking brake cable. The front cable is connected, through an equalizer, to the center of the rear parking brake cable, the ends of which are connected to the two rear wheel brake units. Each cable end is connected to a lever attached to the secondary rear brake shoe. A connecting strut acts against the primary shoe and expands the brake shoes against the brake drum. The amount of brake grip depends on the amount

of tension applied at the actuating hand lever or foot pedal.

A transmission parking brake was available on early Commandos with 3-speed transmissions as an option. The transmission brake is mechanically operated in a manner similar to the conventional parking brake system. The transmission brake is mounted at the rear output bearing housing on the transfer case. It is also very similar to a conventional drum type brake in operation and components. There are two brake shoes mounted inside a brake drum which expand out against the brake drum. The brake drum is attached to the rear driveshaft.

Power brakes were optional on all Commandos and Wagoneers. In 1974, power brakes with front disc brakes were made standard equipment on the Wagoneer and optional on the Cherokee.

Adjustment
DRUM BRAKES ONLY

When the brake linings become worn, effective brake pedal travel is reduced. Adjusting the brake shoes will restore the necessary travel for efficient braking.

Before adjusting the brakes, check the spring nuts, brake dust shield-to-axle flange bolts and wheel bearing adjustments. Any looseness in these parts can cause erratic brake operation.

INITIAL BRAKE SHOE ADJUSTMENT

If the brake assemblies have been disassembled, an initial adjustment must be made before the drum is installed. It may also be necessary to back off the adjustment to remove the drums.

When the brake parts have been installed in their correct position, adjust the adjusting screw assemblies to a point where approximately ⅜ in. of threads are exposed between the starwheel and the starwheel nut.

Non-Self-Adjusting Brakes Adjustment

1. Jack up the vehicle.
2. Remove the adjusting hole dust plug from the back of the brake backing plate.
3. Use a brake adjusting tool to turn the starwheel. Raise the handle of the tool to tighten the shoes against the drum.
4. When the brake shoes are tight against the drum, turn the starwheel in the opposite direction until the wheel rotates freely without brake drag.
5. Repeat the above procedure for all four wheels. Replace the dust plug.

Manually adjusted rear brake assembly

Self-Adjusting Brakes Adjustment

1. Jack up the vehicle.
2. Remove the access slot cover and using

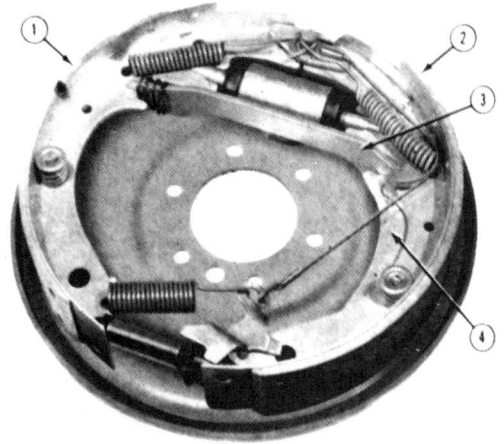

Rear brake assembly with self-adjusters

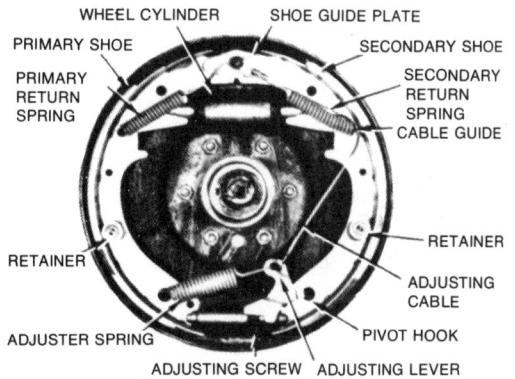

Front drum brake assembly

a brake adjusting tool or screwdriver, rotate the starwheel until the wheel is locked and can't be turned by hand. To tighten, rotate the starwheel in the clockwise direction.

3. Back off the starwheel at least 15 to 20 notches. To back off the starwheel on the brake, insert an ice pick or thin screwdriver in the adjusting screw slot to hold the automatic adjusting lever away from the starwheel. Do not attempt to back off on the adjusting screw without holding the adjusting lever away from the starwheel as the adjuster will be damaged.

HYDRAULIC SYSTEM

When the brake pedal is depressed, the master cylinder piston(s) move forward, displacing the brake fluid. Due to the fact that the fluid volume is constant, the displacement

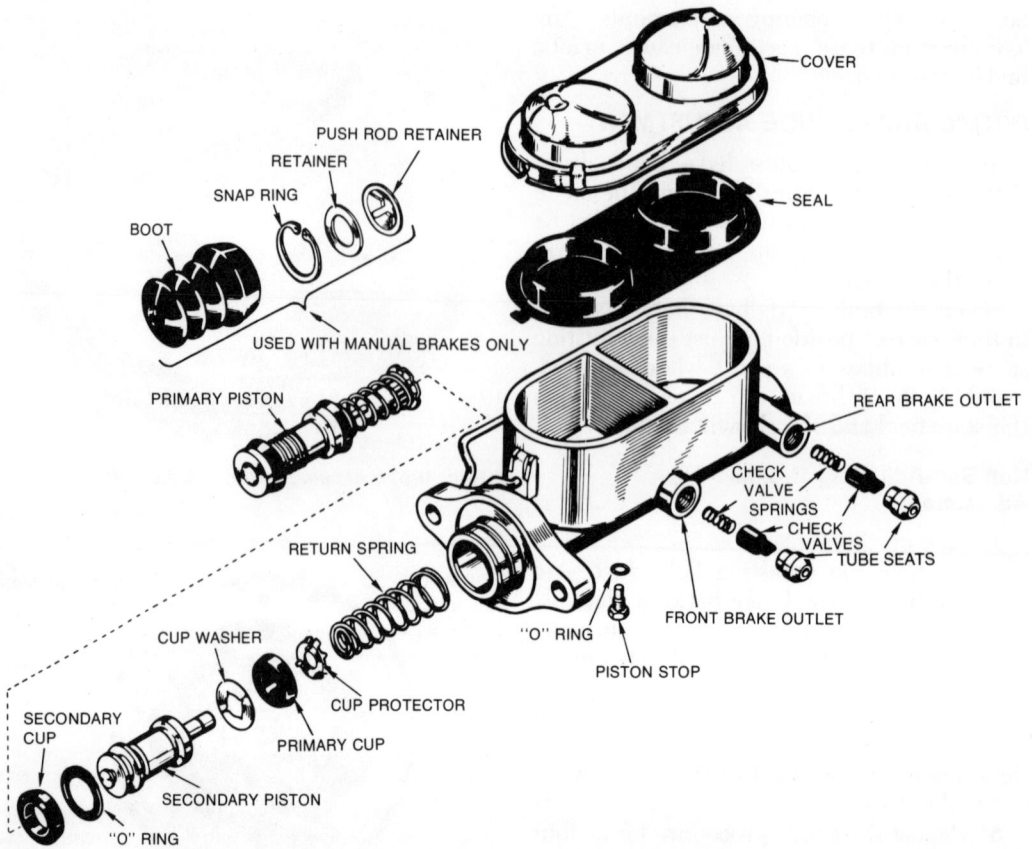

Dual hydraulic master cylinder components

results in increased hydraulic pressure. This pressure is exerted upon the wheel cylinders and/or caliper pistons, thus forcing the brake shoes or friction pads against the drums or discs.

Hydraulic pressure within the brake lines is in direct proportion to the effort applied to the brake pedal. With front disc/rear drum brake applications, a proportioning valve (or metering valve) is utilized. The valve maintains a predetermined front/rear hydraulic pressure ratio which reduces the possibility of premature rear wheel lock-up. The metering valve reduces front brake pressure until rear brake pressure builds up adequately to overcome rear brake shoe return springs.

When the brake pedal is released, hydraulic pressure drops. On drum brakes, the brake return springs, and on disc brakes, the wobbling action of the disc returns the shoes or disc pads to their retracted positions and forces the displaced fluid back into the master cylinder.

Master Cylinder
REMOVAL AND INSTALLATION

To remove the master cylinder, disconnect and plug the brake lines, disconnect the wires from the stoplight switch, remove all attaching bolts and nuts and lift the assembly from the vehicle.

Installation is the reverse of the removal procedure.

OVERHAUL

NOTE: *Do not use any type of mineral oil, gasoline or kerosene to clean any part of any hydraulic brake system. These fluids will cause rubber parts to soften, swell, and distort, resulting in failure. Use only clean brake fluid or alcohol.*

1. Remove the filler cap and empty all the fluid.

2. The stop light switch and primary piston stop, located in the stop light switch outlet hole, must be removed before removing the snap ring from the piston bore. Remove the snap ring, pushrod assembly and

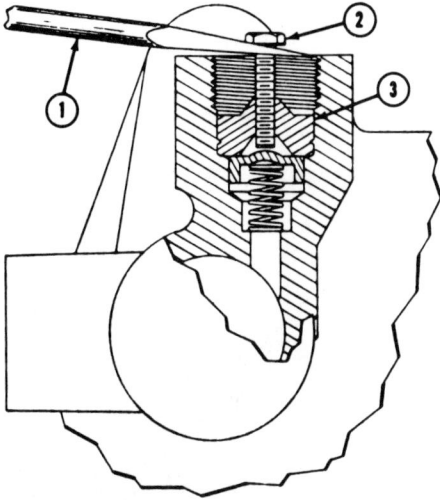

Removing the tube seats from the master cylinder with a screwdriver

the primary and secondary piston assemblies. Air pressure applied in the piston stop hole will help facilitate the removal of the secondary piston assembly.

3. The residual check valves are located under the front and rear fluid outlet tube seats.

4. The tube seats must be removed with self tapping screws to permit the removal of the check valves. Screw the self-tapping screws into the tube seats and place two screwdriver tips under the screw head and force the screw upward.

5. Remove the expander in the rear secondary cup, secondary cups, return spring, cup protector, primary cup, and washer from the secondary piston.

6. Immerse all of the metal parts in clean brake fluid and clean them. Use an air hose to blow out dirt and cleaning solvent from recesses and internal passages.

7. After cleaning, place all of the parts on clean paper in a clean pan.

8. Inspect all parts for damage or excessive wear. Replace any damaged, worn, or chipped parts. Inspect the hydraulic cylinder bore for signs of scoring, rust, pitting, or etching. Any of these will require replacement of the hydraulic cylinder.

9. Prior to assembling the master cylinder, dip all of the components in clean brake fluid and place them on clean paper or in a clean pan.

10. Install the primary cup washer, primary cup, cup protector, and return spring on the secondary piston.

11. Install the piston cups in the double

groove end of the secondary piston, so the flat side of the cups face each other (lip of the cups away from each other). Install the cup expander in the lip groove of the end cup.

12. Coat the cylinder bore and piston assemblies with clean brake fluid before installing any parts in the cylinder.

13. Install the secondary piston assembly first and then the primary piston.

14. Install the pushrod assembly, which includes the pushrod, boot, and rod retainer, and secure with the snap ring. Install the primary piston stop and stop light switch.

15. Place new rubber check valves over the check valve springs and install in the outlet holes, spring first.

16. Install the tube seats, flat side toward the check valve, and press in with tube nuts or the master cylinder brake tube nuts.

17. Before the master cylinder is installed on the vehicle it must be bled. Support the cylinder assembly in a vise and fill both fluid reservoirs with brake fluid.

18. Loosely install a plug in each outlet of the cylinder. Depress the push rod several times until air bubbles cease to appear in the brake fluid.

19. Tighten the plugs and attempt to depress the piston. The piston travel should be restricted after all of the air is expelled.

20. Install the master cylinder in the vehicle and bleed all the hydraulic lines at the wheel cylinders.

BLEEDING THE BRAKES

The hydraulic brake system must be bled whenever a fluid line has been disconnected or air gets into the system. A leak in the system may sometimes be indicated by the presence of a spongy brake pedal. Air trapped in the system is compressible and does not permit the pressure applied to the brake pedal to be transmitted solidly through to the brakes. The system must be absolutely free from air at all times. When bleeding brakes, bleed at the wheel most distant from the master cylinder first, the next most distant second, and so on. During the bleeding operation the master cylinder must be kept at least ¾ full of brake fluid.

NOTE: *On 1973 and later models, a combination differential and proportioning valve is used in the system. It is attached to the inner side of the left frame rail. When bleeding the brakes, the metering section of the valve must be held open. Loosen the front mounting bolt of the valve and insert*

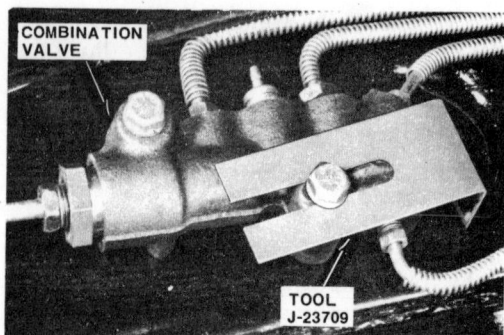

Combination valve during bleeding

Tool J-23709, or its fabricated equivalent, under the bolt. Push in on the metering valve stem to open it and retighten the bolt to hold the tool in place. When bleeding is finished, loosen the bolt, remove the tool and retighten the bolt.

To bleed the brakes, first carefully clean all dirt from around the master cylinder filler plug. If a bleeder tank is used follow the manufacturers instructions. Remove the filler plug and fill the master cylinder to the lower edge of the filler neck. Clean off the bleeder connections at all of the wheel cylinders or disc brake calipers. Attach the bleeder hose and fixture to the right rear wheel cylinder bleeder screw and place the end of the tube in a glass jar, submerged in brake fluid. Open the bleeder valve ½–¾ of a turn.

Have an assistant depress the brake pedal and allow it to return slowly. Continue this pumping action to force any air out of the system. When bubbles cease to appear at the end of the bleeder hose, close the bleeder valve and remove the hose. Check the level of the brake fluid in the master cylinder and add fluid, if necessary.

After the bleeding operation at each caliper or wheel cylinder has been completed, fill the master cylinder reservoir and replace the filler plug.

NOTE: *Never reuse brake fluid which has been removed from the lines through the bleeding process because it contains air bubbles and dirt.*

FRONT DISC BRAKES

The front disc brake consists of 3 assemblies: the caliper assembly, the hub and rotor assembly, and the support and shield assembly.

The caliper is a single-piston sliding type,

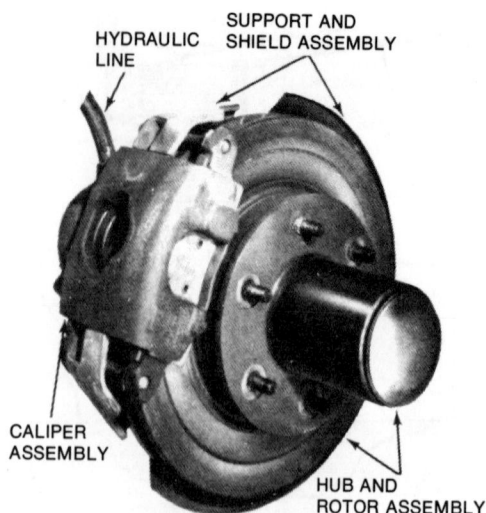

Disc brake assembly

of one-piece casting construction with the inboard side containing the single-piston, piston bore and the bleeder screw and fluid inlet holes. There are two brake pads within the caliper, positioned on either side of the rotor. The brake pads take the place of the brake shoes on drum brakes and the rotor takes the place of brake drums. The pads themselves actually consist of two parts: the metal shoe and the composition lining which is bonded or riveted to the shoe.

The significant operating feature of the single-piston caliper is that it is free to slide laterally on the two mounting bolts which thread into the support bracket. The pressure applied to the piston is transmitted to the inboard brake pad, forcing the lining of the pad against the inboard rotor surface. The pressure applied to the inboard end or

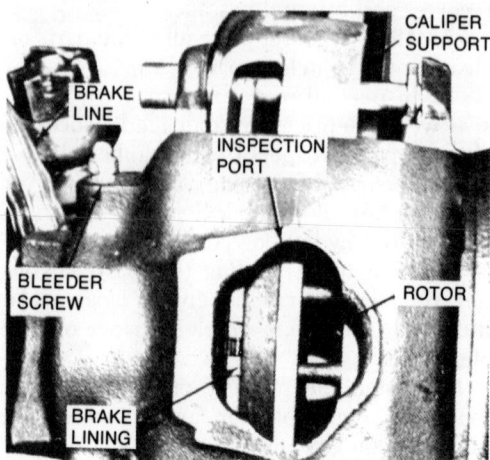

Caliper assembly

bottom of the piston bore forces the caliper to slide on the mounting bolts (which are inserted through sleeves in the caliper body) toward the inboard side. This inward movement of the caliper causes the outboard section of the caliper to apply pressure against the lining of the outboard pad, forcing the lining against the outboard surface of the rotor. As hydraulic pressure builds within the brake lines, due to the increased application of pressure at the brake pedal, the brake pad assemblies press against the rotor surfaces with increasing force, thus slowing the rotation of the rotor.

Disc Brake Calipers

REMOVAL AND INSTALLATION

1. Drain ⅔ of the brake fluid from the front reservoir. Use the bleeder screw at the front outlet port to drain the fluid.
2. Raise the vehicle so that the wheel to be worked on is off the ground. Support the vehicle with jack stands. Remove the front wheels.
3. Remove the front wheels.
4. Place a C-clamp on the caliper so that the solid end contacts the back of the caliper and the screw end contacts the metal part of the outboard brake pad.
5. Tighten the clamp until the caliper moves far enough to force the piston to the bottom of the piston bore. This will back the brake pads off of the rotor surface to facilitate the removal and installation of the caliper assembly.
6. Remove the C-clamp.
NOTE: *Do not push down on the brake pedal or the piston and brake pads will re-*

Removing the caliper mounting bolts

turn to their original positions up against the rotor.
7. Remove both of the allen head mounting bolts and lift the caliper off the rotor.
NOTE: *If just the brake pads are being replaced, it is not necessary to remove the caliper assembly entirely from the vehicle. Do not remove the brake line. Rest the caliper on the front spring or other suitable support. Do not allow the brake hose to support the weight of the caliper.*
8. If the caliper is being removed in order to be rebuilt, then it is necessary to disconnect the brake fluid hose. Clean the brake fluid hose-to-caliper connection thoroughly. Remove the hose-to-caliper bolt. Cap or tape the open ends to keep dirt out. Discard the copper gaskets.
9. Install the caliper in the reverse order of removal.

Compressing the piston with a C-clamp

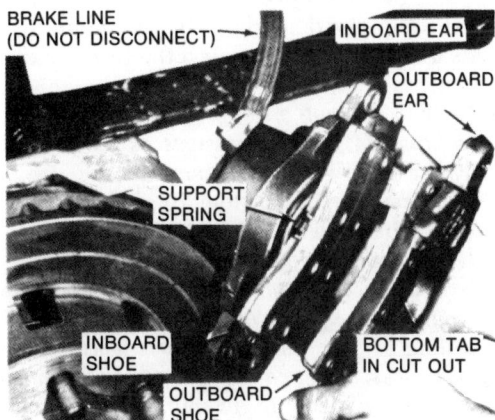
Removing the caliper from the rotor

NOTE: *If the brake fluid hose was disconnected, it will be necessary to bleed the hydraulic system.*

OVERHAUL

1. Remove the caliper assembly and remove the brake pads. If the pads are to be reused, mark their location in the caliper.

2. Clean the caliper exterior with clean brake fluid. Drain any residual fluid from the caliper and place it on a clean work surface.

NOTE: *Removal of the caliper piston requires the use of compressed air. Do not, under any circumstances, place your fingers in front of the piston in an attempt to catch or protect it when applying compressed air to remove the piston.*

3. Pad the interior of the caliper with clean cloths. Use several cloths and pad the interior well to avoid damaging the piston when it comes out of the bore.

4. Insert an air nozzle into the inlet hole in the caliper and gently apply air pressure on the piston to push it out of the bore. Use only enough air pressure to ease the piston out of the bore.

5. Pry the dust boot out of the bore with a screwdriver. Use caution during this operation to prevent scratching the bore. Discard the dust boot.

6. Remove the piston seal from the piston bore and discard the seal. Use only non-scratching implements such as a pencil, wooden stick or a piece of plastic to remove the seal. Do not use a metal tool as it could very easily scratch the bore.

7. Remove the bleeder screw. Remove and discard the sleeves and rubber bushings from the mounting ears.

8. Clean all the parts with clean brake fluid. Blow out all of the passages in the caliper and bleeder valve. Use only dry and filtered compressed air. Replace the mounting bolts if they are corroded or if the threads are damaged.

NOTE: *Do not attempt to clean the attaching bolts with abrasives as their protective plating may be removed.*

9. Examine the piston for defects. Replace the piston if it is nicked, scratched, corroded or the protective plating is worn off.

NOTE: *Do not attempt to refinish the piston in any way. The outside diameter is the sealing surface and is made to very close tolerances. Removal of the nickle-chrome plating will lead to pitting, rusting and eventual cocking of the piston in the piston bore.*

Examine the caliper piston bore for the same defects as the piston. The bore is not plated and minor stains or corrosion can be polished with crocus cloth.

NOTE: *Do not use emery cloth or similar abrasives on the piston bore. If the bore does not clean up with crocus cloth, replace the caliper. Clean the caliper thoroughly with brake fluid if the bore was polished with crocus cloth.*

10. Lubricate the bore and new seal with brake fluid and install the seal in the groove in the bore.

11. Lubricate the piston with brake fluid and install the new dust boot on the piston. Assemble the dust boot into the piston groove so that the fold in the boot faces the open end of the piston. Slide the metal portion of the dust boot over the open end of the piston and push the retainer toward the back of the piston until the lip on the fold seats in the piston groove. Then push the retainer portion of the boot forward until the boot is flush with the rim at the open end of the piston and snaps into place.

12. Insert the piston in the bore, being careful not to unseat the piston seal. Push the piston to the bottom of the bore. It requires 50-100 lbs of force to bottom the piston.

13. Position the dust boot retainer in the counter bore at the top of the piston bore. Seat the dust boot retainer with a flat-ended punch by tapping the metal ring of the dust boot into place. Be careful not to damage the rubber portion of the dust boot. The metal retainer portion of the boot must be evenly seated in the counterbore and fit below the face of the caliper.

14. Install the bleeder screw. Tighten it to 50-140 in. lbs.

15. Connect the brake line to the caliper using new copper gaskets.

16. Install the brake pads, sleeves and rubber bushings.

17. Install the caliper and tighten the mounting bolts to 35 ft lbs. Bleed the hydraulic system.

Disc Brake Pads

REMOVAL AND INSTALLATION

1. Raise and support the vehicle with jack stands and remove the wheel(s) on the side to be worked on.

2. Remove the caliper assembly as previously outlined in this Chapter under "Disc Brake Calipers Removal and Installation". Take heed to the note given after Step 7 of that procedure.

3. Remove the brake pad assemblies. Remove the support spring from the inboard pad. Note the position of the spring before removing it for correct installation later.

4. Remove the sleeves from the inboard ears of the caliper. Remove the rubber bushings from all the holes in the caliper ears. Discard all bushings and sleeves.

5. Clean all the mounting holes and bushing grooves in the caliper ears. Clean the mounting bolts. Replace the bolts if they are corroded or if the threads are damaged.

NOTE: *Do not use abrasives on the bolts in order not to destroy their protective plating.*

Wipe the inside of the caliper clean, including the exterior of the dust boot. Inspect the dust boot for cuts or cracks and for proper seating in the piston bore. If evidence of fluid leakage is noted, the caliper should be rebuilt.

NOTE: *You should not use compressed air to clean the inside of the caliper as it may unseat the dust boot seal.*

6. Lubricate the new bushings, sleeves, bushing grooves and the small ends of the mounting bolts with a silicone lubricant. Install the rubber bushings in all of the caliper mounting ears.

7. Install the sleeves in the inboard mounting ears of the caliper. Position the sleeves so that the sleeve end facing the brake pad lining is flush with the machined surface of the mounting ear.

8. Install the support spring on the inboard brake pad. Place the single tang end of the spring over the notch in the pad. The double tang end of the spring clips onto the bottom of the pad.

9. Install the inboard pad in the caliper. The pad must lay flat against the piston. Be sure that the support spring is fully seated in the piston.

10. Install the outboard pad in the caliper. The ears on the pad should rest on the top of the ears in the caliper. The bottom tab on the pad fits in the cutout in the caliper. Be sure the pad is fully seated.

11. With the pads installed, position the caliper over the rotor. Line up the mounting holes in the caliper and the support bracket

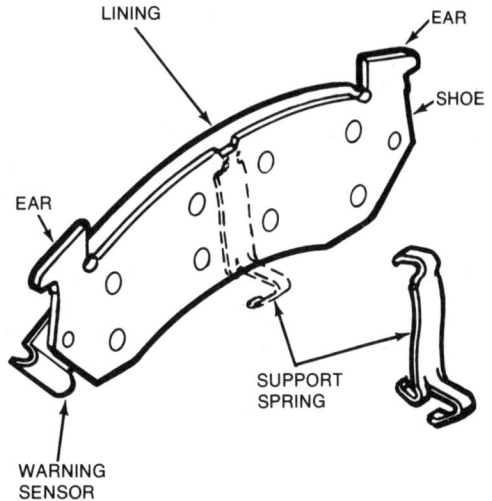

Installing the support spring on the inboard brake pad

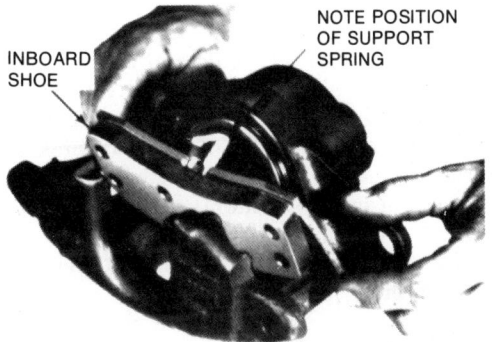

Installing the inboard pad in the caliper

and insert the mounting bolts. Make sure that the bolts pass under the retaining ears on the inboard shoes. Push the bolts through until they engage the holes of the outboard pad and caliper ears. Thread the bolts into the support bracket and tighten them to 35 ft lbs.

12. Fill the master cylinder with brake fluid and pump the brake pedal to seat the pads.

13. Use a pair of channel-lock pliers to bend (clinch) both of the upper ears of the outboard shoe until the radial clearance between the pad and the caliper is eliminated.

NOTE: *Outboard pads with formed ears are designed for original installation only and are fitted to the caliper. The pads should never be relined or reconditioned for installation.*

14. Install the wheel assembly and lower the vehicle. Check the level of the brake fluid in the master cylinder and fill as neces-

sary. Test the operation of the brakes before taking the vehicle onto the road.

Disc Brake Rotor

REMOVAL AND INSTALLATION

The hub and rotor assembly is removed and installed in the same manner as a conventional drum brake hub.

INSPECTION AND MEASUREMENT

Check the rotor for surface cracks, nicks, broken cooling fins and scoring of both contact surfaces. Some scoring of the surfaces may occur during normal use. Scoring that is 0.015 in. deep or less is not detrimental to the operation of the brakes.

If the rotor surface is heavily rusted or scaled, clean both surfaces on a disc brake lathe using flat sanding discs before attempting any measurements.

With the hub and rotor assembly mounted on the spindle of the vehicle or a disc brake lathe and all play removed from the wheel bearings, assemble a dial indicator so that the stem contacts the center of the rotor braking surface. Zero the dial indicator before taking any measurements. Lateral runout must not exceed 0.005 in. with a maximum rate of change not to exceed 0.001 in. in 30 degrees of rotation.

Excessive runout will cause the rotor to wobble and knock the piston back into the caliper causing increased pedal travel, noise and vibration.

After the rotor has been refinished, the minimum thickness of 1.230 in. is acceptable. Discard the rotor if the thickness is 1.215 in. or less.

NOTE: *Remember to adjust the preload on the wheel bearings after the runout measurement has been taken.*

Wheel Bearings

The wheel bearings on vehicles equipped with front disc brakes are serviced in the same manner as those installed on vehicles with front drum brakes. Refer to the "Drum Brakes" section for front wheel bearing service procedures.

NOTE: *There is special high temperature wheel bearing grease available for vehicles equipped with disc brakes.*

DRUM BRAKES

Brake Drums

REMOVAL AND INSTALLATION

On the front wheels, the brake drum is attached to the inside facing of the hub flange. Studs pressed into and through the brake drum protrude through the hub flange, providing mounting points for the wheels as well as securing the brake drum to the hub.

To remove the brake drum, remove the wheel, then the hub and drum assembly from the vehicle. Next, press or drive out the studs to remove the drum from the hub.

When placing the drum on the hub, make sure that the contacting surfaces are clean and flat. Line up the holes in the drum with those in the hub and put the drum over the shoulder on the hub. Insert five new bolts through the drum and hub and drive the bolts into place solidly. Place a round piece of stock approximately the diameter of the head of the bolt, in a vise; next place the hub and drum assembly over it so that the bolt head rests on it. Then flatten the bolt head into the countersunk section of the hub or drum with a punch.

NOTE: *On earlier models the brake drum is mounted behind the hub flange. On later models, its position is reversed.*

The runout of the drum face should be within 0.030 in. If the runout is found to be greater than 0.030 in., it will be necessary to reset the bolts to correct the condition.

The left hand hub bolts have an L stamped on the head of the bolt.

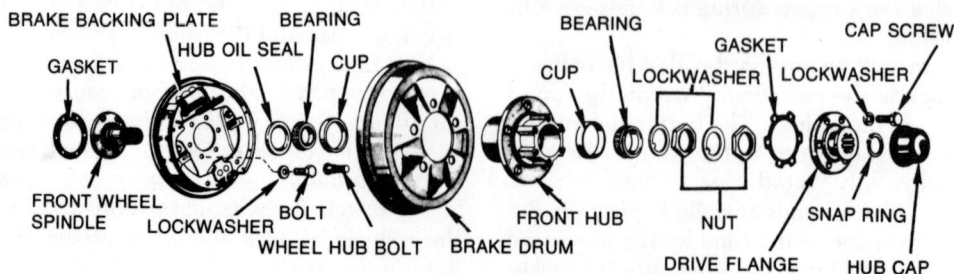

1972 and later front hub and drum

1966–71 front hub and drum

The left hand threaded nuts may have a groove cut around the hexagon faces, or the word LEFT stamped on the face.

Hubs with left hand threaded hub bolts are installed on the left hand side of the vehicle. Later production vehicles are equipped with right hand bolts and nuts on all four hubs.

INSPECTION

Using a brake drum micrometer, check all drums. Should a brake drum be scored or rough, it may be reconditioned by grinding or turning on a lathe. Do not remove more than 0.030 in. thickness of metal. If a drum is reconditioned in this manner, it is recommended that either the correct factory supplied, 0.030 in. oversize lining must be installed, or a shim equal in thickness to the metal removed must be placed between the lining and the brake shoe so that the arc of the lining will be the same as that of the drum.

Brake Shoes

REMOVAL AND INSTALLATION

1. Jack the vehicle up and support it so that the wheels to be worked on are off the ground.

2. Turn the adjustment starwheel so that the brake shoes are retracted from the brake drum.

3. Remove the wheels and the hubs and drums to give access to the brake shoes.

4. Install wheel cylinder clamps to retain the wheel cylinder pistons in place and prevent leakage of brake fluid while replacing the shoes.

5. Remove the return springs with a brake spring remover tool.

6. On models with self adjusters, remove the adjuster cable, cable guide, adjuster lever and adjuster springs.

7. Remove the hold-down washers and springs and remove the brake shoes.

8. Before installing the new shoes, now would be a good time to inspect the oil seals in the hubs. If the condition of the seals is doubtful, replace them. Also check the wheel cylinders for leakage. Pull back the dust covers. If there is fluid present behind the dust cover the wheel cylinder must be rebuilt or replaced.

9. Clean the backing plate with a brush or cloth. Place a dab of Lubriplate on each spot where the brake shoes rub on the backing plate.

CAUTION: *Never clean brake surfaces*

with compressed air. Dirt in the brake area contains asbestos which is hazardous to your health when breathed.

NOTE: *Always replace brake lining in axle sets. Never replace linings on one side or just on one wheel.*

10. Install the brakes in the reverse order of removal.

Wheel Cylinders
OVERHAUL

Wheel cylinder rebuilding kits are available for reconditioning wheel cylinders. The kits usually contain new cup springs, cylinder cups and in some, new boots. The most important factor to keep in mind when rebuilding wheel cylinders is cleanliness. Keep all dirt away from the wheel cylinders when you are reassembling them.

1. To remove the wheel cylinder, jack up the vehicle and remove the wheel, hub, and drum.

2. Disconnect the brake line at the fitting on the brake backing plate.

3. Remove the brake assemblies.

4. Remove the screws or nuts that hold the wheel cylinder to the backing plate and remove the wheel cylinder from the vehicle.

5. Remove the rubber dust covers on the ends of the cylinder. Remove the pistons and piston cups and the spring. Remove the bleeder screw and make sure it is not plugged.

6. Discard all of the parts that the rebuilding kit will replace.

7. Examine the inside of the cylinder. If it is severely rusted, pitted or scratched, then the cylinder must be replaced as the piston cups won't be able to seal against the walls of the cylinder.

8. Using a wheel cylinder hone or emery cloth and crocus cloth, polish the inside of the cylinder. The purpose of this is to put a new surface on the inside of the cylinder. Keep the inside of the cylinder coated with brake fluid while honing.

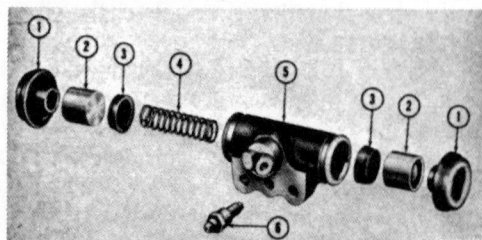

Wheel cylinder

9. Wash out the cylinder with clean brake fluid after honing.

10. When reassembling the cylinder dip all of the parts in clean brake fluid. Reassemble in the reverse order of removal.

Wheel Bearings
REMOVAL AND INSTALLATION

To remove the front wheel bearings, remove the hub cap, snap-ring, drive flange, and the two nuts and lockwashers. The outer wheel bearing can then be removed. To remove the inner wheel bearing, the outer wheel bearing must be removed and the hub removed from the spindle. Turn the hub over and drive out the inner bearing with a hammer and a block of wood, using the block of wood as a drift, having it placed through the center of the hub and up against the inner side of the inner wheel bearing. Discard the grease seal and replace it.

Before handling the bearings there are a few things that you should remember to do and not to do.

Remember to DO the following:

1. Remove all outside dirt from the housing before exposing the bearing.

2. Treat a used bearing as gently as you would a new one.

3. Work with clean tools in clean surroundings.

4. Use clean, dry canvas gloves, or at least clean, dry hands.

5. Clean solvents and flushing fluids are a must.

6. Use clean paper when laying out the bearings to dry.

7. Protect disassembled bearings from rust and dirt. Cover them up.

8. Use clean rags to wipe bearings.

9. Keep the bearings in oil-proof paper when they are to be stored or are not in use.

10. Clean the inside of the housing before replacing the bearing.

Do NOT do the following:

1. Don't work in dirty surroundings.

2. Don't use dirty, chipped, or damaged tools.

3. Try not to work on wooden work benches or use wooden mallets.

4. Don't handle bearings with dirty or moist hands.

5. Do not use gasoline for cleaning; use a safe solvent.

6. Do not spin-dry bearings with compressed air. They will be damaged.

7. Do not spin dirty bearings.

8. Avoid using cotton waste or dirty cloths to wipe bearings.

9. Try not to scratch or nick bearing surfaces.

10. Do not allow the bearing to come in contact with dirt or rust at any time.

Place all of the bearings, nuts, washers, and dust caps in a container of solvent. Cleanliness is basic to wheel bearing maintenance. Use a soft brush to thoroughly clean each part. Make sure that every bit of dirt and grease is rinsed off, then place each cleaned part on an absorbent cloth and let them dry completely.

Inspect the bearings for pitting, flat spots, rust, and rough areas. Check the races on the hub and the spindle for the same defects and rub them clean with a rag that has been soaked in solvent. If the races show hairline cracks or worn, shiny areas, they must be replaced with new parts. Replacement seals, bearings, and other required parts can be bought at an auto parts store. The old parts that are to be replaced should be taken along to be compared with the replacement part to ensure a perfect match.

Pack the wheel bearings with grease. There are special devices made for the specific purpose of greasing bearings, but, if one is not available, pack the wheel bearings by hand. Put a large dab of grease in the palm of your hand and push the bearing through it with a sliding motion. The grease must be forced through the side of the bearing and in between each roller. Continue until the grease begins to ooze out the other side and through the gaps between the rollers; the bearing must be completely packed with grease.

Turn the hub assembly over, making sure that it is perfectly clean, and drop the inner wheel bearing into place. Using a hammer and a block of weed, tap the new seal in place. Do not hit the race with the hammer directly. Move the block of wood around the circumference until it is seated properly.

Slide the hub assembly onto the spindle, and push it as far as it will go, making sure that it has completely covered the brake shoes.

Place the outer bearing in place over the spindle. Press it in until it is snug. Place the washer on the spindle behind the bearing. Put on the spindle adjusting nut. While turning the wheel by hand, turn the nut down until a slight binding is felt, then back off

about $1/6$ of a turn. Bend the lip on the lockwasher over the edge of the nut. Place the second washer and nut on the spindle and tighten them up against the adjusting nut, being careful not to turn the adjusting nut further onto the spindle.

Reassemble the rest of the hub assembly in the reverse order of removal.

If the bearings are correctly adjusted, wheel shake will be just perceptible when the wheel is gripped by hand and shaken from side to side. The wheel will also turn freely with no drag.

If the bearing adjustment is too tight, the rollers may break or become overheated. Loose bearings may cause excessive wear and possible noise.

TRANSMISSION BRAKE

ADJUSTMENT

Make sure that the brake operating handle or foot pedal is fully released. Check the operating linkage and the cable to make sure that they don't bind. If necessary, free the cable and lubricate it. Rotate the brake drum until one pair of the three sets of holes are over the shoe adjusting screw wheels in the brake

Transmission brake assembly

drum as a fulcrum for the brake adjusting tool or a screwdriver. Rotate each notched adjusting screw by moving the handle of the tool away from the center of the driveshaft until the shoes are snug against the drum. Back off seven notches on the adjusting screwwheels to secure the proper running clearance between the shoes and the drum.

PARKING BRAKE
(Rear Wheel Drum Type)

ADJUSTMENT

1. Make sure that the hydraulic brakes are in satisfactory adjustment.

2. Raise the rear wheels off the ground and disengage the parking brake pedal.

3. Loosen the locknut on the brake cable adjusting rod, located directly behind the frame center crossmember.

4. Spin the wheels and tighten the adjustment until the rear wheels drag slightly. Loosen the adjustment until there is no drag and the wheels spin freely.

5. Tighten the locknut to lock the adjusting nut.

Appendix

General Conversion Table

Multiply by	To convert	To	
2.54	Inches	Centimeters	.3937
30.48	Feet	Centimeters	.0328
.914	Yards	Meters	1.094
1.609	Miles	Kilometers	.621
.645	Square inches	Square cm.	.155
.836	Square yards	Square meters	1.196
16.39	Cubic inches	Cubic cm.	.061
28.3	Cubic feet	Liters	.0353
.4536	Pounds	Kilograms	2.2045
4.226	Gallons	Liters	.264
.068	Lbs./sq. in. (psi)	Atmospheres	14.7
.138	Foot pounds	Kg. m.	7.23
1.014	H.P. (DIN)	H.P. (SAE)	.9861
—	To obtain	From	Multiply by

Note: 1 cm. equals 10 mm.; 1 mm. equals .0394″.

Conversion—Common Fractions to Decimals and Millimeters

Common Fractions	Decimal Fractions	Millimeters (approx.)	Common Fractions	Decimal Fractions	Millimeters (approx.)	Common Fractions	Decimal Fractions	Millimeters (approx.)
1/128	.008	0.20	11/32	.344	8.73	43/64	.672	17.07
1/64	.016	0.40	23/64	.359	9.13	11/16	.688	17.46
1/32	.031	0.79	3/8	.375	9.53	45/64	.703	17.86
3/64	.047	1.19	25/64	.391	9.92	23/32	.719	18.26
1/16	.063	1.59	13/32	.406	10.32	47/64	.734	18.65
5/64	.078	1.98	27/64	.422	10.72	3/4	.750	19.05
3/32	.094	2.38	7/16	.438	11.11	49/64	.766	19.45
7/64	.109	2.78	29/64	.453	11.51	25/32	.781	19.84
1/8	.125	3.18	15/32	.469	11.91	51/64	.797	20.24
9/64	.141	3.57	31/64	.484	12.30	13/16	.813	20.64
5/32	.156	3.97	1/2	.500	12.70	53/64	.828	21.03
11/64	.172	4.37	33/64	.516	13.10	27/32	.844	21.43
3/16	.188	4.76	17/32	.531	13.49	55/64	.859	21.83
13/64	.203	5.16	35/64	.547	13.89	7/8	.875	22.23
7/32	.219	5.56	9/16	.563	14.29	57/64	.891	22.62
15/64	.234	5.95	37/64	.578	14.68	29/32	.906	23.02
1/4	.250	6.35	19/32	.594	15.08	59/64	.922	23.42
17/64	.266	6.75	39/64	.609	15.48	15/16	.938	23.81
9/32	.281	7.14	5/8	.625	15.88	61/64	.953	24.21
19/64	.297	7.54	41/64	.641	16.27	31/32	.969	24.61
5/16	.313	7.94	21/32	.656	16.67	63/64	.984	25.00
21/64	.328	8.33						

Conversion—Millimeters to Decimal Inches

mm	inches	mm	inches	mm	inches	mm	inches	mm	inches
1	.039 370	31	1.220 470	61	2.401 570	91	3.582 670	210	8.267 700
2	.078 740	32	1.259 840	62	2.440 940	92	3.622 040	220	8.661 400
3	.118 110	33	1.299 210	63	2.480 310	93	3.661 410	230	9.055 100
4	.157 480	34	1.338 580	64	2.519 680	94	3.700 780	240	9.448 800
5	.196 850	35	1.377 949	65	2.559 050	95	3.740 150	250	9.842 500
6	.236 220	36	1.417 319	66	2.598 420	96	3.779 520	260	10.236 200
7	.275 590	37	1.456 689	67	2.637 790	97	3.818 890	270	10.629 900
8	.314 960	38	1.496 050	68	2.677 160	98	3.858 260	280	11.032 600
9	.354 330	39	1.535 430	69	2.716 530	99	3.897 630	290	11.417 300
10	.393 700	40	1.574 800	70	2.755 900	100	3.937 000	300	11.811 000
11	.433 070	41	1.614 170	71	2.795 270	105	4.133 848	310	12.204 700
12	.472 440	42	1.653 540	72	2.834 640	110	4.330 700	320	12.598 400
13	.511 810	43	1.692 910	73	2.874 010	115	4.527 550	330	12.992 100
14	.551 180	44	1.732 280	74	2.913 380	120	4.724 400	340	13.385 800
15	.590 550	45	1.771 650	75	2.952 750	125	4.921 250	350	13.779 500
16	.629 920	46	1.811 020	76	2.992 120	130	5.118 100	360	14.173 200
17	.669 290	47	1.850 390	77	3.031 490	135	5.314 950	370	14.566 900
18	.708 660	48	1.889 760	78	3.070 860	140	5.511 800	380	14.960 600
19	.748 030	49	1.929 130	79	3.110 230	145	5.708 650	390	15.354 300
20	.787 400	50	1.968 500	80	3.149 600	150	5.905 500	400	15.748 000
21	.826 770	51	2.007 870	81	3.188 970	155	6.102 350	500	19.685 000
22	.866 140	52	2.047 240	82	3.228 340	160	6.299 200	600	23.622 000
23	.905 510	53	2.086 610	83	3.267 710	165	6.496 050	700	27.559 000
24	.944 880	54	2.125 980	84	3.307 080	170	6.692 900	800	31.496 000
25	.984 250	55	2.165 350	85	3.346 450	175	6.889 750	900	35.433 000
26	1.023 620	56	2.204 720	86	3.385 820	180	7.086 600	1000	39.370 000
27	1.062 990	57	2.244 090	87	3.425 190	185	7.283 450	2000	78.740 000
28	1.102 360	58	2.283 460	88	3.464 560	190	7.480 300	3000	118.110 000
29	1.141 730	59	2.322 830	89	3.503 903	195	7.677 150	4000	157.480 000
30	1.181 100	60	2.362 200	90	3.543 300	200	7.874 000	5000	196.850 000

To change decimal millimeters to decimal inches, position the decimal point where desired on either side of the millimeter measurement shown and reset the inches decimal by the same number of digits in the same direction. For example, to convert 0.001 mm to decimal inches, reset the decimal behind the 1 mm (shown on the chart) to 0.001; change the decimal inch equivalent (0.039″ shown) to 0.000039″.

Tap Drill Sizes

National Fine or S.A.E.

Screw & Tap Size	Threads Per Inch	Use Drill Number
No. 5	44	37
No. 6	40	33
No. 8	36	29
No. 10	32	21
No. 12	28	15
1/4	28	3
5/16	24	1
3/8	24	Q
7/16	20	W
1/2	20	29/64
9/16	18	33/64
5/8	18	37/64
3/4	16	11/16
7/8	14	13/16
1 1/8	12	1 3/64
1 1/4	12	1 11/64
1 1/2	12	1 27/64

Tap Drill Sizes

National Coarse or U.S.S.

Screw & Tap Size	Threads Per Inch	Use Drill Number
No. 5	40	39
No. 6	32	36
No. 8	32	29
No. 10	24	25
No. 12	24	17
1/4	20	8
5/16	18	F
3/8	16	5/16
7/16	14	U
1/2	13	27/64
9/16	12	31/64
5/8	11	17/32
3/4	10	21/32
7/8	9	49/64
1	8	7/8
1 1/8	7	63/64
1 1/4	7	1 7/64
1 1/2	6	1 11/32

Decimal Equivalent Size of the Number Drills

Drill No.	Decimal Equivalent	Drill No.	Decimal Equivalent	Drill No.	Decimal Equivalent
80	.0135	53	.0595	26	.1470
79	.0145	52	.0635	25	.1495
78	.0160	51	.0670	24	.1520
77	.0180	50	.0700	23	.1540
76	.0200	49	.0730	22	.1570
75	.0210	48	.0760	21	.1590
74	.0225	47	.0785	20	.1610
73	.0240	46	.0810	19	.1660
72	.0250	45	.0820	18	.1695
71	.0260	44	.0860	17	.1730
70	.0280	43	.0890	16	.1770
69	.0292	42	.0935	15	.1800
68	.0310	41	.0960	14	.1820
67	.0320	40	.0980	13	.1850
66	.0330	39	.0995	12	.1890
65	.0350	38	.1015	11	.1910
64	.0360	37	.1040	10	.1935
63	.0370	36	.1065	9	.1960
62	.0380	35	.1100	8	.1990
61	.0390	34	.1110	7	.2010
60	.0400	33	.1130	6	.2040
59	.0410	32	.1160	5	.2055
58	.0420	31	.1200	4	.2090
57	.0430	30	.1285	3	.2130
56	.0465	29	.1360	2	.2210
55	.0520	28	.1405	1	.2280
54	.0550	27	.1440		

Decimal Equivalent Size of the Letter Drills

Letter Drill	Decimal Equivalent	Letter Drill	Decimal Equivalent	Letter Drill	Decimal Equivalent
A	.234	J	.277	S	.348
B	.238	K	.281	T	.358
C	.242	L	.290	U	.368
D	.246	M	.295	V	.377
E	.250	N	.302	W	.386
F	.257	O	.316	X	.397
G	.261	P	.323	Y	.404
H	.266	Q	.332	Z	.413
I	.272	R	.339		

Anti-Freeze Chart

Temperatures Shown in Degrees Fahrenheit +32 is Freezing

Cooling System Capacity Quarts	1	2	3	4	5	6	7	8	9	10	11	12	13	14
	Quarts of ETHYLENE GLYCOL Needed for Protection to Temperatures Shown Below													
10	+24°	+16°	+ 4°	−12°	−34°	−62°								
11	+25	+18	+ 8	− 6	−23	−47								
12	+26	+19	+10	0	−15	−34	−57°							
13	+27	+21	+13	+ 3	− 9	−25	−45							
14			+15	+ 6	− 5	−18	−34							
15			+16	+ 8	0	−12	−26							
16			+17	+10	+ 2	− 8	−19	−34	−52°					
17			+18	+12	+ 5	− 4	−14	−27	−42					
18			+19	+14	+ 7	0	−10	−21	−34	−50°				
19			+20	+15	+ 9	+ 2	− 7	−16	−28	−42				
20				+16	+10	+ 4	− 3	−12	−22	−34	−48°			
21				+17	+12	+ 6	0	− 9	−17	−28	−41			
22				+18	+13	+ 8	+ 2	− 6	−14	−23	−34	−47°		
23				+19	+14	+ 9	+ 4	− 3	−10	−19	−29	−40		
24				+19	+15	+10	+ 5	0	− 8	−15	−23	−34	−46°	
25				+20	+16	+12	+ 7	+ 1	− 5	−12	−20	−29	−40	−50°
26					+17	+13	+ 8	+ 3	− 3	− 9	−16	−25	−34	−44
27					+18	+14	+ 9	+ 5	− 1	− 7	−13	−21	−29	−39
28					+18	+15	+10	+ 6	+ 1	− 5	−11	−18	−25	−34
29					+19	+16	+12	+ 7	+ 2	− 3	− 8	−15	−22	−29
30					+20	+17	+13	+ 8	+ 4	− 1	− 6	−12	−18	−25

For capacities over 30 quarts divide true capacity by 3. Find quarts Anti-Freeze for the ⅓ and multiply by 3 for quarts to add.

For capacities under 10 quarts multiply true capacity by 3. Find quarts Anti-Freeze for the tripled volume and divide by 3 for quarts to add.

To Increase the Freezing Protection of Anti-Freeze Solutions Already Installed

Cooling System Capacity Quarts	Number of Quarts of ETHYLENE GLYCOL Anti-Freeze Required to Increase Protection													
	From +20° F. to					From +10° F. to					From 0° F. to			
	0°	−10°	−20°	−30°	−40°	0°	−10°	−20°	−30°	−40°	−10°	−20°	−30°	−40°
10	1¾	2¼	3	3½	3¾	¾	1½	2¼	2¾	3¼	¾	1½	2	2½
12	2	2¾	3½	4	4½	1	1¾	2½	3¼	3¾	1	1¾	2½	3¼
14	2¼	3¼	4	4¾	5½	1¼	2	3	3¾	4½	1	2	3	3½
16	2½	3½	4½	5¼	6	1¼	2½	3½	4¼	5¼	1¼	2¼	3¼	4
18	3	4	5	6	7	1½	2¾	4	5	5¾	1½	2½	3¾	4¾
20	3¼	4½	5¾	6¾	7½	1¾	3	4¼	5½	6½	1½	2¾	4¼	5¼
22	3½	5	6¼	7¼	8¼	1¾	3¼	4¾	6	7¼	1¾	3¼	4½	5½
24	4	5½	7	8	9	2	3½	5	6½	7½	1¾	3½	5	6
26	4¼	6	7½	8¾	10	2	4	5½	7	8¼	2	3¾	5½	6¾
28	4½	6¼	8	9½	10½	2¼	4¼	6	7½	9	2	4	5¾	7¼
30	5	6¾	8½	10	11½	2½	4½	6½	8	9½	2¼	4¼	6¼	7¾

Test radiator solution with proper hydrometer. Determine from the table the number of quarts of solution to be drawn off from a full cooling system and replace with undiluted anti-freeze, to give the desired increased protection. For example, to increase protection of a 22-quart cooling system containing Ethylene Glycol (permanent type) anti-freeze, from +20° F. to −20° F. will require the replacement of 6¼ quarts of solution with undiluted anti-freeze.

Index

Chilton's Repair & Tune-Up Guides

The complete line covers domestic cars, imports, trucks, vans, RV's and 4-wheel drive vehicles.

BOOK CODE	TITLE	BOOK CODE	TITLE
#7032	Arrow & D-50 Pick-Ups 79-81	#6980	Honda 73-80
#6637	Aspen & Volare 76-78	#5912	International Scout 67-73
#5902	Audi 70-73	#5998	Jaguar 69-74
#7028	Audi 4000 & 5000 77-81	#7136	Jeep CJ 45-81
#6337	Audi Fox 73-75	#6739	Jeep Wagoneer, Commando, and Cherokee 66-79
#5807	Barracuda and Challenger 65-72		
#6931	Blazer & Jimmy 69-80	#6634	Maverick/Comet 70-77
#6844	BMW 70-79	#6981	Mazda 71-80
#7045	Camaro 67-81	#7031	Mazda RX-7 78-81
#6695	Capri 70-77	#6065	Mercedes-Benz 59-70
#7041	Champ/Arrow/Sapporo 77-81	#5907	Mercedes-Benz 68-73
#6316	Charger, Coronet 71-75	#6809	Mercedes-Benz 74-79
#6836	Chevette 76-80	#6780	MG 61-79
#6840	Chevrolet Mid Size 64-79 Covers Chevelle, Laguna, El Camino, Monte Carlo & Malibu	#6542	Mustang 65-73
		#6812	Mustang II 74-78
		#6963	Mustang & Capri 79-80 Inc. Turbo.
#7135	Chevrolet 68-81 All Full-Size Chevrolet Models	#6845	Omni/Horizon 78-80
#6936	Chevrolet & GMC Pick-Ups 70-80	#5792	Opel 64-70
#6930	Chevrolet & GMC Vans 67-80	#6575	Opel 71-75
#7051	Chevrolet LUV 72-81 Inc. 4 x 4 Models	#6473	Pacer 75-76
#6841	Chevy II, Nova 62-79	#5982	Peugeot 70-74
#7037	Colt & Challenger 71-80	#7027	Pinto & Bobcat 71-80
#6691	Corvair 60-69 All Models and Engines, Inc. Turbo.	#6552	Plymouth 68-76
		#5822	Porsche 69-73
#6576	Corvette 53-62	#6331	Ramcharger & Trail Duster 74-75
#6843	Corvette 63-79	#5985	Rebel/Matador 67-74
#6933	Cutlass 70-80	#5821	Road Runner, Satellite, Belvedere, GTX 68-73
#6962	Dasher, Rabbit, Scirocco, Jetta 74-80		
#5790	Datsun 61-72	#5988	Saab 99 69-75
#6960	Datsun 73-80	#6978	Snowmobiles 76-80
#7050	Datsun Pick-Ups 70-81	#6982	Subaru 70-80
#6932	Datsun Z and ZX 70-80	#5905	Tempest, GTO and Le Mans 68-73
#6554	Dodge 68-77	#5795	Toyota 66-70
#6486	Dodge Charger 67-70	#7036	Toyota Corolla/Carina/Tercel/Starlet 79-81
#6934	Dodge & Plymouth Vans 67-80		
#6320	Fairlane and Torino 62-75	#7043	Toyota Celica & Supra 71-81
#6965	Fairmont & Zephyr 78-80	#7044	Toyota Corona/Cressida/Crown/Mk II 70-81
#6485	Fiat 64-70		
#7042	Fiat 69-81	#6276	Toyota Land Cruiser 66-74
#6846	Fiesta 78-80	#7035	Toyota Pick-Ups 70-81
#5996	Firebird 67-74	#5910	Triumph 69-73
#7140	Ford Bronco 66-81	#6326	Valiant and Duster 68-76
#6983	Ford Courier 72-80	#5796	Volkswagen 49-71
#6842	Ford and Mercury 68-79 All Full-Size Models	#6837	Volkswagen 70-81
		#6529	Volvo 56-69
#6696	Ford and Mercury Mid-Size 71-78 Covers Torino, Gran Torino, Ranchero, Elite, LTD II, Thunderbird, Montego, and Cougar	#7040	Volvo 70-81

AUTOMOTIVE SPECIALITY BOOKS

BOOK CODE	TITLE
#6913	Ford Pick-Ups 65-80
#6849	Ford Vans 61-80
#6935	GM Subcompact 71-80 Covers Vega, Monza, Astre, Sunbird, Starfire, Skyhawk
#7049	GM X-Body 80-81 Covers Citation, Omega, Phoenix and Skylark
#6937	Granada/Monarch 75-80

#6754	Chilton's Diesel Guide
#6942	Chilton's Guide to Consumers' Auto Repairs and Prices
#6940	Chilton's Minor Auto Body Repair
#6908	Chilton's More Miles Per Gallon
#6867	Chilton's Motorcycle Owner's Handbook
#6727	Chilton's Off-Roading Guide
#6811	Chilton's Repair Guide for Small Engines - Covers 2 and 4-stroke air cooled gasoline engines up to 20 hp.

Chilton's Repair & Tune-Up Guides are available at your local retailer or by mailing a check or money order for **$9.95** plus **$1.00** to cover postage and handling to:

Chilton Book Company
Dept. DM,
Radnor, PA 19089

NOTE: When ordering be sure to include name & address, book code & title.